THERMODYNAMICS
AND AN INTRODUCTION TO
THERMOSTATISTICS

THERMODYNAMICS
AND AN INTRODUCTION TO
THERMOSTATISTICS

SECOND EDITION

HERBERT B. CALLEN
University of Pennsylvania

JOHN WILEY & SONS
New York   Chichester   Brisbane   Toronto   Singapore

Library of Congress Cataloging in Publication Data:

Callen, Herbert B.
   Thermodynamics and an Introduction to
Thermostatistics.

   Rev. ed. of: Thermodynamics. 1960.
   Bibliography: p. 485
   Includes index.
   1. Thermodynamics.   2. Statistical Mechanics.
I. Callen, Herbert B.   Thermodynamics.   II. Title.
III. Title: Thermostatistics.
QC311.C25   1985   536'.7   85-6387
ISBN 0-471-86256-8

30 29 28 27 26 25 24 23 22

To Sara
.....and to Jill, Jed,
Zachary and Jessica

PREFACE

Twenty-five years after writing the first edition of *Thermodynamics* I am gratified that the book is now the thermodynamic reference most frequently cited in physics research literature, and that the postulational formulation which it introduced is now widely accepted. Nevertheless several considerations prompt this new edition and extension.

First, thermodynamics advanced dramatically in the 60s and 70s, primarily in the area of critical phenomena. Although those advances are largely beyond the scope of this book, I have attempted to at least describe the nature of the problem and to introduce the critical exponents and scaling functions that characterize the non-analytic behavior of thermodynamic functions at a second-order phase transition. This account is descriptive and simple. It replaces the relatively complicated theory of second-order transitions that, in the view of many students, was the most difficult section of the first edition.

Second, I have attempted to improve the pedagogical attributes of the book for use in courses from the junior undergraduate to the first year graduate level, for physicists, engineering scientists and chemists. This purpose has been aided by a large number of helpful suggestions from students and instructors. Many explanations are simplified, and numerous examples are solved explicitly. The number of problems has been expanded, and partial or complete answers are given for many.

Third, an introduction to the principles of statistical mechanics has been added. Here the spirit of the first edition has been maintained; the emphasis is on the underlying simplicity of principles and on the central train of logic rather than on a multiplicity of applications. For this purpose, and to make the text accessible to advanced undergraduates, I have avoided explicit non-commutivity problems in quantum mechanics. All that is required is familiarity with the fact that quantum mechanics predicts discrete energy levels in finite systems. However, the formulation is designed so that the more advanced student will properly interpret the theory in the non-commutative case.

Fourth, I have long been puzzled by certain conceptual problems lying at the foundations of thermodynamics, and this has led me to an interpretation of the "meaning" of thermodynamics. In the final chapter—an "interpretive postlude" to the main body of the text—I develop the thesis that thermostatistics has its roots in the *symmetries* of the fundamental laws of physics rather than in the quantitative content of those laws. The discussion is qualitative and descriptive, seeking to establish an intuitive framework and to encourage the student to see science as a coherent structure in which thermodynamics has a natural and fundamental role.

Although both statistical mechanics and thermodynamics are included in this new edition, I have attempted neither to separate them completely nor to meld them into the undifferentiated form now popular under the rubric of "thermal physics." I believe that each of these extreme options is misdirected. To divorce thermodynamics completely from its statistical mechanical base is to rob thermodynamics of its fundamental physical origins. Without an insight into statistical mechanics a scientist remains rooted in the macroscopic empiricism of the nineteenth century, cut off from contemporary developments and from an integrated view of science. Conversely, the amalgamation of thermodynamics and statistical mechanics into an undifferentiated "thermal physics" tends to eclipse thermodynamics. The fundamentality and profundity of statistical mechanics are treacherously seductive; "thermal physics" courses almost perforce give short shrift to macroscopic operational principles.* Furthermore the amalgamation of thermodynamics and statistical mechanics runs counter to the "principle of theoretical economy"; the principle that predictions should be drawn from the most general and least detailed assumptions possible. Models, endemic to statistical mechanics, should be eschewed whenever the general methods of macroscopic thermodynamics are sufficient. Such a habit of mind is hardly encouraged by an organization of the subjects in which thermodynamics is little more than a subordinate clause.

The balancing of the two distinct components of the thermal sciences is carried out in this book by introducing the subject at the macroscopic level, by formulating thermodynamics so that its macroscopic postulates are precisely and clearly the theorems of statistical mechanics, and by frequent explanatory allusions to the interrelationships of the two components. Nevertheless, at the option of the instructor, the chapters on statistical mechanics can be interleaved with those on thermodynamics in a sequence to be described. But even in that integrated option the basic macroscopic structure of thermodynamics is established *before* statistical reasoning is introduced. Such a separation and sequencing of the subjects

---

*The American Physical Society Committee on Applications of Physics reported [*Bulletin of the APS*, Vol. 22 #10, 1233 (1971)] that a survey of industrial research leaders designated thermodynamics above all other subjects as requiring increased emphasis in the undergraduate curriculum. That emphasis subsequently has *decreased*.

preserves and emphasizes the hierarchical structure of science, organizing physics into coherent units with clear and easily remembered interrelationships. Similarly, classical mechanics is best understood as a self-contained postulatory structure, only *later* to be validated as a limiting case of quantum mechanics.

Two primary curricular options are listed in the "menu" following. In one option the chapters are followed in sequence (Column A alone, or followed by all or part of column B). In the "integrated" option the menu is followed from top to bottom. Chapter 15 is a short and elementary statistical interpretation of entropy; it can be inserted immediately after Chapter 1, Chapter 4, or Chapter 7.

The chapters listed below the first dotted line are freely flexible with respect to sequence, or to inclusion or omission. To balance the concrete and particular against more esoteric sections, instructors may choose to insert parts of Chapter 13 (Properties of Materials) at various stages, or to insert the Postlude (Chapter 21, Symmetry and Conceptual Foundations) at any point in the course.

The minimal course, for junior year undergraduates, would involve the first seven chapters, with Chapter 15 and 16 optionally included as time permits.

Philadelphia, Pennsylvania   **Herbert B. Callen**

CONTENTS

xi

PART II
STATISTICAL MECHANICS

PART III
FOUNDATIONS

PART I

GENERAL PRINCIPLES OF CLASSICAL THERMODYNAMICS

INTRODUCTION

The Nature of Thermodynamics and the Basis of ThermoStatistics

Whether we are physicists, chemists, biologists, or engineers, our primary interface with nature is through the properties of macroscopic matter. Those properties are subject to universal regularities and to stringent limitations. Subtle relationships exist among apparently unconnected properties.

The existence of such an underlying order has far reaching implications. Physicists and chemists familiar with that order need not confront each new material as a virgin puzzle. Engineers are able to anticipate limitations to device designs predicated on creatively imagined (but yet undiscovered) materials with the requisite properties. And the specific form of the underlying order provides incisive clues to the structure of fundamental physical theory.

Certain primal concepts of thermodynamics are intuitively familiar. A metallic block released from rest near the rim of a smoothly polished metallic bowl oscillates within the bowl, approximately conserving the sum of potential and kinetic energies. But the block eventually comes to rest at the bottom of the bowl. Although the mechanical energy appears to have vanished, an observable effect is wrought upon the material of the bowl and block; they are very slightly, but perceptibly, "warmer." Even before studying thermodynamics, we are qualitatively aware that the mechanical energy has merely been converted to another form, that the fundamental principle of energy conservation is preserved, and that the physiological sensation of "warmth" is associated with the thermodynamic concept of "temperature."

Vague and undefined as these observations may be, they nevertheless reveal a notable dissimilarity between thermodynamics and the other branches of classical science. Two prototypes of the classical scientific paradigm are mechanics and electromagnetic theory. The former addresses itself to the dynamics of particles acted upon by forces, the latter to the dynamics of the fields that mediate those forces. In each of these cases a new "law" is formulated—for mechanics it is Newton's Law (or Lagrange or Hamilton's more sophisticated variants); for electromagnetism it is the Maxwell equations. In either case it remains only to explicate the consequences of the law.

Thermodynamics is quite different. It neither claims a unique domain of systems over which it asserts primacy, nor does it introduce a new fundamental law analogous to Newton's or Maxwell's equations. In contrast to the specificity of mechanics and electromagnetism, the hallmark of thermodynamics is generality. Generality first in the sense that thermodynamics applies to all types of systems in macroscopic aggregation, and second in the sense that thermodynamics does not predict specific numerical values for observable quantities. Instead, thermodynamics sets limits (*in*equalities) on permissible physical processes, and it establishes relationships among apparently unrelated properties.

The contrast between thermodynamics and its counterpart sciences raises fundamental questions which we shall address directly only in the final chapter. There we shall see that whereas thermodynamics is not based on a new and particular law of nature, it instead reflects a commonality or universal feature of *all* laws. In brief, *thermodynamics is the study of the restrictions on the possible properties of matter that follow from the symmetry properties of the fundamental laws of physics.*

The connection between the symmetry of fundamental laws and the macroscopic properties of matter is not trivially evident, and we do not attempt to derive the latter from the former. Instead we follow the postulatory formulation of thermodynamics developed in the first edition of this text, returning to an interpretive discussion of symmetry origins in Chapter 21. But even the preliminary assertion of this basis of thermodynamics may help to prepare the reader for the somewhat uncommon form of thermodynamic theory. Thermodynamics inherits its universality, it nonmetric nature, and its emphasis on relationships from its symmetry parentage.

THE PROBLEM AND THE POSTULATES

1-1 THE TEMPORAL NATURE OF MACROSCOPIC MEASUREMENTS

Perhaps the most striking feature of macroscopic matter is the incredible simplicity with which it can be characterized. We go to a pharmacy and request one liter of ethyl alcohol, and that meager specification is pragmatically sufficient. Yet from the atomistic point of view, we have specified remarkably little. A complete mathematical characterization of the system would entail the specification of coordinates and momenta for each molecule in the sample, plus sundry additional variables descriptive of the internal state of each molecule—altogether at least 10^{23} numbers to describe the liter of alcohol! A computer printing one coordinate each microsecond would require 10 billion years—the age of the universe—to list the atomic coordinates. Somehow, among the 10^{23} atomic coordinates, or linear combinations of them, all but a few are macroscopically irrelevant. The pertinent few emerge as *macroscopic coordinates*, or "thermodynamic coordinates."

Like all sciences, thermodynamics is a description of the results to be obtained in particular types of measurements. The character of the contemplated measurements dictates the appropriate descriptive variables; these variables, in turn, ordain the scope and structure of thermodynamic theory.

The key to the simplicity of macroscopic description, and the criterion for the choice of thermodynamic coordinates, lies in two attributes of macroscopic measurement. *Macroscopic measurements are extremely slow on the atomic scale of time, and they are extremely coarse on the atomic scale of distance.*

While a macroscopic measurement is being made, the atoms of a system go through extremely rapid and complex motions. To measure the length of a bar of metal we might choose to calibrate it in terms of the wavelength of yellow light, devising some arrangement whereby reflection from the end of the bar produces interference fringes. These fringes are then to be photographed and counted. The duration of the measurement is determined by the shutter speed of the camera—typically on the order of one hundredth of a second. But the characteristic period of vibration of the atoms at the end of the bar is on the order of 10^{-15} seconds!

A macroscopic observation cannot respond to those myriads of atomic coordinates which vary in time with typical atomic periods. *Only those few particular combinations of atomic coordinates that are essentially time independent are macroscopically observable.*

The word *essentially* is an important qualification. In fact we are able to observe macroscopic processes that are almost, but not quite, time independent. With modest difficulty we might observe processes with time scales on the order of 10^{-7} s or less. Such observable processes are still enormously slow relative to the atomic scale of 10^{-15} s. It is rational then to first consider the *limiting* case and to erect a theory of time-independent phenomena. Such a theory is thermodynamics.

By definition, suggested by the nature of macroscopic observations, thermodynamics describes only static states of macroscopic systems.

Of all the 10^{23} atomic coordinates, or combinations thereof, only a few are time independent.

Quantities subject to conservation principles are the most obvious candidates as time-independent thermodynamic coordinates: the energy, each component of the total momentum, and each component of the total angular momentum of the system. But there are other time-independent thermodynamic coordinates, which we shall enumerate after exploring the *spatial* nature of macroscopic measurement.

1-2   THE SPATIAL NATURE OF MACROSCOPIC MEASUREMENTS

Macroscopic measurements are not only extremely slow on the atomic scale of time, but they are correspondingly coarse on the atomic scale of distance. We probe our system always with "blunt instruments." Thus an optical observation has a resolving power defined by the wavelength of light, which is on the order of 1000 interatomic distances. The smallest resolvable volume contains approximately 10^9 atoms! *Macroscopic observations sense only coarse spatial averages of atomic coordinates.*

The two types of averaging implicit in macroscopic observations together effect the enormous reduction in the number of pertinent variables, from the initial 10^{23} atomic coordinates to the remarkably small number of thermodynamic coordinates. The manner of reduction can be illustrated schematically by considering a simple model system, as shown in Fig. 1.1. The model system consists not of 10^{23} atoms, but of only 9. These atoms are spaced along a one-dimensional line, are constrained to

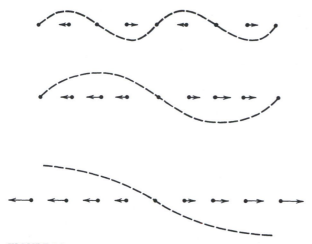

FIGURE 1.1

Three normal modes of oscillation in a nine-atom model system. The wave lengths of the three modes are four, eight and sixteen interatomic distances. The dotted curves are a transverse representation of the longitudinal displacements.

move only along that line, and interact by linear forces (as if connected by springs).

The motions of the individual atoms are strongly coupled, so the atoms tend to move in organized patterns called *normal modes*. Three such normal modes of motion are indicated schematically in Fig. 1.1. The arrows indicate the displacements of the atoms at a particular moment; the atoms oscillate back and forth, and half a cycle later all the arrows would be reversed.

Rather than describe the atomic state of the system by specifying the position of each atom, it is more convenient (and mathematically equivalent) to specify the instantaneous amplitude of each normal mode. These amplitudes are called *normal coordinates*, and the number of normal coordinates is exactly equal to the number of atomic coordinates.

In a "macroscopic" system composed of only nine atoms there is no precise distinction between "macroscopic" and "atomic" observations. For the purpose of illustration, however, we think of a macroscopic observation as a kind of "blurred" observation with low resolving power; the spatial coarseness of macroscopic measurements is qualitatively analogous to visual observation of the system through spectacles that are somewhat out of focus. Under such observation the fine structure of the first two modes in Fig. 1.1 is unresolvable, and these modes are rendered unobservable and macroscopically irrelevant. The third mode, however, corresponds to a relatively *homogeneous net expansion* (or contraction) of the whole system. Unlike the first two modes, it is easily observable through "blurring spectacles." The amplitude of this mode describes the length (or volume, in three dimensions) of the system. *The length (or volume*) *remains as a thermodynamic variable, undestroyed by the spatial averaging, because of its spatially homogeneous* (*long wavelength*) *structure.*

The *time* averaging associated with macroscopic measurements augments these considerations. Each of the normal modes of the system has a characteristic frequency, the frequency being smaller for modes of longer wavelength. The frequency of the third normal mode in Fig. 1.1 is the lowest of those shown, and if we were to consider systems with very large numbers of atoms, the frequency of the longest wavelength mode would approach zero (for reasons to be explored more fully in Chapter 21). Thus all the short wavelength modes are lost in the time averaging, but *the long wavelength mode corresponding to the "volume" is so slow that it survives the time averaging as well as the spatial averaging.*

This simple example illustrates a very general result. Of the enormous number of atomic coordinates, a very few, with unique symmetry properties, survive the statistical averaging associated with a transition to a macroscopic description. Certain of these surviving coordinates are mechanical in nature—they are volume, parameters descriptive of the shape (components of elastic strain), and the like. Other surviving coordinates are electrical in nature—they are electric dipole moments, magnetic dipole moments, various multipole moments, and the like. *The study of mechanics* (*including elasticity*) *is the study of one set of surviving coordinates. The subject of electricity* (*including electrostatics, magnetostatics, and ferromagnetism*) *is the study of another set of surviving coordinates.*

Thermodynamics, in contrast, is concerned with the macroscopic consequences of the myriads of atomic coordinates that, by virtue of the coarseness of macroscopic observations, do not appear explicitly in a macroscopic description of a system.

Among the many consequences of the "hidden" atomic modes of motion, the most evident is the ability of these modes to act as a repository for energy. Energy transferred via a "mechanical mode" (i.e., one associated with a mechanical macroscopic coordinate) is called *mechanical work*. Energy transferred via an "electrical mode" is called *electrical work*. Mechanical work is typified by the term $-P\,dV$ (P is pressure, V is volume), and electrical work is typified by the term $-E_e\,d\mathscr{P}$ (E_e is electric field, $\mathscr{P}$ is electric dipole moment). These energy terms and various other mechanical and electrical work terms are treated fully in the standard mechanics and electricity references. *But it is equally possible to transfer energy via the hidden atomic modes of motion as well as via those that happen to be macroscopically observable.* An energy transfer via the hidden atomic modes is called *heat*. Of course this descriptive characterization of heat is not a sufficient basis for the formal development of thermodynamics, and we shall soon formulate an appropriate operational definition.

With this contextual perspective we proceed to certain definitions and conventions needed for the theoretical development.

1-3 THE COMPOSITION OF THERMODYNAMIC SYSTEMS

Thermodynamics is a subject of great generality, applicable to systems of elaborate structure with all manner of complex mechanical, electrical, and thermal properties. We wish to focus our chief attention on the thermal properties. Therefore it is convenient to idealize and simplify the mechanical and electrical properties of the systems that we shall study initially. Similarly, in mechanics we consider uncharged and unpolarized systems; whereas in electricity we consider systems with no elastic compressibility or other mechanical attributes. The generality of either subject is not essentially reduced by this idealization, and after the separate content of each subject has been studied it is a simple matter to combine the theories to treat systems of simultaneously complicated electrical and mechanical properties. Similarly, in our study of thermodynamics we idealize our systems so that their mechanical and electrical properties are almost trivially simple. When the essential content of thermodynamics has thus been developed, it again is a simple matter to extend the analysis to systems with relatively complex mechanical and electrical structure. The essential point to be stressed is that the restrictions on the types of systems considered in the following several chapters are *not* basic limitations on the generality of thermodynamic theory but are adopted merely for simplicity of exposition.

We (temporarily) restrict our attention to *simple systems*, defined as *systems that are macroscopically homogeneous, isotropic, and uncharged, that are large enough so that surface effects can be neglected, and that are not acted on by electric, magnetic, or gravitational fields*.

For such a simple system there are no macroscopic electric coordinates whatsoever. The system is uncharged and has neither electric nor magnetic dipole, quadrupole, or higher-order moments. All elastic shear components and other such mechanical parameters are zero. The volume V does remain as a relevant mechanical parameter. Furthermore, a simple system has a definite *chemical composition* which must be described by an appropriate set of parameters. One reasonable set of composition parameters is the numbers of molecules in each of the chemically pure components of which the system is a mixture. Alternatively, to obtain numbers of more convenient size, we adopt the *mole numbers*, defined as the actual number of each type of molecule divided by Avogadro's number ($N_A = 6.02217 \times 10^{23}$).

This definition of the mole number refers explicitly to the "number of molecules," and it therefore lies outside the boundary of purely macroscopic physics. An equivalent definition which avoids the reference to molecules simply designates 12 grams as the molar mass of the isotope ^{12}C. The molar masses of other isotopes are then defined to stand in the same ratio as the conventional "atomic masses," a partial list of which is given in Table 1.1.

TABLE 1.1
**Atomic Masses (g) of Some Naturally
Occurring Elements (Mixtures of Isotopes)**[a]

H	1.0080	F	18.9984
Li	6.941	Na	22.9898
C	12.011	Al	26.9815
N	14.0067	S	32.06
O	15.9994	Cl	35.453

[a] As adopted by the International Union of Pure and Applied Chemistry, 1969.

If a system is a mixture of r chemical components, the r ratios $N_k/(\sum_{j=1}^r N_j)$ ($k = 1, 2, \ldots, r$) are called the *mole fractions*. The sum of all r mole fractions is unity. The quantity $V/(\sum_{j=1}^r N_j)$ is called the *molar volume*.

The macroscopic parameters $V, N_1, N_2, \ldots, N_r$ have a common property that will prove to be quite significant. Suppose that we are given two identical systems and that we now regard these two systems taken together as a single system. The value of the volume for the composite system is then just twice the value of the volume for a single subsystem. Similarly, each of the mole numbers of the composite system is twice that for a single subsystem. Parameters that have values in a composite system equal to the sum of the values in each of the subsystems are called *extensive* parameters. Extensive parameters play a key role throughout thermodynamic theory.

PROBLEMS

1.3-1. One tenth of a kilogram of NaCl and 0.15 kg of sugar ($C_{12}H_{22}O_{11}$) are dissolved in 0.50 kg of pure water. The volume of the resultant thermodynamic system is 0.55×10^{-3} m³. What are the mole numbers of the three components of the system? What are the mole fractions? What is the molar volume of the system? It is sufficient to carry the calculations only to two significant figures.

Answer:
Mole fraction of NaCl = 0.057;
molar volume = 18×10^{-6} m³/mole.

1.3-2. Naturally occurring boron has an atomic mass of 10.811 g. It is a mixture of the isotopes ^{10}B with an atomic mass of 10.0129 g and ^{11}B with an atomic mass of 11.0093 g. What is the mole fraction of ^{10}B in the mixture?

1.3-3. Twenty cubic centimeters each of ethyl alcohol (C_2H_5OH; density = 0.79 g/cm³), methyl alcohol (CH_3OH; density = 0.81 g/cm³), and water (H_2O;

density = 1 g/cm³) are mixed together. What are the mole numbers and mole fractions of the three components of the system?

> *Answer:*
> mole fractions = 0.17, 0.26, 0.57

1.3-4. A 0.01 kg sample is composed of 50 molecular percent H_2, 30 molecular percent HD (hydrogen deuteride), and 20 molecular percent D_2. What additional mass of D_2 must be added if the mole fraction of D_2 in the final mixture is to be 0.3?

1.3-5. A solution of sugar $(C_{12}H_{22}O_{11})$ in water is 20% sugar by weight. What is the mole fraction of sugar in the solution?

1.3-6. An aqueous solution of an unidentified solute has a total mass of 0.1029 kg. The mole fraction of the solute is 0.1. The solution is diluted with 0.036 kg of water, after which the mole fraction of the solute is 0.07. What would be a reasonable guess as to the chemical identity of the solute?

1.3-7. One tenth of a kg of an aqueous solution of HCl is poured into 0.2 kg of an aqueous solution of NaOH. The mole fraction of the HCl solution was 0.1, whereas that of the NaOH solution was 0.25. What are the mole fractions of each of the components in the solution after the chemical reaction has come to completion?

> *Answer:*
> $x_{H_2O} = N_{H_2O}/N = 0.84$

1-4 THE INTERNAL ENERGY

The development of the principle of conservation of energy has been one of the most significant achievements in the evolution of physics. The present form of the principle was not discovered in one magnificent stroke of insight but was slowly and laboriously developed over two and a half centuries. The first recognition of a conservation principle, by Leibniz in 1693, referred only to the sum of the kinetic energy $(\frac{1}{2} mv^2)$ and the potential energy (mgh) of a simple mechanical mass point in the terrestrial gravitational field. As additional types of systems were considered the established form of the conservation principle repeatedly failed, but in each case it was found possible to revive it by the addition of a new mathematical term—a "new kind of energy." Thus consideration of charged systems necessitated the addition of the *Coulomb interaction energy* (Q_1Q_2/r) and eventually of the energy of the electromagnetic field. In 1905 Einstein extended the principle to the relativistic region, adding such terms as the relativistic rest-mass energy. In the 1930s Enrico Fermi postulated the existence of a new particle called the *neutrino* solely for the purpose of retaining the energy conservation principle in nuclear reactions. The principle of energy conservation is now seen as a reflection of the (presumed) fact that the fundamental laws of physics are the same today as they were eons ago, or as they will be in the remote future; the laws of physics are unaltered by a shift in the scale of time ($t \rightarrow t +$ constant). Of this basis for energy conservation we shall have more to say in Chapter 21. Now we simply note that the energy conservation principle is one of the most fundamental, general, and significant principles of physical theory.

Viewing a macroscopic system as an agglomerate of an enormous number of electrons and nuclei, interacting with complex but definite forces to which the energy conservation principle applies, we conclude that *macroscopic systems have definite and precise energies, subject to a definite conservation principle*. That is, we now accept the existence of a well-defined energy of a thermodynamic system as a macroscopic manifestation of a conservation law, highly developed, tested to an extreme precision, and apparently of complete generality at the atomic level.

The foregoing justification of the existence of a thermodynamic energy function is quite different from the historical thermodynamic method. Because thermodynamics was developed largely before the atomic hypothesis was accepted, the existence of a conservative macroscopic energy function had to be demonstrated by purely macroscopic means. A significant step in that direction was taken by Count Rumford in 1798 as he observed certain thermal effects associated with the boring of brass cannons. Sir Humphry Davy, Sadi Carnot, Robert Mayer, and, finally (between 1840 and 1850), James Joule carried Rumford's initial efforts to their logical fruition. The history of the concept of heat as a form of energy transfer is unsurpassed as a case study in the tortuous development of scientific theory, as an illustration of the almost insuperable inertia presented by accepted physical doctrine, and as a superb tale of human ingenuity applied to a subtle and abstract problem. The interested reader is referred to *The Early Development of the Concepts of Temperature and Heat* by D. Roller (Harvard University Press, 1950) or to any standard work on the history of physics.

Although we shall not have recourse explicitly to the experiments of Rumford and Joule in order to justify our postulate of the *existence* of an energy function, we make reference to them in Section 1.7 in our discussion of the *measurability* of the thermodynamic energy.

Only differences of energy, rather than absolute values of the energy, have physical significance, either at the atomic level or in macroscopic systems. It is conventional therefore to adopt some particular state of a system as a fiducial state, the energy of which is arbitrarily taken as zero. The energy of a system in any other state, relative to the energy of the system in the fiducial state, is then called the thermodynamic *internal energy* of the system in that state and is denoted by the symbol U. *Like the volume and the mole numbers, the internal energy is an extensive parameter.*

1-5 THERMODYNAMIC EQUILIBRIUM

Macroscopic systems often exhibit some "memory" of their recent history. A stirred cup of tea continues to swirl within the cup. Cold-worked steel maintains an enhanced hardness imparted by its mechanical treatment. But memory eventually fades. Turbulences damp out, internal strains yield to plastic flow, concentration inhomogeneities diffuse to uniformity. Systems tend to subside to very simple states, independent of their specific history.

In some cases the evolution toward simplicity is rapid; in other cases it can proceed with glacial slowness. But *in all systems there is a tendency to evolve toward states in which the properties are determined by intrinsic factors and not by previously applied external influences. Such simple terminal states are, by definition, time independent. They are called equilibrium states.*

Thermodynamics seeks to describe these simple, static "equilibrium" states to which systems eventually evolve.

To convert this statement to a formal and precise postulate we first recognize that an appropriate criterion of simplicity is the possibility of description in terms of a small number of variables. It therefore seems plausible to adopt the following postulate, suggested by experimental observation and formal simplicity, and to be verified ultimately by the success of the derived theory:

Postulate I. *There exist particular states (called equilibrium states) of simple systems that, macroscopically, are characterized completely by the internal energy U, the volume V, and the mole numbers $N_1, N_2, \ldots, N_r$ of the chemical components.*

As we expand the generality of the systems to be considered, eventually permitting more complicated mechanical and electrical properties, the number of parameters required to characterize an equilibrium state increases to include, for example, the electric dipole moment and certain elastic strain parameters. These new variables play roles in the formalism which are completely analogous to the role of the volume V for a simple system.

A persistent problem of the experimentalist is to determine somehow whether a given system actually is in an equilibrium state, to which thermodynamic analysis can be applied. He or she can, of course, observe whether the system is static and quiescent. But quiescence is not sufficient. As the state is assumed to be characterized completely by the extensive parameters, $U, V, N_1, N_2, \ldots, N_r$, it follows that the properties of the system must be independent of the past history. This is hardly an operational prescription for the recognition of an equilibrium state, but in certain cases this independence of the past history is obviously *not* satisfied, and these cases give some insight into the significance of equilibrium. Thus two pieces of chemically identical commercial steel may have very different properties imparted by cold-working, heat treatment, quenching, and annealing in the manufacturing process. Such systems are clearly not in equilibrium. Similarly, the physical characteristics of glass depend upon the cooling rate and other details of its manufacture; hence glass is not in equilibrium.

If a system that is not in equilibrium is analyzed on the basis of a thermodynamic formalism predicated on the supposition of equilibrium, inconsistencies appear in the formalism and predicted results are at variance with experimental observations. This failure of the theory is used by the experimentalist as an a posteriori criterion for the detection of nonequilibrium states.

In those cases in which an unexpected inconsistency arises in the thermodynamic formalism a more incisive quantum statistical theory usually provides valid reasons for the failure of the system to attain equilibrium. The occasional theoretical discrepancies that arise are therefore of great heuristic value in that they call attention to some unsuspected complication in the molecular mechanisms of the system. Such circumstances led to the discovery of ortho- and parahydrogen,[1] and to the understanding of the molecular mechanism of conversion between the two forms.

From the atomic point of view, the macroscopic equilibrium state is associated with incessant and rapid transitions among all the atomic states consistent with the given boundary conditions. If the transition mechanism among the atomic states is sufficiently effective, the system passes rapidly through all representative atomic states in the course of a macroscopic observation; such a system is in equilibrium. However, under certain unique conditions, the mechanism of atomic transition may be ineffective and the system may be trapped in a small subset of atypical atomic states. Or even if the system is not completely trapped the rate of transition may be so slow that a macroscopic measurement does not yield a proper average over all possible atomic states. In these cases the system is not in equilibrium. It is readily apparent that such situations are most likely to occur in solid rather than in fluid systems, for the comparatively high atomic mobility in fluid systems and the random nature of the

---

[1] If the two nuclei in a H_2 molecule have parallel angular momentum, the molecule is called ortho-H_2; if antiparallel, para-H_2. The ratio of ortho-H_2 to para-H_2 in a gaseous H_2 system should have a definite value in equilibrium, but this ratio may not be obtained under certain conditions. The resultant failure of H_2 to satisfy certain thermodynamic equations motivated the investigations of the ortho- and para-forms of H_2.

interatomic collisions militate strongly against any restrictions of the atomic transition probabilities.

In actuality, few systems are in absolute and true equilibrium. In absolute equilibrium all radioactive materials would have decayed completely and nuclear reactions would have transmuted all nuclei to the most stable of isotopes. Such processes, which would take cosmic times to complete, generally can be ignored. A system that has completed the relevant processes of spontaneous evolution, and that can be described by a reasonably small number of parameters, can be considered to be in *metastable equilibrium*. Such a limited equilibrium is sufficient for the application of thermodynamics.

In practice the criterion for equilibrium is circular. *Operationally, a system is in an equilibrium state if its properties are consistently described by thermodynamic theory*!

It is important to reflect upon the fact that the circular character of thermodynamics is *not* fundamentally different from that of mechanics. A particle of known mass in a known gravitational field might be expected to move in a specific trajectory; if it does not do so we do not reject the theory of mechanics, but we simply conclude that some additional force acts on the particle. Thus the existence of an electrical charge on the particle, and the associated relevance of an electrical force, cannot be known a priori. It is inferred only by circular reasoning, in that dynamical predictions are incorrect unless the electric contribution to the force is included. Our model of a mechanical system (including the assignment of its mass, moment of inertia, charge, dipole moment, etc.) is "correct" if it yields successful predictions.

1-6 WALLS AND CONSTRAINTS

A description of a thermodynamic system requires the specification of the "walls" that separate it from the surroundings and that provide its boundary conditions. It is by means of manipulations of the walls that the extensive parameters of the system are altered and processes are initiated.

The processes arising by manipulations of the walls generally are associated with a redistribution of some quantity among various systems or among various portions of a single system. A formal classification of thermodynamic walls accordingly can be based on the property of the walls in permitting or preventing such redistributions. As a particular illustration, consider two systems separated by an internal piston within a closed, rigid cylinder. If the position of the piston is rigidly fixed the "wall" prevents the redistribution of volume between the two systems, but if the piston is left free such a redistribution is permitted. The cylinder and the rigidly fixed piston may be said to constitute a wall *restrictive* with respect to the volume, whereas the cylinder and the movable piston may be said to constitute a wall *nonrestrictive* with respect to the volume. In general, a wall that constrains an extensive parameter of a system to have a definite and particular value is said to be restrictive with respect to that parameter, whereas a wall that permits the parameter to change freely is said to be nonrestrictive with respect to that parameter.

A wall that is impermeable to a particular chemical component is restrictive with respect to the corresponding mole number; whereas a permeable membrane is nonrestrictive with respect to the mole number. Semipermeable membranes are restrictive with respect to certain mole numbers and nonrestrictive with respect to others. A wall with holes in it is nonrestrictive with respect to all mole numbers.

The existence of walls that are restrictive with respect to the energy is associated with the larger problem of measurability of the energy, to which we now turn our attention.

1-7 MEASURABILITY OF THE ENERGY

On the basis of atomic considerations, we have been led to accept the existence of a macroscopic conservative energy function. In order that this energy function may be meaningful in a practical sense, however, we must convince ourselves that it is macroscopically *controllable* and *measurable*. We shall now show that practical methods of measurement of the energy do exist, and in doing so we shall also be led to a quantitative operational definition of heat.

An essential prerequisite for the measurability of the energy is the existence of walls that do not permit the transfer of energy in the form of heat. We briefly examine a simple experimental situation that suggests that such walls do indeed exist.

Consider a system of ice and water enclosed in a container. We find that the ice can be caused to melt rapidly by stirring the system vigorously. By stirring the system we are clearly transferring energy to it mechanically, so that we infer that the melting of the ice is associated with an input of energy to the system. If we now observe the system on a summer day, we find that the ice spontaneously melts despite the fact that no work is done on the system. It therefore seems plausible that energy is being transferred to the system in the form of heat. We further observe that the rate of melting of the ice is progressively decreased by changing the wall surrounding the system from thin metal sheet, to thick glass, and thence to a Dewar wall (consisting of two silvered glass sheets separated by an evacuated interspace). This observation strongly suggests that the metal, glass, and Dewar walls are progressively less permeable to the flow of heat. The ingenuity of experimentalists has produced walls that are able to reduce the melting rate of the ice to a negligible value, and such walls are correspondingly excellent approximations to the limiting idealization of a wall that is truly impermeable to the flow of heat.

It is conventional to refer to a wall that is impermeable to the flow of heat as *adiabatic*; whereas a wall that permits the flow of heat is termed *diathermal*. If a wall allows the flux of neither work nor heat, it is *restrictive with respect to the energy*. A system enclosed by a wall that is restrictive with respect to the energy, volume, and all the mole numbers is said to be *closed*.[2]

The existence of these several types of walls resolves the first of our concerns with the thermodynamic energy. That is, these walls demonstrate that the energy is macroscopically *controllable*. It can be trapped by restrictive walls and manipulated by diathermal walls. If the energy of a system is measured today, and if the system is enclosed by a wall restrictive with respect to the energy, we can be certain of the energy of the system tomorrow. Without such a wall the concept of a macroscopic thermodynamic energy would be purely academic.

We can now proceed to our second concern—that of *measurability* of the energy. More accurately, we are concerned with the measurability of energy *differences*, which alone have physical significance. Again we invoke the existence of adiabatic walls, and we note that for a simple system enclosed by an impermeable adiabatic wall the only type of permissible energy transfer is in the form of work. The theory of mechanics provides us with quantitative formulas for its measurement. If the work is done by compression, displacing a piston in a cylinder, the work is the product of force times displacement; or if the work is done by stirring, it is the product of the torque times the angular rotation of the stirrer shaft. In either case, the work is well defined and measurable by the theory of mechanics. We conclude that we are able to measure the energy difference of two states *provided that one state can be reached from the other by some mechanical process while the system is enclosed by an adiabatic impermeable wall.*

The entire matter of controllability and measurability of the energy can be succinctly stated as follows: *There exist walls, called adiabatic, with the property that the work done in taking an adiabatically enclosed system between two given states is determined entirely by the states, independent of all external conditions. The work done is the difference in the internal energy of the two states.*

As a specific example suppose we are given an equilibrium system composed of ice and water enclosed in a rigid adiabatic impermeable wall. Through a small hole in this wall we pass a thin shaft carrying a propellor blade at the inner end and a crank handle at the outer end. By turning the crank handle we can do work on the system. The work done is equal to the angular rotation of the shaft multiplied by the viscous torque. After turning the shaft for a definite time the system is allowed to come to a new equilibrium state in which some definite amount of the ice is observed

---

[2] This definition of closure differs from a usage common in chemistry, in which closure implies only a wall restrictive with respect to the transfer of matter.

to have been melted. The difference in energy of the final and initial states is equal to the work that we have done in turning the crank.

We now inquire about the possibility of starting with some arbitrary given state of a system, of enclosing the system in an adiabatic impermeable wall, and of then being able to contrive some mechanical process that will take the system to another arbitrarily specified state. To determine the existence of such processes, we must have recourse to experimental observation, and it is here that the great classical experiments of Joule are relevant. His work can be interpreted as demonstrating that *for a system enclosed by an adiabatic impermeable wall any two equilibrium states with the same set of mole numbers $N_1, N_2, \ldots, N_r$ can be joined by some possible mechanical process*. Joule discovered that if two states (say A and B) are specified it may *not* be possible to find a mechanical process (consistent with an adiabatic impermeable wall) to take the system *from A to B* but that it is always possible to find *either* a process to take the system from A to B *or* a process to take the system from B to A. That is, for any states A and B with equal mole numbers, either the adiabatic mechanical process $A \to B$ or $B \to A$ exists. For our purposes either of these processes is satisfactory. Experiment thus shows that *the methods of mechanics permit us to measure the energy difference of any two states with equal mole numbers*.

Joule's observation that only one of the processes $A \to B$ or $B \to A$ may exist is of profound significance. This asymmetry of two given states is associated with the concept of *irreversibility*, with which we shall subsequently be much concerned.

The only remaining limitation to the measurability of the energy difference of any two states is the requirement that the states must have equal mole numbers. This restriction is easily eliminated by the following observation. Consider two simple subsystems separated by an impermeable wall and assume that the energy of each subsystem is known (relative to appropriate fiducial states, of course). If the impermeable wall is removed, the subsystems will intermix, but the total energy of the composite system will remain constant. Therefore the energy of the final mixed system is known to be the sum of the energies of the original subsystems. This technique enables us to relate the energies of states with different mole numbers.

In summary, we have seen that *by employing adiabatic walls and by measuring only mechanical work, the energy of any thermodynamic system, relative to an appropriate reference state, can be measured*.

1-8 QUANTITATIVE DEFINITION OF HEAT—UNITS

The fact that the energy difference of any two equilibrium states is measurable provides us directly with a quantitative definition of the heat: *The heat flux to a system in any process (at constant mole numbers) is simply the difference in internal energy between the final and initial states, diminished by the work done in that process.*

Consider some specified process that takes a system from the initial state A to the final state B. We wish to know the amount of energy transferred to the system in the form of work and the amount transferred in the form of heat in that particular process. The work is easily measured by the method of mechanics. Furthermore, the total energy difference $U_B - U_A$ is measurable by the procedures discussed in Section 1.7. Subtracting the work from the total energy difference gives us the heat flux in the specified process.

It should be noted that the amount of work associated with different processes may be different, even though each of the processes initiates in the same state A and each terminates in the same state B. Similarly, the heat flux may be different for each of the processes. But the sum of the work and heat fluxes is just the total energy difference $U_B - U_A$ and is the same for each of the processes. In referring to the total energy flux we therefore need specify only the initial and terminal states, but in referring to heat or work fluxes we must specify in detail the process considered.

Restricting our attention to thermodynamic simple systems, the quasi-static work is associated with a change in volume and is given quantitatively by

$$dW_M = -P\,dV \tag{1.1}$$

where P is the pressure. In recalling this equation from mechanics, we stress that the equation applies only to *quasi-static processes*. A precise definition of quasi-static processes will be given in Section 4.2, but now we merely indicate the essential qualitative idea of such processes. Let us suppose that we are discussing, as a particular system, a gas enclosed in a cylinder fitted with a moving piston. If the piston is pushed in very rapidly, the gas immediately behind the piston acquires kinetic energy and is set into turbulent motion and the pressure is not well defined. In such a case the work done on the system is not quasi-static and is not given by equation 1.1. If, however, the piston is pushed in at a vanishingly slow rate (quasi-statically), the system is at every moment in a quiescent equilibrium state, and equation 1.1 then applies. The "infinite slowness" of the process is, roughly, the essential feature of a quasi-static process.

A second noteworthy feature of equation 1.1 is the sign convention. The work is taken to be positive if it increases the energy of the system. If the volume of the system is decreased, work is done on the system, increasing its energy; hence the negative sign in equation 1.1.

With the quantitative expression $dW_M = -P\,dV$ for the quasi-static work, we can now give a quantitative expression for the heat flux. In an infinitesimal quasi-static process at constant mole numbers the *quasi-static heat dQ* is defined by the equation

$$dQ = dU - dW_M \quad \text{at constant mole numbers} \tag{1.2}$$

or

$$dQ = dU + P \, dV \quad \text{at constant mole numbers} \qquad (1.3)$$

It will be noted that we use the terms *heat* and *heat flux* interchangeably. Heat, like work, is only a form of energy *transfer*. Once energy is transferred to a system, either as heat or as work, it is indistinguishable from energy that might have been transferred differently. Thus, although dQ and dW_M add together to give dU, the energy U of a state *cannot* be considered as the sum of "work" and "heat" components. To avoid this implication we put a stroke through the symbol d: infinitesimals such as dW_M and dQ are called *imperfect differentials*. The integrals of dW_M and dQ for a particular process are the work and heat fluxes *in that process*; the sum is the energy difference ΔU, which alone is independent of the process.

The concepts of heat, work, and energy may possibly be clarified in terms of a simple analogy. A certain farmer owns a pond, fed by one stream and drained by another. The pond also receives water from an occasional rainfall and loses it by evaporation, which we shall consider as "negative rain." In this analogy the pond is our system, the water within it is the internal energy, water transferred by the streams is work, and water transferred as rain is heat.

The first thing to be noted is that no examination of the pond at any time can indicate how much of the water within it came by way of the stream and how much came by way of rain. The term *rain* refers only to a method of water *transfer*.

Let us suppose that the owner of the pond wishes to measure the amount of water in the pond. He can purchase flow meters to be inserted in the streams, and with these flow meters he can measure the amount of stream water entering and leaving the pond. But he cannot purchase a rain meter. However, he can throw a tarpaulin over the pond, enclosing the pond in a wall impermeable to rain (an *adiabatic wall*). The pond owner consequently puts a vertical pole into the pond, covers the pond with his tarpaulin, and inserts his flow meters into the streams. By damming one stream and then the other, he varies the level in the pond at will, and by consulting his flow meters he is able to calibrate the pond level, as read on his vertical stick, with total water content (U). Thus, by carrying out processes on the system enclosed by an adiabatic wall, he is able to measure the total water content of any state of his pond.

Our obliging pond owner now removes his tarpaulin to permit rain as well as stream water to enter and leave the pond. He is then asked to evaluate the amount of rain entering his pond during a particular day. He proceeds simply; he reads the difference in water content from his vertical stick, and from this he deducts the total flux of stream water as registered by his flow meters. The difference is a quantitative measure of the rain. The strict analogy of each of these procedures with its thermodynamic counterpart is evident.

Since work and heat refer to particular modes of energy transfer, each is measured in energy units. In the cgs system the unit of energy, and hence of work and heat, is the erg. In the mks system the unit of energy is the joule, or 10^7 ergs.

A practical unit of energy is the calorie,[3] or 4.1858 J. Historically, the calorie was introduced for the measurement of heat flux before the relationship of heat and work was clear, and the prejudice toward the use of the calorie for heat and of the joule for work still persists. Nevertheless, the calorie and the joule are simply alternative units of energy, either of which is acceptable whether the energy flux is work, heat, or some combination of both.

Other common units of energy are the British thermal unit (Btu), the liter–atmosphere, the foot–pound and the watt–hour. Conversion factors among energy units are given inside the back cover of this book.

Example 1

A particular gas is enclosed in a cylinder with a moveable piston. It is observed that if the walls are adiabatic, a quasi-static increase in volume results in a decrease in pressure according to the equation

$$P^3 V^5 = \text{constant} \qquad (\text{for } Q = 0)$$

a) Find the quasi-static work done on the system and the net heat transfer to the system in each of the three processes (*ADB*, *ACB*, and the direct linear process *AB*) as shown in the figure.

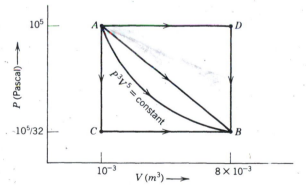

In the process *ADB* the gas is heated at constant pressure ($P = 10^5$ Pa) until its volume increases from its initial value of 10^{-3} m^3 to its final value of 8×10^{-3} m^3. The gas is then cooled at constant volume until its pressure decreases to $10^5/32$ Pa. The other processes (*ACB* and *AB*) can be similarly interpreted, according to the figure.

---

[3] Nutritionists refer to a kilocalorie as a "Calorie"—presumably to spare calorie counters the trauma of large numbers. To compound the confusion the initial capital C is often dropped, so that a kilocalorie becomes a "calorie"!

b) A small paddle is installed inside the system and is driven by an external motor (by means of a magnetic coupling through the cylinder wall). The motor exerts a torque, driving the paddle at an angular velocity ω, and the pressure of the gas (at constant volume) is observed to increase at a rate given by

$$\frac{dP}{dt} = \frac{2}{3}\frac{\omega}{V} \times \text{torque}$$

Show that the energy difference of any two states of equal volumes can be determined by this process. In particular, evaluate $U_C - U_A$ and $U_D - U_B$.

Explain why this process can proceed only in one direction (vertically upward rather than downward in the $P-V$ plot).

c) Show that *any* two states (any two points in the $P-V$ plane) can be connected by a combination of the processes in (*a*) and (*b*). In particular, evaluate $U_D - U_A$.

d) Calculate the work W_{AD} in the process $A \rightarrow D$. Calculate the heat transfer Q_{AD}. Repeat for $D \rightarrow B$, and for $C \rightarrow A$. Are these results consistent with those of (*a*)?

The reader should attempt to solve this problem *before* reading the following solution!

Solution

a) Given the equation of the "adiabat" (for which $Q = 0$ and $\Delta U = W$), we find

$$U_B - U_A = W_{AB} = -\int_{V_A}^{V_B} P\, dV = -P_A \int_{V_A}^{V_B}\left(\frac{V_A}{V}\right)^{5/3} dV$$

$$= \frac{3}{2}P_A V_A^{5/3}\left(V_B^{-2/3} - V_A^{-2/3}\right)$$

$$= \frac{3}{2}(25 - 100) = -112.5 \text{ J}$$

Now consider process ADB:

$$W_{ADB} = -\int P\, dV = -10^5 \times (8 \times 10^{-3} - 10^{-3}) = -700 \text{ J}$$

But

$$U_B - U_A = W_{ADB} + Q_{ADB}$$

$$Q_{ADB} = -112.5 + 700 = 587.5 \text{ J}$$

Note that we are able to calculate Q_{ADB}, but not Q_{AD} and Q_{DB} separately, for we do not (yet) know $U_D - U_A$.

Similarly we find $W_{ACB} = -21.9$ J and $Q_{ACB} = -90.6$ J. Also $W_{AB} = -360.9$ J and $Q_{AB} = 248.4$ J.

b) As the motor exerts a torque, and turns through an angle $d\theta$, it delivers an energy[4] dU = torque $\times$ $d\theta$ to the system. But $d\theta = \omega\, dt$, so that

$$dP = \frac{2}{3}\frac{1}{V}\,(\text{torque})\,\omega\,dt$$

$$= \frac{2}{3}\frac{1}{V}\,dU$$

or

$$dU = \frac{3}{2}V\,dP$$

This process is carried out at constant V and furthermore $dU \geq 0$ (and consequently $dP \geq 0$). The condition $dU \geq 0$ follows from $dU =$ torque $\times$ $d\theta$, for the sign of the rotation $d\theta$ is the same as the sign of the torque that induces that rotation. In particular

$$U_A - U_C = \frac{3}{2}V(P_A - P_C) = \frac{3}{2}\times 10^{-3}\times\left(10^5 - \frac{1}{32}\times 10^5\right) = 145.3\text{ J}$$

and

$$U_D - U_B = \frac{3}{2}V(P_D - P_B) = \frac{3}{2}\times 8\times 10^{-3}\times\left(10^5 - \frac{1}{32}\times 10^5\right) = 1162.5\text{ J}$$

c) To connect any two points in the plane we draw an adiabat through one and an isochor (V = constant) through the other. These two curves intersect, thereby connecting the two states. Thus we have found (using the adiabatic process) that $U_B - U_A = -112.5$ J and (using the irreversible stirrer process) that $U_D - U_B = 1162.5$ J. Therefore $U_D - U_A = 1050$ J. Equivalently, if we assign the value zero to U_A then

$$U_A = 0, \qquad U_B = -112.5\text{ J}, \qquad U_C = -145.3\text{ J}, \qquad U_D = 1050\text{ J}$$

and similarly every state can be assigned a value of U.

d) Now having $U_D - U_A$ and W_{AD} we can calculate Q_{AD}.

$$U_D - U_A = W_{AD} + Q_{AD}$$

$$1050 = -700 + Q_{AD}$$

$$Q_{AD} = 1750\text{ J}$$

Also

$$U_B - U_D = W_{DB} + Q_{DB}$$

or

$$-1162.5 = 0 + Q_{DB}$$

To check, we note that $Q_{AD} + Q_{DB} = 587.5$ J, which is equal to Q_{ADB} as found in (a).

---

[4] Note that the energy output of the motor is delivered to the system as energy that cannot be classified either as work or as heat—it is a *non-quasi-static* transfer of energy.

PROBLEMS

1.8-1. For the system considered in Example 1, calculate the energy of the state with $P = 5 \times 10^4$ Pa and $V = 8 \times 10^{-3}$ m³.

1.8-2. Calculate the heat transferred to the system considered in Example 1 in the process in which it is taken in a straight line (on the $P-V$ diagram) from the state A to the state referred to in the preceding problem.

1.8-3. For a particular gaseous system it has been determined that the energy is given by

$$U = 2.5PV + \text{constant}$$

The system is initially in the state $P = 0.2$ MPa (mega-Pascals), $V = 0.01$ m³; designated as point A in the figure. The system is taken through the cycle of three processes ($A \rightarrow B$, $B \rightarrow C$, and $C \rightarrow A$) shown in the figure. Calculate Q and W for each of the three processes. Calculate Q and W for a process from A to B along the parabola $P = 10^5 + 10^9 \times (V - .02)^2$.

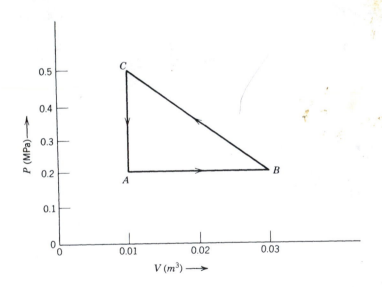

Answer:
$$W_{BC} = 7 \times 10^3 \text{ J}; \quad Q_{BC} = -9.5 \times 10^3 \text{ J}$$

1.8-4. For the system of Problem 1.8-3 find the equation of the adiabats in the $P-V$ plane (i.e., find the form of the curves $P = P(V)$ such that $dQ = 0$ along the curves).

Answer:
$$V^7 P^5 = \text{constant}$$

1.8-5. The energy of a particular system, of one mole, is given by

$$U = AP^2V$$

where A is a positive constant of dimensions $[P]^{-1}$. Find the equation of the adiabats in the $P-V$ plane.

1.8-6. For a particular system it is found that if the volume is kept constant at the value V_0 and the pressure is changed from P_0 to an arbitrary pressure P', the heat transfer to the system is

$$Q' = A(P' - P_0) \quad (A > 0)$$

In addition it is known that the adiabats of the system are of the form

$$PV^\gamma = \text{constant} \ (\gamma \text{ a positive constant})$$

Find the energy $U(P, V)$ for an arbitrary point in the $P-V$ plane, expressing $U(P, V)$ in terms of P_0, V_0, A, $U_0 \equiv U(P_0, V_0)$ and γ (as well as P and V).

Answer:
$$U - U_0 = A(Pr^\gamma - P_0) + [PV/(\gamma - 1)](1 - r^{\gamma - 1}) \quad \text{where } r \equiv V/V_0$$

1.8-7. Two moles of a particular single-component system are found to have a dependence of internal energy U on pressure and volume given by

$$U = APV^2 \quad (\text{for } N = 2)$$

Note that doubling the system doubles the volume, energy, and mole number, but leaves the pressure unaltered. Write the complete dependence of U on P, V, and N for arbitrary mole number.

1-9   THE BASIC PROBLEM OF THERMODYNAMICS

The preliminaries thus completed, we are prepared to formulate first the seminal problem of thermodynamics and then its solution.

Surveying those preliminaries retrospectively, it is remarkable how far reaching and how potent have been the consequences of the mere choice of thermodynamic coordinates. Identifying the criteria for those coordinates revealed the role of measurement. The distinction between the macroscopic coordinates and the incoherent atomic coordinates suggested the distinction between work and heat. The completeness of the description by the thermodynamic coordinates defined equilibrium states. The thermodynamic coordinates will now provide the framework for the solution of the central problem of thermodynamics.

There is, in fact, one central problem that defines the core of thermodynamic theory. All the results of thermodynamics propagate from its solution.

The single, all-encompassing problem of thermodynamics is the determination of the equilibrium state that eventually results after the removal of internal constraints in a closed, composite system.

Let us suppose that two simple systems are contained within a closed cylinder, separated from each other by an internal piston. Assume that the cylinder walls and the piston are rigid, impermeable to matter, and adiabatic and that the position of the piston is firmly fixed. Each of the systems is closed. If we now free the piston, it will, in general, seek some new position. Similarly, if the adiabatic coating is stripped from the fixed piston, so that heat can flow between the two systems, there will be a redistribution' of energy between the two systems. Again, if holes are punched in the piston, there will be a redistribution of matter (and also of energy) between the two systems. The removal of a constraint in each case results in the onset of some spontaneous process, and when the systems finally settle into new equilibrium states they do so with new values of the parameters $U^{(1)}, V^{(1)}, N_1^{(1)} \cdots$ and $U^{(2)}, V^{(2)}, N_1^{(2)} \cdots$. The basic problem of thermodynamics is the calculation of the equilibrium values of these parameters.

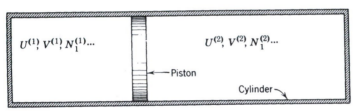

FIGURE 1.2

Before formulating the postulate that provides the means of solution of the problem, we rephrase the problem in a slightly more general form without reference to such special devices as cylinders and pistons. Given two or more simple systems, they may be considered as constituting a single *composite* system. The composite system is termed *closed* if it is surrounded by a wall that is restrictive with respect to the total energy, the total volume, and the total mole numbers of each component of the composite system. The individual simple systems within a closed composite system need not themselves be closed. Thus, in the particular example referred to, the composite system is closed even if the internal piston is free to move or has holes in it. Constraints that prevent the flow of energy, volume, or matter among the simple systems constituting the composite system are known as *internal* constraints. If a closed composite system is in equilibrium with respect to internal constraints, and if some of these constraints are then removed, certain previously disallowed processes become permissible. These processes bring the system to a new equilibrium state. Prediction of the new equilibrium state is the central problem of thermodynamics.

1.10 THE ENTROPY MAXIMUM POSTULATES

The induction from experimental observation of the central principle that provides the solution of the basic problem is subtle indeed. The historical method, culminating in the analysis of Caratheodory, is a tour de force of delicate and formal logic. The statistical mechanical approach pioneered by Josiah Willard Gibbs required a masterful stroke of inductive inspiration. The symmetry-based foundations to be developed in Chapter 21 will provide retrospective understanding and interpretation, but they are not yet formulated as a deductive basis. We therefore merely formulate the solution to the basic problem of thermodynamics in a set of postulates depending upon a posteriori rather than a priori justification. These postulates are, in fact, the most natural guess that we might make, providing the *simplest conceivable formal solution* to the basic problem. On this basis alone the problem *might* have been solved; the tentative postulation of the simplest formal solution of a problem is a conventional and frequently successful mode of procedure in theoretical physics.

What then is the simplest criterion that reasonably can be imagined for the determination of the final equilibrium state? From our experience with many physical theories we might expect that the most economical form for the equilibrium criterion would be in terms of an extremum principle. That is, we might anticipate the values of the extensive parameters in the final equilibrium state to be simply those that maximize[5] some function. And, straining our optimism to the limit, we might hope that this hypothetical function would have several particularly simple mathematical properties, designed to guarantee simplicity of the derived theory. We develop this proposed solution in a series of postulates.

Postulate II. *There exists a function (called the entropy S) of the extensive parameters of any composite system, defined for all equilibrium states and having the following property: The values assumed by the extensive parameters in the absence of an internal constraint are those that maximize the entropy over the manifold of constrained equilibrium states.*

It must be stressed that we postulate the existence of the entropy only for equilibrium states and that our postulate makes no reference whatsoever to nonequilibrium states. In the absence of a constraint the system is free to select any one of a number of states, *each of which might also be realized in the presence of a suitable constraint.* The entropy of each of these constrained equilibrium states is definite, and the entropy is largest in some particular state of the set. In the absence of the constraint this state of maximum entropy is selected by the system.

---

[5] Or minimize the function, this being purely a matter of convention in the choice of the sign of the function, having no consequence whatever in the logical structure of the theory.

In the case of two systems separated by a diathermal wall we might wish to predict the manner in which the total energy U distributes between the two systems. We then consider the composite system with the internal diathermal wall replaced by an adiabatic wall and with particular values of $U^{(1)}$ and $U^{(2)}$ (consistent, of course, with the restriction that $U^{(1)} + U^{(2)} = U$). For each such constrained equilibrium state there is an entropy of the composite system, and for some particular values of $U^{(1)}$ and $U^{(2)}$ this entropy is maximum. These, then, are the values of $U^{(1)}$ and $U^{(2)}$ that obtain in the presence of the diathermal wall, or in the absence of the adiabatic constraint.

All problems in thermodynamics are derivative from the basic problem formulated in Section 1.9. The basic problem can be completely solved with the aid of the extremum principle if the entropy of the system is known as a function of the extensive parameters. The relation that gives the entropy as a function of the extensive parameters is known as a *fundamental relation*. It therefore follows that *if the fundamental relation of a particular system is known all conceivable thermodynamic information about the system is ascertainable from it.*

The importance of the foregoing statement cannot be overemphasized. The information contained in a fundamental relation is all-inclusive—it is equivalent to all conceivable numerical data, to all charts, and to all imaginable types of descriptions of thermodynamic properties. If the fundamental relation of a system is known, every thermodynamic attribute is completely and precisely determined.

Postulate III. *The entropy of a composite system is additive over the constituent subsystems. The entropy is continuous and differentiable and is a monotonically increasing function of the energy.*

Several mathematical consequences follow immediately. The additivity property states that the entropy S of the composite system is merely the sum of the entropies $S^{(\alpha)}$ of the constituent subsystems:

$$S = \sum_\alpha S^{(\alpha)} \tag{1.4}$$

The entropy of each subsystem is a function of the extensive parameters of that subsystem alone

$$S^{(\alpha)} = S^{(\alpha)}\left(U^{(\alpha)}, V^{(\alpha)}, N_1^{(\alpha)}, \ldots, N_r^{(\alpha)}\right) \tag{1.5}$$

The additivity property applied to spatially separate subsystems requires the following property: *The entropy of a simple system is a homogeneous first-order function of the extensive parameters.* That is, if all the extensive parameters of a system are multiplied by a constant λ, the entropy is multiplied by this same constant. Or, omitting the superscript (α),

$$S(\lambda U, \lambda V, \lambda N_1, \ldots, \lambda N_r) = \lambda S(U, V, N_1, \ldots, N_r) \qquad (1.6)$$

The monotonic property postulated implies that *the partial derivative* $(\partial S/\partial U)_{V, N_1, \ldots, N_r}$ is a positive quantity,

$$\left(\frac{\partial S}{\partial U}\right)_{V, N_1, \ldots, N_r} > 0 \qquad (1.7)$$

As the theory develops in subsequent sections, we shall see that the reciprocal of this partial derivative is taken as the definition of the temperature. Thus the temperature is postulated to be nonnegative.[6]

The continuity, differentiability, and monotonic property imply that the entropy function can be inverted with respect to the energy and that *the energy is a single-valued, continuous, and differentiable function of* $S, V, N_1, \ldots, N_r$. The function

$$S = S(U, V, N_1, \ldots, N_r) \qquad (1.8)$$

can be solved uniquely for U in the form

$$U = U(S, V, N_1, \ldots, N_r) \qquad (1.9)$$

Equations 1.8 and 1.9 are alternative forms of the fundamental relation, and each contains *all* thermodynamic information about the system.

We note that the extensivity of the entropy permits us to scale the properties of a system of N moles from the properties of a system of 1 mole. The fundamental equation is subject to the identity

$$S(U, V, N_1, N_2, \ldots, N_r) = NS(U/N, V/N, N_1/N, \ldots, N_r/N) \qquad (1.10)$$

in which we have taken the scale factor λ of equation 1.6 to be equal to $1/N \equiv 1/\sum_k N_k$. For a single-component simple system, in particular,

$$S(U, V, N) = NS(U/N, V/N, 1) \qquad (1.11)$$

But U/N is the energy per mole, which we denote by u.

$$u \equiv U/N \qquad (1.12)$$

---

[6] The possibility of negative values of this derivative (i.e., of negative temperatures) has been discussed by N. F. Ramsey, *Phys. Rev.* **103**, 20 (1956). Such states are not equilibrium states in real systems, and they do not invalidate equation 1.7. They can be produced only in certain very unique systems (specifically in isolated spin systems) and they spontaneously decay away. Nevertheless the study of these states is of statistical mechanical interest, elucidating the statistical mechanical concept of temperature.

Also, V/N is the volume per mole, which we denote by v.

$$v \equiv V/N \qquad (1.13)$$

Thus $S(U/N, V/N, 1) \equiv S(u, v, 1)$ is the entropy of a system of a single mole, to be denoted by $s(u, v)$.

$$s(u, v) \equiv S(u, v, 1) \qquad (1.14)$$

Equation 1.11 now becomes

$$S(U, V, N) = Ns(u, v) \qquad (1.15)$$

Postulate IV. *The entropy of any system vanishes in the state for which*

$$(\partial U/\partial S)_{V, N_1, \ldots, N_r} = 0 \qquad (\textit{that is, at the zero of temperature})$$

We shall see later that the vanishing of the derivative $(\partial U/\partial S)_{V, N_1, \ldots, N_2}$ is equivalent to the vanishing of the temperature, as indicated. Hence the fourth postulate is that zero temperature implies zero entropy.

It should be noted that an immediate implication of postulate IV is that S (like V and N, but unlike U) has a uniquely defined zero.

This postulate is an extension, due to Planck, of the so-called *Nernst postulate or third law of thermodynamics*. Historically, it was the latest of the postulates to be developed, being inconsistent with classical statistical mechanics and requiring the prior establishment of quantum statistics in order that it could be properly appreciated. The bulk of thermodynamics does not require this postulate, and I make no further reference to it until Chapter 10. Nevertheless, I have chosen to present the postulate at this point to close the postulatory basis.

The foregoing postulates are the logical bases of our development of thermodynamics. In the light of these postulates, then, it may be wise to reiterate briefly the method of solution of the standard type of thermodynamic problem, as formulated in Section 1.9. We are given a composite system and we assume the fundamental equation of each of the constituent systems to be known in principle. These fundamental equations determine the individual entropies of the subsystems when these systems are in equilibrium. If the total composite system is in a constrained equilibrium state, with particular values of the extensive parameters of each constituent system, the total entropy is obtained by addition of the individual entropies. This total entropy is known as a function of the various extensive parameters of the subsystems. By straightforward differentiation we compute the extrema of the total entropy function, and then, on the basis of the sign of the second derivative, we classify these extrema as minima, maxima, or as horizontal inflections. In an appropriate physical terminology we first find the *equilibrium states* and we then classify them on the basis of *stability*. It should be noted that in the adoption of this conventional terminology we augment our previous definition of equilibrium; that which was previously termed *equilibrium* is now termed *stable equilibrium*, whereas *unstable equilibrium* states are newly defined in terms of extrema other than maxima.

It is perhaps appropriate at this point to acknowledge that although all applications of thermodynamics are equivalent in principle to the procedure outlined, there are several alternative procedures that frequently prove more convenient. These alternate procedures are developed in subsequent chapters. Thus we shall see that under appropriate conditions the energy $U(S, V, N_1, \ldots)$ may be minimized rather than the entropy $S(U, V, N_1, \ldots)$, maximized. That these two procedures determine the same final state is analogous to the fact that a circle may be characterized either as the closed curve of minimum perimeter for a given area or as the closed curve of maximum area for a given perimeter. In later chapters we shall encounter several new functions, the minimization of which is logically equivalent to the minimization of the energy or to the maximization of the entropy.

The inversion of the fundamental equation and the alternative statement of the basic extremum principle in terms of a minimum of the energy (rather than a maximum of the entropy) suggests another viewpoint from which the extremum postulate perhaps may appear plausible. In the theories of electricity and mechanics, ignoring thermal effects, the energy is a function of various mechanical parameters, and the condition of equilibrium is that the energy shall be a minimum. Thus a cone is stable lying on its side rather than standing on its point because the first position is of lower energy. If thermal effects are to be included the energy ceases to be a function simply of the mechanical parameters. According to the inverted fundamental equation, however, the energy is a function of the mechanical parameters and of one additional parameter (the entropy). By the introduction of this additional parameter the form of the energy-minimum principle is extended to the domain of thermal effects as well as to pure mechanical phenomena. In this manner we obtain a sort of correspondence principle between thermodynamics and mechanics— ensuring that the thermodynamic equilibrium principle reduces to the mechanical equilibrium principle when thermal effects can be neglected.

We shall see that the mathematical condition that a maximum of $S(U, V, N_1, \ldots)$ implies a minimum of $U(S, V, N_1, \ldots)$ is that the derivative $(\partial S / \partial U)_{V, N_1, \ldots}$ be positive. The motivation for the introduction of this statement in postulate III may be understood in terms of our desire to ensure that the entropy-maximum principle will go over into an energy-minimum principle on inversion of the fundamental equation.

In Parts II and III the concept of the entropy will be more deeply explored, both in terms of its symmetry roots and in terms of its statistical mechanical interpretation. Pursuing those inquires now would take us too far afield. In the classical spirit of thermodynamics we temporarily defer such interpretations while exploring the far-reaching consequences of our simple postulates.

PROBLEMS

1.10-1. The following ten equations are purported to be fundamental equations of various thermodynamic systems. However, five are inconsistent with one or more of postulates II, III, and IV and consequently are not physically acceptable.
  In each case qualitatively sketch the fundamental relationship between S and U (with N and V constant). Find the five equations that are not physically permissible and indicate the postulates violated by each.
  The quantities v_0, θ, and R are positive constants, and in all cases in which fractional exponents appear only the real positive root is to be taken.

a) $S = \left(\dfrac{R^2}{v_0\theta}\right)^{1/3}(NVU)^{1/3}$

b) $S = \left(\dfrac{R}{\theta^2}\right)^{1/3}\left(\dfrac{NU}{V}\right)^{2/3}$

c) $S = \left(\dfrac{R}{\theta}\right)^{1/2}\left(NU + \dfrac{R\theta V^2}{v_0^2}\right)^{1/2}$

d) $S = \left(\dfrac{R^2\theta}{v_0^3}\right)V^3/NU$

e) $S = \left(\dfrac{R^3}{v_0\theta^2}\right)^{1/5}[N^2VU^2]^{1/5}$

f) $S = NR\ln(UV/N^2R\theta v_0)$

g) $S = \left(\dfrac{R}{\theta}\right)^{1/2}[NU]^{1/2}\exp(-V^2/2N^2v_0^2)$

h) $S = \left(\dfrac{R}{\theta}\right)^{1/2}(NU)^{1/2}\exp\left(-\dfrac{UV}{NR\theta v_0}\right)$

i) $U = \left(\dfrac{v_0\theta}{R}\right)\dfrac{S^2}{V}\exp(S/NR)$

j) $U = \left(\dfrac{R\theta}{v_0}\right)NV\left(1 + \dfrac{S}{NR}\right)\exp(-S/NR)$

1.10-2. For each of the five physically acceptable fundamental equations in problem 1.10-1 find U as a function of S, V, and N.

1.10-3. The fundamental equation of system A is

$$S = \left(\frac{R^2}{v_0\theta}\right)^{1/3} (NVU)^{1/3}$$

and similarly for system B. The two systems are separated by a rigid, imperme-able, adiabatic wall. System A has a volume of 9×10^{-6} m^3 and a mole number of 3 moles. System B has a volume of 4×10^{-6} m^3 and a mole number of 2 moles. The total energy of the composite system is 80 J. Plot the entropy as a function of $U_A/(U_A + U_B)$. If the internal wall is now made diathermal and the system is allowed to come to equilibrium, what are the internal energies of each of the individual systems? (As in Problem 1.10-1, the quantities v_0, θ, and R are positive constants.)

$$\frac{2}{}$$

THE CONDITIONS
OF EQUILIBRIUM

2-1   INTENSIVE PARAMETERS

By virtue of our interest in processes, and in the associated changes of the extensive parameters, we anticipate that we shall be concerned with the differential form of the fundamental equation. Writing the fundamental equation in the form

$$U = U(S, V, N_1, N_2, \ldots, N_r) \tag{2.1}$$

we compute the first differential:

$$dU = \left(\frac{\partial U}{\partial S}\right)_{V, N_1, \ldots, N_r} dS + \left(\frac{\partial U}{\partial V}\right)_{S, N_1, \ldots, N_r} dV + \sum_{j=1}^{r} \left(\frac{\partial U}{\partial N_j}\right)_{S, V, \ldots, N_r} dN_j \tag{2.2}$$

The various partial derivatives appearing in the foregoing equation recur so frequently that it is convenient to introduce special symbols for them. They are called *intensive parameters*, and the following notation is conventional:

$$\left(\frac{\partial U}{\partial S}\right)_{V, N_1, \ldots, N_r} \equiv T, \text{ the } \textit{temperature} \tag{2.3}$$

$$-\left(\frac{\partial U}{\partial V}\right)_{S, N_1, \ldots, N_r} \equiv P, \text{ the } \textit{pressure} \tag{2.4}$$

$$\left(\frac{\partial U}{\partial N_j}\right)_{S, V, \ldots, N_k, \ldots} \equiv \mu_j, \quad \begin{array}{l}\text{the } \textit{electrochemical potential of}\\ \textit{the } j\text{th component}\end{array} \tag{2.5}$$

35

With this notation, equation 2.2 becomes

$$dU = T\, dS - P\, dV + \mu_1\, dN_1 + \cdots + \mu_r\, dN_r \qquad (2.6)$$

The formal definition of the temperature soon will be shown to agree with our intuitive qualitative concept, based on the physiological sensations of "hot" and "cold." We certainly would be reluctant to adopt a definition of the temperature that would contradict such strongly entrenched although qualitative notions. For the moment, however, we merely introduce the concept of temperature by the formal definition (2.3).

Similarly, we shall soon corroborate that the pressure defined by equation 2.4 agrees in every respect with the pressure defined in mechanics. With respect to the several electrochemical potentials, we have no prior definitions or concepts and we are free to adopt the definition (equation 2.5) forthwith.

For brevity, the electrochemical potential is often referred to simply as the *chemical potential*, and we shall use these two terms interchangeably[1].

The term $-P\, dV$ in equation 2.6 is identified as the quasi-static work dW_M, as given by equation 1.1.

In the special case of constant mole numbers equation 2.6 can then be written as

$$T\, dS = dU - dW_M \qquad \text{if} \quad dN_1 = dN_2 = dN_r = 0 \qquad (2.7)$$

Recalling the definition of the quasi-static heat, or comparing equation 2.7 with equation 1.2, we now recognize $T\, dS$ as the quasi-static heat flux.

$$dQ = T\, dS \qquad (2.8)$$

A quasi-static flux of heat into a system is associated with an increase of entropy of that system.

The remaining terms in equation 2.6 represent an increase of internal energy associated with the addition of matter to a system. This type of energy flux, although intuitively meaningful, is not frequently discussed outside thermodynamics and does not have a familiar distinctive name. We shall call $\sum_j \mu_j\, dN_j$ the *quasi-static chemical work*.

$$dW_c \equiv \sum_{j=1}^{r} \mu_j\, dN_j \qquad (2.9)$$

---

[1] However it should be noted that occasionally, and particularly in the theory of solids, the "chemical potential" is defined as the electrochemical potential μ minus the molar electrostatic energy.

Therefore

$$dU = dQ + dW_M + dW_c \qquad (2.10)$$

Each of the terms $T\,dS, - P\,dV$, $\mu_j\,dN_j$, in equation 2.6 has the dimensions of energy. The matter of units will be considered in Section 2.6. We can observe here, however, that having not yet specified the units (nor even the dimensions) of entropy, the units and dimensions of temperature remain similarly undetermined. The units of μ are the same as those of energy (as the mole numbers are dimensionless). The units of pressure are familiar, and conversion factors are listed inside the back cover of this book.

2-2 EQUATIONS OF STATE

The temperature, pressure, and electrochemical potentials are partial derivatives of functions of $S, V, N_1, \ldots, N_r$ and consequently are also functions of $S, V, N_1, \ldots, N_r$. We thus have a set of functional relationships

$$T = T(S, V, N_1, \ldots, N_r) \qquad (2.11)$$

$$P = P(S, V, N_1, \ldots, N_r) \qquad (2.12)$$

$$\mu_j = \mu_j(S, V, N_1, \ldots, N_r) \qquad (2.13)$$

Such relationships, expressing intensive parameters in terms of the independent extensive parameters, are called *equations of state*.

Knowledge of a single equation of state does *not* constitute complete knowledge of the thermodynamic properties of a system. We shall see, subsequently, that knowledge of *all* the equations of state of a system is equivalent to knowledge of the fundamental equation and consequently is thermodynamically complete.

The fact that the fundamental equation must be homogeneous first order has direct implications for the functional form of the equations of state. It follows immediately that the equations of state are *homogeneous zero order*. That is, multiplication of each of the independent extensive parameters by a scalar λ leaves the function unchanged.

$$T(\lambda S, \lambda V, \lambda N_1, \ldots, \lambda N_r) = T(S, V, N_1, \ldots, N_r) \qquad (2.14)$$

It therefore follows that the temperature of a portion of a system is equal to the temperature of the whole system. This is certainly in agreement with the intuitive concept of temperature. The pressure and the electrochemical potentials also have the property (2.14), and together with the temperature are said to be *intensive*.

To summarize the foregoing considerations it is convenient to adopt a condensed notation. We denote the extensive parameters $V, N_1, \ldots, N_r$ by the symbols $X_1, X_2, \ldots, X_t$, so that the fundamental relation takes the form

$$U = U(S, X_1, X_2, \ldots, X_t) \tag{2.15}$$

The intensive parameters are denoted by

$$\left(\frac{\partial U}{\partial S} \right)_{X_1, X_2, \ldots} \equiv T = T(S, X_1, X_2, \ldots, X_t) \tag{2.16}$$

$$\left(\frac{\partial U}{\partial X_j} \right)_{S, \ldots, X_k, \ldots} \equiv P_j = P_j(S, X_1, X_2, \ldots, X_t) \qquad j = 1, 2, \ldots, t \tag{2.17}$$

whence

$$dU = T\,dS + \sum_{j=1}^{t} P_j\,dX_j \tag{2.18}$$

It should be noted that a negative sign appears in equation 2.4, but does not appear in equation 2.17. The formalism of thermodynamics is uniform if the *negative pressure*, $-P$, is considered as an intensive parameter analogous to T and $\mu_1, \mu_2, \ldots$. Correspondingly one of the general intensive parameters P_j of equation 2.17 is $-P$.

For single-component simple systems the energy differential is frequently written in terms of molar quantities. Analogous to equations 1.11 through 1.15, the fundamental equation per mole is

$$u = u(s, v) \tag{2.19}$$

where

$$s = S/N, \qquad v = V/N \tag{2.20}$$

and

$$u(s, v) = \frac{1}{N} U(S, V, N) \tag{2.21}$$

Taking an infinitesimal variation of equation 2.19

$$du = \frac{\partial u}{\partial s} \, ds + \frac{\partial u}{\partial v} \, dv \tag{2.22}$$

However

$$\left(\frac{\partial u}{\partial s} \right)_v = \left(\frac{\partial u}{\partial s} \right)_{V,N} = \left(\frac{\partial U}{\partial S} \right)_{V,N} = T \tag{2.23}$$

and similarly

$$\left(\frac{\partial u}{\partial v} \right)_s = -P \tag{2.24}$$

Thus

$$du = T \, ds - P \, dv \tag{2.25}$$

PROBLEMS

2.2-1. Find the three equations of state for a system with the fundamental equation

$$U = \left(\frac{v_0 \theta}{R^2} \right) \frac{S^3}{NV}$$

Corroborate that the equations of state are homogeneous zero order (i.e., that T, P, and μ are intensive parameters).

2.2-2. For the system of problem 2.2-1 find μ as a function of T, V, and N.

2.2-3. Show by a diagram (drawn to arbitrary scale) the dependence of pressure on volume for fixed temperature for the system of problem 2.2-1. Draw two such "isotherms," corresponding to two values of the temperature, and indicate which isotherm corresponds to the higher temperature.

2.2-4. Find the three equations of state for a system with the fundamental equation

$$u = \left(\frac{\theta}{R} \right) s^2 - \left(\frac{R\theta}{v_0^2} \right) v^2$$

and show that, for this system, $\mu = -u$.

2.2-5. Express μ as a function of T and P for the system of problem 2.2-4.

2.2-6. Find the three equations of state for a system with the fundamental equation

$$u = \left(\frac{v_0 \theta}{R} \right) \frac{s^2}{v} e^{s/R}$$

2.2-7. A particular system obeys the relation

$$u = Av^{-2}\exp(s/R)$$

N moles of this substance, initially at temperature T_0 and pressure P_0, are expanded isentropically (s = constant) until the pressure is halved. What is the final temperature?

Answer:
$$T_f = 0.63\ T_0$$

2.2-8. Show that, in analogy with equation 2.25, for a system with r components

$$du = T\,ds - P\,dv + \sum_{j=1}^{r-1} (\mu_j - \mu_r)\,dx_j$$

where the x_j are the mole fractions ($= N_j/N$).

2.2-9. Show that if a single-component system is such that PV^k is constant in an adiabatic process (k is a positive constant) the energy is

$$U = \frac{1}{k-1}PV + Nf(PV^k/N^k)$$

where f is an arbitrary function.

Hint: PV^k must be a function of S, so that $(\partial U/\partial V)_S = g(S) \cdot V^{-k}$, where $g(S)$ is an unspecified function.

2-3 ENTROPIC INTENSIVE PARAMETERS

If, instead of considering the fundamental equation in the form $U = U(S, \ldots, X_j, \ldots)$ with U as dependent, we had considered S as dependent, we could have carried out all the foregoing formalism in an inverted but equivalent fashion. Adopting the notation X_0 for U, we write

$$S = S(X_0, X_1, \ldots, X_t) \tag{2.26}$$

We take an infinitesimal variation to obtain

$$dS = \sum_{k=0}^{t} \frac{\partial S}{\partial X_k}\,dX_k \tag{2.27}$$

The quantities $\partial S / \partial X_k$ are denoted by F_k.

$$F_k \equiv \frac{\partial S}{\partial X_k} \tag{2.28}$$

By carefully noting which variables are kept constant in the various partial derivatives (and by using the calculus of partial derivatives as reviewed in Appendix A) the reader can demonstrate that

$$F_0 = \frac{1}{T}, \qquad F_k = \frac{-P_k}{T} \qquad (k = 1, 2, 3, \ldots) \tag{2.29}$$

These equations also follow from solving equation 2.18 for dS and comparing with equation 2.27.

Despite the close relationship between the F_k and the P_k, there is a very important difference in principle. Namely, the P_k are obtained by differentiating a function of $S, \ldots, X_j, \ldots$ and are considered as functions of these variables, whereas the F_k are obtained by differentiating a function of $U, \ldots, X_j, \ldots$ and are considered as functions of these latter variables. That is, in one case the entropy is a member of the set of independent parameters, and in the second case the energy is such a member. In performing formal manipulations in thermodynamics it is extremely important to make a definite commitment to one or the other of these choices and to adhere rigorously to that choice. A great deal of confusion results from a vacillation between these two alternatives within a single problem.

If the entropy is considered dependent and the energy independent, as in $S = S(U, \ldots, X_k, \ldots)$, we shall refer to the analysis as being in the *entropy representation*. If the energy is dependent and the entropy is independent, as in $U = U(S, \ldots, X_k, \ldots)$, we shall refer to the analysis as being in the *energy representation*.

The formal development of thermodynamics can be carried out in either the energy or entropy representations alone, but for the solution of a particular problem either one or the other representation may prove to be by far the more convenient. Accordingly, we shall develop the two representations in parallel, although a discussion presented in one representation generally requires only a brief outline in the alternate representation.

The relation $S = S(X_0, \ldots, X_j, \ldots)$ is said to be the *entropic fundamental relation*, the set of variables $X_0, \ldots, X_j, \ldots$ is called the *entropic extensive parameters*, and the set of variables $F_0, \ldots, F_j, \ldots$ is called the *entropic intensive parameters*. Similarly, the relation $U = U(S, X_1, \ldots, X_j, \ldots)$ is said to be the *energetic fundamental relation*; the set of variables $S, X_1, \ldots, X_j, \ldots$ is called the *energetic extensive parameters*; and the set of variables $T, P_1, \ldots, P_j, \ldots$ is called the *energetic intensive parameters*.

PROBLEMS

2.3-1. Find the three equations of state in the entropy representation for a system with the fundamental equation

$$u = \left(\frac{v_0^{1/2}\theta}{R^{3/2}} \right) \frac{s^{5/2}}{v^{1/2}}$$

Answer:

$$\frac{1}{T} = \frac{2}{5} \left(\frac{v_0^{1/2}\theta}{R^{3/2}} \right)^{-2/5} \frac{v^{1/5}}{u^{3/5}}$$

$$\frac{\mu}{T} = -\frac{2}{5} \left(\frac{v_0^{1/2}\theta}{R^{3/2}} \right)^{-2/5} u^{2/5}v^{1/5}$$

2.3-2. Show by a diagram (drawn to arbitrary scale) the dependence of temperature on volume for fixed pressure for the system of problem 2.3-1. Draw two such "isobars" corresponding to two values of the pressure, and indicate which isobar corresponds to the higher pressure.

2.3-3. Find the three equations of state in the entropy representation for a system with the fundamental equation

$$u = \left(\frac{\theta}{R} \right) s^2 e^{-v^2/v_0^2}$$

2.3-4. Consider the fundamental equation

$$S = AU^n V^m N^r$$

where A is a positive constant. Evaluate the permissible values of the three constants n, m, and r if the fundamental equation is to satisfy the thermodynamic postulates and if, in addition, we wish to have P increase with U/V, at constant N. (This latter condition is an intuitive substitute for stability requirements to be studied in Chapter 8.) For definiteness, the zero of energy is to be taken as the energy of the zero-temperature state.

2.3-5. Find the three equations of state for a system with the fundamental relation

$$\frac{S}{R} = \frac{UV}{N} - \frac{N^3}{UV}$$

a) Show that the equations of state in entropy representation are homogeneous zero-order functions.

b) Show that the temperature is intrinsically positive.

c) Find the "mechanical equation of state" $P = P(T, v)$.

d) Find the form of the adiabats in the $P-v$ plane. (An "adiabat" is a locus of constant entropy, or an "isentrope").

2-4 THERMAL EQUILIBRIUM—TEMPERATURE

We are now in a position to illustrate several interesting implications of the extremum principle which has been postulated for the entropy. Consider a closed composite system consisting of two simple systems separated by a wall that is rigid and impermeable to matter but that does allow the flow of heat. The volumes and mole numbers of each of the simple systems are fixed, but the energies $U^{(1)}$ and $U^{(2)}$ are free to change, subject to the conservation restriction

$$U^{(1)} + U^{(2)} = \text{constant} \qquad (2.30)$$

imposed by the closure of the composite system as a whole. Assuming that the system has come to equilibrium, we seek the values of $U^{(1)}$ and $U^{(2)}$. According to the fundamental postulate, the values of $U^{(1)}$ and $U^{(2)}$ are such as to maximize the entropy. Therefore, by the usual mathematical condition for an extremum, it follows that in the equilibrium state a virtual infinitesimal transfer of energy from system 1 to system 2 will produce no change in the entropy of the whole system. That is,

$$dS = 0 \qquad (2.31)$$

The additivity of the entropy for the two subsystems gives the relation

$$S = S^{(1)}\left(U^{(1)}, V^{(1)}, \ldots, N_j^{(1)}, \ldots\right) + S^{(2)}\left(U^{(2)}, V^{(2)}, \ldots, N_j^{(2)}, \ldots\right).$$

$$(2.32)$$

As $U^{(1)}$ and $U^{(2)}$ are changed by the virtual energy transfer, the entropy change is

$$dS = \left(\frac{\partial S^{(1)}}{\partial U^{(1)}}\right)_{V^{(1)}, \ldots, N_j^{(1)}, \ldots} dU^{(1)} + \left(\frac{\partial S^{(2)}}{\partial U^{(2)}}\right)_{V^{(2)}, \ldots, N_j^{(2)}, \ldots} dU^{(2)} \qquad (2.33)$$

or, employing the definition of the temperature

$$dS = \frac{1}{T^{(1)}} \, dU^{(1)} + \frac{1}{T^{(2)}} \, dU^{(2)} \tag{2.34}$$

By the conservation condition (equation 2.30), we have

$$dU^{(2)} = -dU^{(1)} \tag{2.35}$$

whence

$$dS = \left(\frac{1}{T^{(1)}} - \frac{1}{T^{(2)}} \right) dU^{(1)} \tag{2.36}$$

The condition of equilibrium (equation 2.31) demands that dS vanish for arbitrary values of $dU^{(1)}$, whence

$$\frac{1}{T^{(1)}} = \frac{1}{T^{(2)}} \tag{2.37}$$

This is the condition of equilibrium. If the fundamental equations of each of the subsystems were known, then $1/T^{(1)}$ would be a known function of $U^{(1)}$ (and of $V^{(1)}$ and $N_k^{(1)}, \ldots$, which, however, are merely constants). Similarly, $1/T^{(2)}$ would be a known function of $U^{(2)}$, and the equation $1/T^{(1)} = 1/T^{(2)}$ would be one equation in $U^{(1)}$ and $U^{(2)}$. The conservation condition $U^{(1)} + U^{(2)} = $ constant provides a second equation, and these two equations completely determine, in principle, the values of $U^{(1)}$ and of $U^{(2)}$. To proceed further and actually to obtain the values of $U^{(1)}$ and $U^{(2)}$ would require knowledge of the explicit forms of the fundamental equations of the systems. In thermodynamic theory, however, we accept the existence of the fundamental equations, but we do not assume explicit forms for them, and we therefore do not obtain explicit answers. In practical applications of thermodynamics the fundamental equations may be known, either by empirical observations (in terms of measurements to be described later) or on the basis of statistical mechanical calculations based on simple models. In this way applied thermodynamics is able to lead to explicit numerical answers.

   Equation 2.37 could also be written as $T^{(1)} = T^{(2)}$. We write it in the form $1/T^{(1)} = 1/T^{(2)}$ to stress the fact that the analysis is couched in the entropy representation. By writing $1/T^{(1)}$, we indicate a function of $U^{(1)}, V^{(1)}, \ldots$, whereas $T^{(1)}$ would imply a function of $S^{(1)}, V^{(1)}, \ldots$. The *physical* significance of equation 2.37, however, remains the equality of the temperatures of the two subsystems.

   A second phase of the problem is the investigation of the stability of the predicted final state. In the solution given we have not exploited fully the basic postulate that the entropy is a maximum in equilibrium; rather, we merely have investigated the consequences of the fact that it is an extremum. The condition that it be a maximum requires, in addition to the condition $dS = 0$, that

$$d^2S < 0 \tag{2.38}$$

The consequences of this condition lead to considerations of stability, to which we shall give explicit attention in Chapter 8.

2-5 AGREEMENT WITH INTUITIVE CONCEPT OF TEMPERATURE

In the foregoing example we have seen that if two systems are separated by a diathermal wall, heat will flow until each of the system attains the same temperature. This prediction is in agreement with our intuitive notion of temperature, and it is the first of several observations that corroborate the plausibility of the formal definition of the temperature.

Inquiring into the example in slightly more detail, we suppose that the two subsystems initially are separated by an adiabatic wall and that the temperatures of the two subsystems are almost, but not quite, equal. In particular we assume that

$$T^{(1)} > T^{(2)} \tag{2.39}$$

The system is considered initially to be in equilibrium with respect to the internal adiabatic constraint. If the internal adiabatic constraint now is removed, the system is no longer in equilibrium, heat flows across the wall, and the entropy of the composite system *increases*. Finally the system comes to a new equilibrium state, determined by the condition that the final values of $T^{(1)}$ and $T^{(2)}$ are equal, and with the maximum possible value of the entropy that is consistent with the remaining constraints. Compare the initial and the final states. If ΔS denotes the entropy difference between the final and initial states

$$\Delta S > 0 \tag{2.40}$$

But, as in equation 2.36,

$$\Delta S \simeq \left(\frac{1}{T^{(1)}} - \frac{1}{T^{(2)}} \right) \Delta U^{(1)} \tag{2.41}$$

where $T^{(1)}$ and $T^{(2)}$ are the initial values of the temperatures. By the condition that $T^{(1)} > T^{(2)}$, it follows that

$$\Delta U^{(1)} < 0 \tag{2.42}$$

This means that the spontaneous process that occurred was one in which heat flowed *from* subsystem 1 *to* subsystem 2. We conclude therefore that heat tends to flow *from* a system with a *high* value of T *to* a system with a *low* value of T. This is again in agreement with the intuitive notion of temperature. It should be noted that these conclusions do not depend on the assumption that $T^{(1)}$ is approximately equal to $T^{(2)}$; this assumption was made merely for the purpose of obtaining mathematical simplicity in equation 2.41, which otherwise would require a formulation in terms of integrals.

If we now take stock of our intuitive notion of temperature, based on the physiological sensations of hot and cold, we realize that it is based upon two essential properties. First, we expect temperature to be an intensive parameter, having the same value in a part of a system as it has in the entire system. Second, we expect that heat should tend to flow from regions of high temperature toward regions of low temperature. These properties imply that thermal equilibrium is associated with equality and homogeneity of the temperature. Our formal definition of the temperature possesses each of these properties.

2-6 TEMPERATURE UNITS

The physical dimensions of temperature are those of energy divided by those of entropy. But we have not yet committed ourselves on the dimensions of entropy; in fact its dimensions can be selected quite arbitrarily. If the entropy is multiplied by any positive dimensional constant we obtain a new function of different dimensions but with exactly the same extremum properties—and therefore equally acceptable as the entropy. We summarily resolve the arbitrariness simply by adopting the convention that the entropy is dimensionless (from the more incisive viewpoint of statistical mechanics this *is* a physically reasonable choice). Consequently the dimensions of temperature are identical to those of energy. However, just as torque and work have the same dimensions, but are different types of quantities and are measured in different units (the meter–Newton and the joule, respectively), so the temperature and the energy should be carefully distinguished. The *dimensions* of both energy and temperature are [mass · (length)^2/(time)^2]. The *units* of energy are joules, ergs, calories, and the like. The *units* of temperature remain to be discussed.

In our later discussion of thermodynamic "Carnot" engines, in Chapter 4, we shall find that the optimum performance of an engine in contact with two thermodynamic systems is completely determined by the ratio of the temperatures of those two systems. That is, *the principles of thermodynamics provide an experimental procedure that unambiguously determines the ratio of the temperatures of any two given systems.*

The fact that the *ratio* of temperatures is measurable has immediate consequences. First the zero of temperature is uniquely determined and cannot be arbitrarily assigned or "shifted." Second we are free to assign the value of unity (or some other value) to *one* arbitrary chosen state. All other temperatures are thereby determined.

Equivalently, the single arbitrary aspect of the temperature scale is the size of the temperature unit, determined by assigning a specific temperature to some particular state of a standard system.

The assignment of different temperature values to standard states leads to different thermodynamic temperature scales, but all thermodynamic temperature scales coincide at $T = 0$. Furthermore, according to equation 1.7 no system can have a temperature lower than zero. Needless to say, this essential positivity of the temperature is in full agreement with all measurements of thermodynamic temperatures.

The Kelvin scale of temperature, which is the official Système International (SI) system, is defined by assigning the number 273.16 to the temperature of a mixture of pure ice, water, and water vapor in mutual equilibrium; a state which we show in our later discussion of "triple points" determines a unique temperature. The corresponding unit of temperature is called a *kelvin*, designated by the notation K.

The ratio of the kelvin and the joule, two units with the same dimensions, is 1.3806×10^{-23} joules/kelvin. This ratio is known as Boltzmann's constant and is generally designated as k_B. Thus $k_B T$ is an energy.

The Rankine scale is obtained by assigning the temperature $(\frac{9}{5}) \times 273.16 = 491.688°R$ to the ice–water–water vapor system just referred to. The unit, denoted by °R, is called the *degree Rankine*. Rankine temperatures are merely $\frac{9}{5}$ times the corresponding Kelvin temperature.

Closely related to the "absolute" Kelvin scale of temperature is the *International Kelvin* scale, which is a "practical" scale, defined in terms of the properties of particular systems in various temperature ranges and contrived to coincide as closely as possible with the (absolute) Kelvin scale. The practical advantage of the International Kelvin scale is that it provides reproducible laboratory standards for temperature measurement throughout the temperature range. However, from the thermodynamic point of view, it is not a true temperature scale, and to the extent that it deviates from the absolute Kelvin scale it will not yield temperature ratios that are consistent with those demanded by the thermodynamic formalism.

The values of the temperature of everyday experiences are large numbers on both the Kelvin and the Rankine scales. Room temperatures are in the region of 300 K, or 540°R. For common usage, therefore, two derivative scales are in common use. The Celsius scale is defined as

$$T(°C) = T(K) - 273.15 \qquad (2.43)$$

where $T(°C)$ denotes the "Celsius temperature," for which the unit is called the *degree Celsius*, denoted by °C. The zero of this scale is displaced relative to the true zero of temperature, so *the Celsius temperature scale is not a thermodynamic temperature scale at all*. Negative temperatures appear, the zero is incorrect, and ratios of temperatures are not in agreement with thermodynamic principles. Only temperature differences are correctly given.

On the Celsius scale the "temperature" of the triple point (ice, water, and water vapor in mutual equilibrium) is 0.01°C. The Celsius temperature of an equilibrium mixture of ice and water, maintained at a pressure of 1 atm, is even closer to 0°C, with the difference appearing only in the third decimal place. Also the Celsius temperature of boiling water at 1 atm pressure is very nearly 100°C. These near equalities reveal the historical origin[2] of the Celsius scale; before it was recognized that the zero of temperature is unique it was thought that two points, rather than one, could be arbitrarily assigned and these were taken (by Anders Celsius, in 1742) as the 0°C and 100°C just described.

The Fahrenheit scale is a similar "practical" scale. It is now defined by

$$T(°F) \equiv T(°R) - 459.67 = \tfrac{9}{5}T(°C) + 32 \qquad (2.44)$$

The Fahrenheit temperature of ice and water at 1 atm pressure is roughly 32°F; the temperature of boiling water at 1 atm pressure is about 212°F; and room temperatures are in the vicinity of 70°F. More suggestive of the presumptive origins of this scale are the facts that ice, salt, and water coexist in equilibrium at 1 atm pressure at a temperature in the vicinity of 0°F, and that the body (i.e., rectal) temperature of a cow is roughly 100°F.

Although we have defined the temperature formally in terms of a partial derivative of the fundamental relation, we briefly note the conventional method of introduction of the temperature concept, as developed by Kelvin and Caratheodory. The heat flux dQ is first defined very much as we have introduced it in connection with the energy conservation principle. From the consideration of certain cyclic processes it is then inferred that there exists an integrating factor $(1/T)$ such that the product of this integrating factor with the imperfect differential dQ is a perfect differential (dS).

$$dS = \frac{1}{T}dQ \qquad (2.45)$$

[2]A very short but fascinating review of the history of temperature scales is given by E. R. Jones, Jr., *The Physics Teacher* **18**, 594 (1980).

The temperature and the entropy thereby are introduced by analysis of the existence of integrating factors in particular types of differential equations called *Pfaffian forms*.

PROBLEMS

2.6-1. The temperature of a system composed of ice, water, and water vapor in mutual equilibrium has a temperature of *exactly* 273.16 K, by definition. The temperature of a system of ice and water at 1 atm of pressure is then measured as 273.15 K, with the third and later decimal places uncertain. The temperature of a system of water and water vapor (i.e., boiling water) at 1 atm is measured as 373.15 K ± 0.01 K. Compute the temperature of water–water vapor at 1 atm, with its probable error, on the Celsius, absolute Fahrenheit, and Fahrenheit scales.

2.6-2. The "gas constant" R is defined as the product of Avogadro's number ($N_A = 6.0225 \times 10^{23}$/mole) and Boltzmann's constant $R \equiv N_A k_B$. Correspondingly $R \simeq 8.314$ J/mole K. Since the size of the Celsius degree is the same as the size of Kelvin degree, it has the value 8.314 J/mole°C. Express R in units of J/mole°F.

2.6-3. Two particular systems have the following equations of state:

$$\frac{1}{T^{(1)}} = \frac{3}{2} R \frac{N^{(1)}}{U^{(1)}}$$

and

$$\frac{1}{T^{(2)}} = \frac{5}{2} R \frac{N^{(2)}}{U^{(2)}}$$

where R is the gas constant (Problem 2.6-2). The mole number of the first system is $N^{(1)} = 2$ and that of the second is $N^{(2)} = 3$. The two systems are separated by a diathermal wall, and the total energy in the composite system is 2.5×10^3 J. What is the internal energy of each system in equilibrium?

Answer:
$$U^{(1)} = 714.3 \text{ J}$$

2.6-4. Two systems with the equations of state given in Problem 2.6-3 are separated by a diathermal wall. The respective mole numbers are $N^{(1)} = 2$ and $N^{(2)} = 3$. The initial temperatures are $T^{(1)} = 250$ K and $T^{(2)} = 350$ K. What are the values of $U^{(1)}$ and $U^{(2)}$ after equilibrium has been established? What is the equilibrium temperature?

2-7 MECHANICAL EQUILIBRIUM

A second application of the extremum principle for the entropy yields an even simpler result and therefore is useful in making the procedure clear. We consider a closed composite system consisting of two simple systems separated by a movable diathermal wall that is impervious to the flow of matter. The values of the mole numbers are fixed and constant, but the values of $U^{(1)}$ and $U^{(2)}$ can change, subject only to the closure condition

$$U^{(1)} + U^{(2)} = \text{constant} \tag{2.46}$$

and the values of $V^{(1)}$ and $V^{(2)}$ can change, subject only to the closure condition

$$V^{(1)} + V^{(2)} = \text{constant} \tag{2.47}$$

The extremum principle requires that no change in entropy result from infinitesimal virtual processes consisting of transfer of heat across the wall or of displacement of the wall.

Then

$$dS = 0 \tag{2.48}$$

where

$$dS = \left(\frac{\partial S^{(1)}}{\partial U^{(1)}} \right)_{V^{(1)}, \dots N_k^{(1)}, \dots} dU^{(1)} + \left(\frac{\partial S^{(1)}}{\partial V^{(1)}} \right)_{U^{(1)}, \dots N_k^{(1)}, \dots} dV^{(1)}$$

$$+ \left(\frac{\partial S^{(2)}}{\partial U^{(2)}} \right)_{V^{(2)}, \dots N_k^{(2)}, \dots} dU^{(2)} + \left(\frac{\partial S^{(2)}}{\partial V^{(2)}} \right)_{U^{(2)}, \dots N_k^{(2)}, \dots} dV^{(2)} \tag{2.49}$$

By the closure conditions

$$dU^{(2)} = -dU^{(1)} \tag{2.50}$$

and

$$dV^{(2)} = -dV^{(1)} \tag{2.51}$$

whence

$$dS = \left(\frac{1}{T^{(1)}} - \frac{1}{T^{(2)}} \right) dU^{(1)} + \left(\frac{P^{(1)}}{T^{(1)}} - \frac{P^{(2)}}{T^{(2)}} \right) dV^{(1)} = 0 \tag{2.52}$$

As this expression must vanish for arbitrary and independent values of $dU^{(1)}$ and $dV^{(1)}$, we must have

$$\frac{1}{T^{(1)}} - \frac{1}{T^{(2)}} = 0 \tag{2.53}$$

and

$$\frac{P^{(1)}}{T^{(1)}} - \frac{P^{(2)}}{T^{(2)}} = 0 \qquad (2.54)$$

Although these two equations are the equilibrium conditions in the proper form appropriate to the entropy representation, we note that they imply the physical conditions of equality of both temperature and pressure.

$$T^{(1)} = T^{(2)} \qquad (2.55)$$

$$P^{(1)} = P^{(2)} \qquad (2.56)$$

The equality of the temperatures is just our previous result for equilibrium with a diathermal wall. The equality of the pressures is the new feature introduced by the fact that the wall is movable. Of course, the equality of the pressures is precisely the result that we would expect on the basis of mechanics, and this result corroborates the identification of the function P as the mechanical pressure.

Again we stress that this result is a formal solution of the given problem. In the entropy representation, $1/T^{(1)}$ is a function of $U^{(1)}$, $V^{(1)}$, and $N^{(1)}$ (an entropic equation of state), so that equation 2.53 is formally a relationship among $U^{(1)}$, $V^{(1)}$, $U^{(2)}$, and $V^{(2)}$ (with $N^{(1)}$ and $N^{(2)}$ each held fixed). Similarly $P^{(1)}/T^{(1)}$ is a function of $U^{(1)}$, $V^{(1)}$, and $N^{(1)}$, so that equation 2.54 is a second relationship among $U^{(1)}$, $V^{(1)}$, $U^{(2)}$, and $V^{(2)}$. The two conservation equations 2.46 and 2.47 complete the four equations required to determine the four sought-for variables. Again thermodynamics provides the methodology, which becomes explicit when applied to a concrete system with a definite fundamental relation, or with known equations of state.

The case of a moveable adiabatic (rather than diathermal) wall presents a unique problem with subtleties that are best discussed after the formalism is developed more fully; we shall return to that case in Problem 2.7-3 and in Problem 5.1-2.

Example 1

Three cylinders of identical cross-sectional areas are fitted with pistons, and each contains a gaseous system (not necessarily of the same composition). The pistons are connected to a rigid bar hinged on a fixed fulcrum, as indicated in Fig. 2.1. The "moment arms," or the distances from the fulcrum, are in the ratio of $1:2:3$. The cylinders rest on a heat conductive table of negligible mass; the table makes no contribution to the physics of the problem except to ensure that the three cylinders are in diathermal contact. The entire system is isolated and no pressure acts on the external surfaces of the pistons. Find the ratio of pressures and of temperatures in the three cylinders.

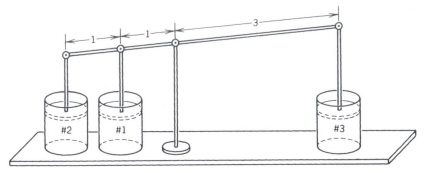

FIGURE 2.1
Three volume-coupled systems (Example 2.7-1).

Solution

The closure condition for the total energy is

$$\delta U^{(1)} + \delta U^{(2)} + \delta U^{(3)} = 0$$

and the coupling of the pistons imposes the conditions that

$$\delta V^{(2)} = 2\,\delta V^{(1)}$$

and

$$\delta V^{(3)} = -3\,\delta V^{(1)}$$

Then the extremal property of the entropy is

$$\delta S = \frac{1}{T^{(1)}}\,\delta U^{(1)} + \frac{1}{T^{(2)}}\,\delta U^{(2)} + \frac{1}{T^{(3)}}\,\delta U^{(3)} + \frac{P^{(1)}}{T^{(1)}}\,\delta V^{(1)}$$

$$+ \frac{P^{(2)}}{T^{(2)}}\,\delta V^{(2)} + \frac{P^{(3)}}{T^{(3)}}\,\delta V^{(3)} = 0$$

Eliminating $U^{(3)}$, $V^{(2)}$, and $V^{(3)}$

$$\delta S = \left(\frac{1}{T^{(1)}} - \frac{1}{T^{(3)}}\right)\delta U^{(1)} + \left(\frac{1}{T^{(2)}} - \frac{1}{T^{(3)}}\right)\delta U^{(2)}$$

$$+ \left(\frac{P^{(1)}}{T^{(1)}} + 2\frac{P^{(2)}}{T^{(2)}} - 3\frac{P^{(3)}}{T^{(3)}}\right)\delta V^{(1)} = 0$$

The remaining three variations $\delta U^{(1)}$, $\delta U^{(2)}$, and $\delta V^{(1)}$ are arbitrary and unconstrained, so that the coefficient of each must vanish separately. From the coefficient of $\delta U^{(1)}$ we find $T^{(1)} = T^{(3)}$, and from the coefficient of $\delta U^{(2)}$ we find $T^{(2)} = T^{(3)}$. Hence all three systems come to a common final temperature. From the coefficient of $\delta V^{(1)}$, and using the equality of the temperatures, we find

$$P^{(1)} + 2P^{(2)} = 3P^{(3)}$$

This is the expected result, embodying the familiar mechanical principle of the lever. Explicit knowledge of the equations of state would enable us to convert this into a solution for the volumes of the three systems.

PROBLEMS

2.7-1. Three cylinders are fitted with four pistons, as shown in Fig. 2.2. The cross-sectional areas of the cylinders are in the ratio $A_1 : A_2 : A_3 = 1 : 2 : 3$. Pairs of pistons are coupled so that their displacements (linear motions) are equal. The walls of the cylinders are diathermal and are connected by a heat conducting bar (crosshatched in the figure). The entire system is isolated (so that, for instance, there is no pressure exerted on the outer surfaces of the pistons). Find the ratios of pressures in the three cylinders.

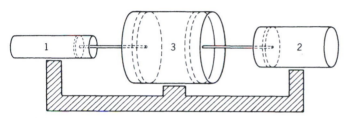

FIGURE 2.2
Three volume-coupled systems. (Problem 2.7-1)

2.7-2. Two particular systems have the following equations of state:

$$\frac{1}{T^{(1)}} = \frac{3}{2} R \frac{N^{(1)}}{U^{(1)}}, \qquad \frac{P^{(1)}}{T^{(1)}} = R \frac{N^{(1)}}{V^{(1)}}$$

and

$$\frac{1}{T^{(2)}} = \frac{5}{2} R \frac{N^{(2)}}{U^{(2)}}, \qquad \frac{P^{(2)}}{T^{(2)}} = R \frac{N^{(2)}}{V^{(2)}}$$

The mole number of the first system is $N^{(1)} = 0.5$ and that of the second is $N^{(2)} = 0.75$. The two systems are contained in a closed cylinder, separated by a fixed, adiabatic, and impermeable piston. The initial temperatures are $T^{(1)} = 200$ K and $T^{(2)} = 300$ K, and the total volume is 20 liters. The "setscrew" which prevents the motion of the piston is then removed, and simultaneously the adiabatic insulation of the piston is stripped off, so that the piston becomes moveable, diathermal, and impermeable. What is the energy, volume, pressure, and temperature of each subsystem when equilibrium is established?

It is sufficient to take $R \simeq 8.3$ J/mole K and to assume the external pressure to be zero.

Answer:
$U^{(1)} = 1700$ J

2.7-3. The hypothetical problem of equilibrium in a closed composite system with an internal moveable *adiabatic* wall is a unique *indeterminate* problem. Physically, release of the piston would lead it to perpetual oscillation in the absence of viscous damping. With viscous damping the piston would eventually come to rest at such a position that the pressures on either side would be equal, but the temperatures in each subsystem would then depend on the relative viscosity in each subsystem. The solution of this problem depends on *dynamical* considerations. Show that the application of the entropy maximum formalism is correspondingly indeterminate with respect to the temperatures (but determinate with respect to the pressures).

Hint: First show that with $dU^{(1)} = -P^{(1)}dV^{(1)}$, and similarly for subsystem 2, energy conservation gives $P^{(1)} = P^{(2)}$. Then show that the entropy maximum condition vanishes identically, giving no solution for $T^{(1)}$ or $T^{(2)}$.

2-8 EQUILIBRIUM WITH RESPECT TO MATTER FLOW

Consideration of the flow of matter provides insight into the nature of the chemical potential. We consider the equilibrium state of two simple systems connected by a rigid and diathermal wall, permeable to one type of material (N_1) and impermeable to all others ($N_2, N_3, \ldots, N_r$). We seek the equilibrium values of $U^{(1)}$ and $U^{(2)}$ and of $N_1^{(1)}$ and $N_1^{(2)}$. The virtual change in entropy in the appropriate virtual process is

$$dS = \frac{1}{T^{(1)}} dU^{(1)} - \frac{\mu_1^{(1)}}{T^{(1)}} dN_1^{(1)} + \frac{1}{T^{(2)}} dU^{(2)} - \frac{\mu_1^{(2)}}{T^{(2)}} dN_1^{(2)} \quad (2.57)$$

and the closure conditions demand

$$dU^{(2)} = -dU^{(1)} \quad (2.58)$$

and

$$dN_1^{(2)} = -dN_1^{(1)} \quad (2.59)$$

whence

$$dS = \left(\frac{1}{T^{(1)}} - \frac{1}{T^{(2)}} \right) dU^{(1)} - \left(\frac{\mu_1^{(1)}}{T^{(1)}} - \frac{\mu_1^{(2)}}{T^{(2)}} \right) dN_1^{(1)} \quad (2.60)$$

As dS must vanish for arbitrary values of both $dU^{(1)}$ and $dN_1^{(1)}$, we find as the conditions of equilibrium

$$\frac{1}{T^{(1)}} = \frac{1}{T^{(2)}} \quad (2.61)$$

and

$$\frac{\mu_1^{(1)}}{T^{(1)}} = \frac{\mu_1^{(2)}}{T^{(2)}} \qquad \left(\text{hence also } \mu_1^{(1)} = \mu_1^{(2)} \right) \quad (2.62)$$

Thus, just as the temperature can be looked upon as a sort of "potential" for heat flux and the pressure can be looked upon as a sort of "potential" for volume changes, so the chemical potential can be looked upon as a sort of "potential" for matter flux. A difference in chemical potential provides a "generalized force" for matter flow.

The direction of the matter flow can be analyzed by the same method used in Section 2.5 to analyze the direction of the heat flow. If we assume that the temperatures $T^{(1)}$ and $T^{(2)}$ are equal, equation 2.60 becomes

$$dS = \frac{\mu_1^{(2)} - \mu_1^{(1)}}{T} \, dN_1^{(1)} \tag{2.63}$$

If $\mu_1^{(1)}$ is greater than $\mu_1^{(2)}$, $dN_1^{(1)}$ will be negative, since dS must be positive. Thus matter tends to flow from regions of high chemical potential to regions of low chemical potential.

In later chapters we shall see that the chemical potential provides the generalized force not only for the flow of matter from point to point but also for its changes of phase and for chemical reactions. The chemical potential thus plays a dominant role in theoretical chemistry.

The units of chemical potential are joules per mole (or any desired energy unit per mole).

PROBLEMS

2.8-1. The fundamental equation of a particular type of two-component system is

$$S = NA + NR \ln \frac{U^{3/2}V}{N^{5/2}} - N_1 R \ln \frac{N_1}{N} - N_2 R \ln \frac{N_2}{N}$$

$$N \equiv N_1 + N_2$$

where A is an unspecified constant. A closed rigid cylinder of total volume 10 liters is divided into two chambers of equal volume by a diathermal rigid membrane, permeable to the first component but impermeable to the second. In one chamber is placed a sample of the system with original parameters $N_1^{(1)} = 0.5$, $N_2^{(1)} = 0.75$, $V^{(1)} = 5$ liters, and $T^{(1)} = 300$ K. In the second chamber is placed a sample with original parameters $N_1^{(2)} = 1$, $N_2^{(2)} = 0.5$, $V^{(2)} = 5$ liters, and $T^{(2)} = 250$ K. After equilibrium is established, what are the values of $N_1^{(1)}$, $N_1^{(2)}$, T, $P^{(1)}$, and $P^{(2)}$?

Answer:
$T = 272.7$ K

2.8-2. A two-component gaseous system has a fundamental equation of the form

$$S = AU^{1/3}V^{1/3}N^{1/3} + \frac{BN_1 N_2}{N}, \qquad N = N_1 + N_2$$

where A and B are positive constants. A closed cylinder of total volume $2V_0$ is separated into two equal subvolumes by a rigid diathermal partition permeable only to the first component. One mole of the first component, at a temperature T_ℓ, is introduced in the left-hand subvolume, and a mixture of $\frac{1}{2}$ mole of each component, at a temperature T_r, is introduced into the right-hand subvolume.

Find the equilibrium temperature T_e and the mole numbers in each subvolume when the system has come to equilibrium, assuming that $T_r = 2T_\ell = 400$ K and that $37B^2 = 100A^3V_0$. Neglect the heat capacity of the walls of the container!

Answer:

$$N_{1e} = 0.9$$

2-9 CHEMICAL EQUILIBRIUM

Systems that can undergo chemical reactions bear a strong formal similarity to the diffusional systems considered in the preceding section. Again they are governed by equilibrium conditions expressed in terms of the chemical potential μ—whence derives its name *chemical potential*.

In a chemical reaction the mole numbers of the system change, some increasing at the expense of a decrease in others. The relationships among the changing mole numbers are governed by chemical reaction equations such as

$$2H_2 + O_2 \rightleftharpoons 2H_2O \tag{2.64}$$

or

$$2O \rightleftharpoons O_2 \tag{2.65}$$

The meaning of the first of these equations is that the changes in the mole numbers of hydrogen, oxygen, and water stand in the ratio of $-2 : -1 : +2$. More generally one writes a chemical reaction equation, for a system with r components, in the form

$$0 \rightleftharpoons \sum_j \nu_j A_j \tag{2.66}$$

The ν_j are the "stoichiometric coefficients" ($-2, -1, +2$ for the reaction of hydrogen and oxygen to form water), and the A_j are the symbols for the chemical components ($A_1 = H_2$, $A_2 = O_2$, and $A_3 = H_2O$ for the preceding reaction). If the reaction is viewed in the reverse sense (for instance, as the *dissociation* of water to hydrogen plus oxygen) the opposite signs would be assigned to each of the ν_j; this is a matter of arbitrary choice and only the relative signs of the ν_j are significant.

The fundamental equation of the system is

$$S = S(U, V, N_1, N_2, \ldots, N_r) \qquad (2.67)$$

In the course of the chemical reaction both the total energy U and the total volume V remain fixed, the system being considered to be enclosed in an adiabatic and rigid "reaction vessel." This is not the most common boundary condition for chemical reactions, which are more often carried out in open vessels, free to interchange energy and volume with the ambient atmosphere; we shall return to these open boundary conditions in Section 6.4.

The change in entropy in a virtual chemical process is then

$$dS = -\sum_{j=1}^{r} \frac{\mu_j}{T} \, dN_j \qquad (2.68)$$

However, the changes in the mole numbers are proportional to the stoichiometric coefficients ν_j. Let the factor of proportionality be denoted by $d\tilde{N}$, so that

$$dS = -\frac{d\tilde{N}}{T} \sum_{j=1}^{r} \mu_j \nu_j \qquad (2.69)$$

Then the extremum principle dictates that, in equilibrium

$$\sum_{j=1}^{r} \mu_j \nu_j = 0 \qquad (2.70)$$

If the equations of state of the mixture are known, the equilibrium condition (2.70) permits explicit solution for the final mole numbers.

It is of interest to examine this "solution in principle" in a slightly richer case. If hydrogen, oxygen, and carbon dioxide are introduced into a vessel the following chemical reactions may occur.

$$H_2 + \tfrac{1}{2} O_2 \rightleftharpoons H_2O$$

$$CO_2 + H_2 \rightleftharpoons CO + H_2O \qquad (2.71)$$

$$CO + \tfrac{1}{2} O_2 \rightleftharpoons CO_2$$

In equilibrium we then have

$$\mu_{H_2} + \tfrac{1}{2}\mu_{O_2} = \mu_{H_2O}$$

$$\mu_{CO_2} + \mu_{H_2} = \mu_{CO} + \mu_{H_2O} \qquad (2.72)$$

$$\mu_{CO} + \tfrac{1}{2}\mu_{O_2} = \mu_{CO_2}$$

These constitute *two* independent equations, for the first equation is simply the sum of the two following equations (just as the first chemical reaction is the net result of the two succeeding reactions). The amounts of hydrogen, oxygen, and carbon introduced into the system (in whatever chemical combinations) specify three additional constraints. There are thus five constraints, and there are precisely five mole numbers to be found (the quantities of H_2, O_2, H_2O, CO_2, and CO). The problem is thereby solved in principle.

As we observed earlier, chemical reactions more typically occur in open vessels with only the final pressure and temperature determined. The number of variables is then increased by two (the energy and the volume) but the specification of T and P provides two additional constraints. Again the problem is determinate.

We shall return to a more thorough discussion of chemical reactions in Section 6.4. For now it is sufficient to stress that the chemical potential plays a role in matter transfer or chemical reactions fully analogous to the role of temperature in heat transfer or pressure in volume transfer.

PROBLEMS

2.9-1. The hydrogenation of propane (C_3H_8) to form methane (CH_4) proceeds by the reaction

$$C_3H_8 + 2H_2 \rightleftharpoons 3CH_4$$

Find the relationship among the chemical potentials and show that both the problem and the solution are formally identical to Example 1 on mechanical equilibrium.

3

SOME FORMAL RELATIONSHIPS, AND SAMPLE SYSTEMS

3-1 THE EULER EQUATION

Having seen how the fundamental postulates lead to a solution of the equilibrium problem, we now pause to examine in somewhat greater detail the mathematical properties of fundamental equations.

The homogeneous first-order property of the fundamental relation permits that equation to be written in a particularly convenient form, called the Euler form.

From the definition of the homogeneous first-order property we have, for any λ

$$U(\lambda S, \lambda X_1, \ldots, \lambda X_t) = \lambda U(S, X_1, \ldots, X_t) \tag{3.1}$$

Differentiating with respect to λ

$$\frac{\partial U(\ldots, \lambda X_k, \ldots)}{\partial(\lambda S)} \frac{\partial(\lambda S)}{\partial \lambda} + \frac{\partial U(\ldots, \lambda X_k, \ldots)}{\partial(\lambda X_j)} \frac{\partial(\lambda X_j)}{\partial \lambda}$$

$$+ \cdots = U(S, X_1, \ldots, X_t) \tag{3.2}$$

or

$$\frac{\partial U(\ldots, \lambda X_k, \ldots)}{\partial(\lambda S)} S + \sum_{j=1}^{t} \frac{\partial U(\ldots, \lambda X_k, \ldots)}{\partial(\lambda X_j)} X_j$$

$$= U(S, X_1, \ldots, X_t) \tag{3.3}$$

This equation is true for any λ and in particular for $\lambda = 1$, in which case it takes the form

$$\frac{\partial U}{\partial S}S + \sum_{j=1}^{t}\frac{\partial U}{\partial X_j}X_j + \cdots = U \tag{3.4}$$

$$U = TS + \sum_{j=1}^{t}P_jX_j \tag{3.5}$$

For a simple system in particular we have

$$U = TS - PV + \mu_1 N_1 + \cdots + \mu_r N_r \tag{3.6}$$

The relation 3.5 or 3.6 is the particularization to thermodynamics of the Euler theorem on homogeneous first-order forms. The foregoing development merely reproduces the standard mathematical derivation. We refer to equation 3.5 or 3.6 as the Euler relation.

In the entropy representation the Euler relation takes the form

$$S = \sum_{j=0}^{t}F_jX_j \tag{3.7}$$

or

$$S = \left(\frac{1}{T}\right)U + \left(\frac{P}{T}\right)V - \sum_{k=1}^{r}\left(\frac{\mu_k}{T}\right)N_k \tag{3.8}$$

PROBLEMS

3.1-1. Write each of the five physically acceptable fundamental equations of Problem 1.10-1 in the Euler form.

3-2   THE GIBBS–DUHEM RELATION

In Chapter 2 we arrived at equilibrium criteria involving the temperature, pressure, and chemical potentials. Each of the intensive parameters entered the theory in a similar way, and the formalism is, in fact, symmetric in the several intensive parameters. Despite this symmetry, however, the reader is apt to feel an intuitive response to the concepts of temperature and pressure, which is lacking, at least to some degree, in the case of the chemical potential. It is of interest, then, to note that the intensive parameters are not all independent. There is a relation among the intensive parameters, and for a single-component system μ is a function of T and P.

The existence of a relationship among the various intensive parameters is a consequence of the homogeneous first-order property of the fundamental relation. For a single-component system this property permits the fundamental relation to be written in the form $u = u(s, v)$, as in equation 2.19; each of the three intensive parameters is then also a function of s and v. Elimination of s and v from among the three equations of state yields a relation among T, P, and μ.

The argument can easily be extended to the more general case, and it again consists of a straightforward counting of variables. Suppose we have a fundamental equation in $(t + 1)$ extensive variables

$$U = U(S, X_1, X_2, \ldots, X_t) \tag{3.9}$$

yielding, in turn, $t + 1$ equations of state

$$P_k = P_k(S, X_1, X_2, \ldots, X_t) \tag{3.10}$$

If we choose the parameter λ of equation 2.14 as $\lambda = 1/X_t$, we then have

$$P_k = P_k(S/X_t, X_1/X_t, \ldots, X_{t-1}/X_t, 1) \tag{3.11}$$

Thus each of the $(t + 1)$ intensive parameters is a function of just t variables. Elimination of these t variables among the $(t + 1)$ equations yields the desired relation among the intensive parameters.

To find the explicit functional relationship that exists among the set of intensive parameters would require knowledge of the explicit fundamental equation of the system. That is, the analytic form of the relationship varies from system to system. Given the fundamental relation, the procedure is evident and follows the sequence of steps indicated by equations 3.9 through 3.11.

A differential form of the relation among the intensive parameters can be obtained directly from the Euler relation and is known as the Gibbs–Duhem relation. Taking the infinitesimal variation of equation 3.5, we find

$$dU = T\,dS + S\,dT + \sum_{j=1}^{t} P_j\,dX_j + \sum_{j=1}^{t} X_j\,dP_j \tag{3.12}$$

But, in accordance with equation 2.6, we certainly know that

$$dU = T\,dS + \sum_{j=1}^{t} P_j\,dX_j \tag{3.13}$$

whence, by subtraction we find the Gibbs–Duhem relation

$$S\,dT + \sum_{j=1}^{t} X_j\,dP_j = 0 \tag{3.14}$$

For a single-component simple system, in particular, we have

$$S\,dT - V\,dP + N\,d\mu = 0 \tag{3.15}$$

or

$$d\mu = -s\,dT + v\,dP \tag{3.16}$$

The variation in chemical potential is not independent of the variations in temperature and pressure, but the variation of any one can be computed in terms of the variations of the other two.

The Gibbs–Duhem relation presents the relationship among the intensive parameters in differential form. Integration of this equation yields the relation in explicit form, and this is a procedure alternative to that presented in equations 3.9 through 3.11. In order to integrate the Gibbs–Duhem relation, one must know the equations of state that enable one to write the X_j's in terms of the P_j's, or vice versa.

The number of intensive parameters capable of independent variation is called the number of *thermodynamic degrees of freedom* of a given system. *A simple system of r components has r + 1 thermodynamic degrees of freedom.*

In the entropy representation the Gibbs–Duhem relation again states that the sum of products of the extensive parameters and the differentials of the corresponding intensive parameters vanishes.

$$\sum_{j=0}^{t} X_j\,dF_j = 0 \tag{3.17}$$

or

$$U d\left(\frac{1}{T}\right) + V d\left(\frac{P}{T}\right) - \sum_{k=1}^{r} N_k d\left(\frac{\mu_k}{T}\right) = 0 \tag{3.18}$$

PROBLEMS

3.2-1. Find the relation among T, P, and μ for the system with the fundamental equation

$$U = \left(\frac{v_0^2 \theta}{R^3}\right)\frac{S^4}{NV^2}$$

3-3 SUMMARY OF FORMAL STRUCTURE

Let us now summarize the structure of the thermodynamic formalism in the energy representation. For the sake of clarity, and in order to be explicit, we consider a single-component simple system. The fundamental equation

$$U = U(S, V, N) \tag{3.19}$$

contains *all* thermodynamic information about a system. With the definitions $T = \partial U/\partial S$, and so forth, the fundamental equation implies three equations of state

$$T = T(S, V, N) = T(s, v) \tag{3.20}$$

$$P = P(S, V, N) = P(s, v) \tag{3.21}$$

$$\mu = \mu(S, V, N) = \mu(s, v) \tag{3.22}$$

If *all three* equations of state are known, they may be substituted into the Euler relation, thereby recovering the fundamental equation. *Thus the totality of all three equations of state is equivalent to the fundamental equation* and contains all thermodynamic information about a system. Any single equation of state contains less thermodynamic information than the fundamental equation.

If two equations of state are known, the Gibbs–Duhem relation can be integrated to obtain the third. The equation of state so obtained will contain an undetermined integration constant. Thus two equations of state are sufficient to determine the fundamental equation, except for an undetermined constant.

A logically equivalent but more direct and generally more convenient method of obtaining the fundamental equation when two equations of state are given is by direct integration of the molar relation

$$du = T\,ds - P\,dv \tag{3.23}$$

Clearly, knowledge of $T = T(s, v)$ and $P = P(s, v)$ yields a differential equation in the three variables u, s, and v, and integration gives

$$u = u(s, v) \tag{3.24}$$

which is a fundamental equation. Again, of course, we have an undetermined constant of integration.

It is always possible to express the internal energy as a function of parameters other than S, V, and N. Thus we could eliminate S from $U = U(S, V, N)$ and $T = T(S, V, N)$ to obtain an equation of the form $U = U(T, V, N)$. However, I stress that such an equation is *not* a fundamental relation and does not contain all possible thermodynamic informa-

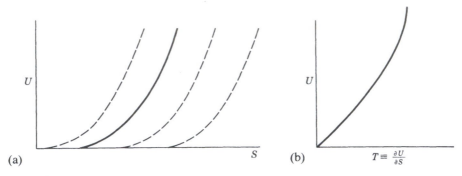

(a)

(b)

S

$T \equiv \frac{\partial U}{\partial S}$

FIGURE 3.1

tion about the system. In fact, recalling the definition of T as $\partial U/\partial S$, we see that $U = U(T, V, N)$ actually is a partial differential equation. Even if this equation were integrable, it would yield a fundamental equation with undetermined functions. Thus knowledge of the relation $U = U(S, V, N)$ allows one to compute the relation $U = U(T, V, N)$, but knowledge of $U = U(T, V, N)$ does not permit one inversely to compute $U = U(S, V, N)$. Associated with every equation there is both a truth value and an informational content. Each of the equations $U = U(S, V, N)$ and $U = U(T, V, N)$ may be true, but only the former has the optimum informational content.

These statements are graphically evident if we focus, for instance, on the dependence of U on S at constant V and N. Let that dependence be as shown in the solid curve in Fig. 3.1(a). This curve uniquely determines the dependence of U on T, shown in Fig. 3.1(b); for each point on the $U(S)$ curve there is a definite U and a definite slope $T = \partial U/\partial S$, determining a point on the $U(T)$ curve. Suppose, however, that we are given the $U(T)$ curve (an equation of state) and we seek to recover the fundamental $U(S)$ curve. Each of the dotted curves in Fig. 3.1(a) is *equally* compatible with the given $U(T)$ curve, for all have the same slope T at a given U. The curves differ by an arbitrary displacement, corresponding to the arbitrary "constant of integration" in the solution of the differential equation $U = U(\partial U/\partial S)$. Thus, Fig. 3.1(a) implies Fig. 3.1(b), but the reverse is not true. Equivalently stated, only $U = U(S)$ is a fundamental relation. The formal structure is illustrated by consideration of several specific and explicit systems in the following Sections of this book.

Example

A particular system obeys the equations

$$U = \tfrac{1}{2}PV$$

and

$$T^2 = \frac{AU^{3/2}}{VN^{1/2}}$$

where A is a positive constant. Find the fundamental equation.

Solution
Writing the two equations in the form of equations of state in the entropy representation (which is suggested by the appearance of U, V, and N as independent parameters)

$$\frac{1}{T} = A^{-1/2}u^{-3/4}v^{1/2}$$

$$\frac{P}{T} = 2A^{-1/2}u^{1/4}v^{-1/2}$$

Then the differential form of the molar fundamental equation (the analogue of equation 3.23) is

$$ds = \frac{1}{T}\,du + \frac{P}{T}\,dv$$

$$= A^{-1/2}(u^{-3/4}v^{1/2}\,du + 2u^{1/4}v^{-1/2}\,dv)$$

$$= 4A^{-1/2}d(u^{1/4}v^{1/2})$$

so that

$$s = 4A^{-1/2}u^{1/4}v^{1/2} + s_0$$

and

$$S = 4A^{-1/2}U^{1/4}V^{1/2}N^{1/4} + Ns_0$$

The reader should compare this method with the alternative technique of first integrating the Gibbs–Duhem relation to obtain $\mu(u, v)$, and then inserting the three equations of state into the Euler equation.

Particular note should be taken of the manner in which ds is integrated to obtain s. The equation for ds in terms of du and dv is a *partial differential equation*—it certainly can*not* be integrated term by term, nor by any of the familiar methods for ordinary differential equations in one independent variable. We have integrated the equation by "inspection"; simply "recognizing" that $u^{-3/4}v^{1/2}\,du + 2u^{1/4}v^{-1/2}\,dv$ is the differential of $u^{1/4}v^{1/2}$.

PROBLEMS

3.3-1. A particular system obeys the two equations of state

$$T = \frac{3As^2}{v}, \qquad \text{the } \textit{thermal equation of state}$$

and

$$P = \frac{As^3}{v^2}, \qquad \text{the } mechanical \text{ } equation \text{ } of \text{ } state$$

where A is constant.

a) Find μ as a function of s and v, and then find the fundamental equation.

b) Find the fundamental equation of this system by direct integration of the molar form of the equation.

3.3-2. It is found that a particular system obeys the relations

$$U = PV$$

and

$$P = BT^2$$

where B is constant. Find the fundamental equation of this system.

3.3-3. A system obeys the equations

$$P = - \frac{NU}{NV - 2AVU}$$

and

$$T = 2C\frac{U^{1/2}V^{1/2}}{N - 2AU}e^{AU/N}$$

Find the fundamental equation.

Hint: To integrate, let

$$s = Du^n v^m e^{-Au}$$

where D, n, and m are constants to be determined.

3.3-4. A system obeys the two equations $u = \frac{3}{2}Pv$ and $u^{1/2} = BTv^{1/3}$. Find the fundamental equation of this system.

3-4   THE SIMPLE IDEAL GAS AND MULTICOMPONENT SIMPLE IDEAL GASES

A "simple ideal gas" is characterized by the two equations

$$PV = NRT \qquad (3.25)$$

and

$$U = cNRT \qquad (3.26)$$

where c is a constant and R is the "universal gas constant" ($R = N_A k_B = 8.3144$ J/mole K).

Gases composed of noninteracting monatomic atoms (such as He, Ar, Ne) are observed to satisfy equations 3.25 and 3.26 at temperatures such that $k_B T$ is small compared to electronic excitation energies (i.e., $T \lesssim 10^4$ K), and at low or moderate pressures. All such "monatomic ideal gases" have a value of $c = \frac{3}{2}$.

Under somewhat more restrictive conditions of temperature and pressure other real gases may conform to the simple ideal gas equations 3.25 and 3.26, but with other values of the constant c. For diatomic molecules (such as O_2 or NO) there tends to be a considerable region of temperature for which $c \simeq \frac{5}{2}$ and another region of higher temperature for which $c \simeq \frac{7}{2}$ (with the boundary between these regions generally occurring at temperatures on the order of 10^3 K).

Equations 3.25 and 3.26 permit us to determine the fundamental equation. The explicit appearance of the energy U in one equation of state (equation 3.26) suggests the entropy representation. Rewriting the equations in the correspondingly appropriate form

$$\frac{1}{T} = cR\left(\frac{N}{U}\right) = \frac{cR}{u} \tag{3.27}$$

and

$$\frac{P}{T} = R\left(\frac{N}{V}\right) = \frac{R}{v} \tag{3.28}$$

From these two entropic equations of state we find the third equation of state

$$\frac{\mu}{T} = \text{function of } u, v \tag{3.29}$$

by integration of the Gibbs–Duhem relation

$$d\left(\frac{\mu}{T}\right) = ud\left(\frac{1}{T}\right) + vd\left(\frac{P}{T}\right) \tag{3.30}$$

Finally, the three equations of state will be substituted into the Euler equation

$$S = \left(\frac{1}{T}\right)U + \left(\frac{P}{T}\right)V - \left(\frac{\mu}{T}\right)N \tag{3.31}$$

Proceeding in this way the Gibbs–Duhem relation (3.30) becomes

$$d\left(\frac{\mu}{T}\right) = u \times \left(-\frac{cR}{u^2}\right)du + v \times \left(-\frac{R}{v^2}\right)dv = -cR\frac{du}{u} - R\frac{dv}{v} \tag{3.32}$$

and integrating

$$\frac{\mu}{T} - \left(\frac{\mu}{T}\right)_0 = -cR\ln\frac{u}{u_0} - R\ln\frac{v}{v_0} \tag{3.33}$$

Here u_0 and v_0 are the parameters of a fixed reference state, and $(\mu/T)_0$ arises as an undetermined constant of integration. Then, from the Euler relation (3.31)

$$S = Ns_0 + NR \ln\left[\left(\frac{U}{U_0}\right)^c\left(\frac{V}{V_0}\right)\left(\frac{N}{N_0}\right)^{-(c+1)}\right] \tag{3.34}$$

where

$$s_0 = (c + 1)R - \left(\frac{\mu}{T}\right)_0 \tag{3.35}$$

Equation 3.34 is the desired fundamental equation; if the integration constant s_0 were known equation 3.34 would contain all possible thermodynamic information about a simple ideal gas.

This procedure is neither the sole method, nor even the preferred method. Alternatively, and more directly, we could integrate the molar equation

$$ds = \left(\frac{1}{T}\right)du + \left(\frac{P}{T}\right)dv \tag{3.36}$$

which, in the present case, becomes

$$ds = c\left(\frac{R}{u}\right)du + \left(\frac{R}{v}\right)dv \tag{3.37}$$

giving, on integration,

$$s = s_0 + cR \ln\left(\frac{u}{u_0}\right) + R \ln\left(\frac{v}{v_0}\right) \tag{3.38}$$

This equation is equivalent to equation 3.34.

It should, perhaps, be noted that equation 3.37 is integrable term by term, despite our injunction (in Example 3) that such an approach generally is not possible. The segregation of the independent variables u and v in separate terms in equation 3.37 is a fortunate but unusual simplification which permits term by term integration in this special case.

A mixture of two or more simple ideal gases—a "multicomponent simple ideal gas"—is characterized by a fundamental equation which is most simply written in parametric form, with the temperature T playing the role of the parametric variable.

$$S = \sum_j N_j s_{j0} + \left(\sum_j N_j c_j\right) R \ln\frac{T}{T_0} + \sum_j N_j R \ln\left(\frac{V}{N_j v_0}\right)$$

$$U = \left(\sum_j N_j c_j\right) RT \tag{3.39}$$

Elimination of T between these equations gives a single equation of the standard form $S = S(U, V, N_1, N_2, \ldots)$.

Comparison of the individual terms of equations 3.39 with the expression for the entropy of a single-component ideal gas leads to the following interpretation (often referred to as Gibbs's Theorem). *The entropy of a mixture of ideal gases is the sum of the entropies that each gas would have if it alone were to occupy the volume V at temperature T.* The theorem is, in fact, true for all ideal gases (Chapter 13).

It is also of interest to note that the first of equations 3.39 can be written in the form

$$S = \sum_j N_j s_{j0} + \left(\sum_j N_j c_j\right) R \ln\frac{T}{T_0} + NR \ln\frac{V}{Nv_0} - R\sum_j N_j \ln\frac{N_j}{N}$$

$$(3.40)$$

and the last term is known as the "entropy of mixing." *It represents the difference in entropies between that of a mixture of gases and that of a collection of separate gases each at the same temperature and the same density as the original mixture* $N_j/V_j = N/V$, (and hence at the same pressure as the original mixture; see Problem 3.4-15. The close similarity, and the important distinction, between Gibbs's theorem and the interpretation of the entropy of mixing of ideal gases should be noted carefully by the reader. An application of the entropy of mixing to the problem of isotope separation will be given in Section 4.4 (Example 4).

Gibbs's theorem is demonstrated very neatly by a simple "thought experiment." A cylinder (Fig. 3.2) of total volume $2V_0$ is divided into four chambers (designated as $\alpha, \beta, \gamma, \delta$) by a fixed wall in the center and by two sliding walls. The two sliding walls are coupled together so that their distance apart is always one half the length of the cylinder ($V_\alpha = V_\gamma$ and $V_\beta = V_\delta$). Initially, the two sliding walls are coincident with the left end and the central fixed partition, respectively, so that $V_\alpha = V_\gamma = 0$. The chamber β, of volume V_0, is filled with a mixture of N_0 moles of a simple ideal gas A and N_0 moles of a simple ideal gas B. Chamber δ is initially evacuated. The entire system is maintained at temperature T.

The left-hand sliding wall is permeable to component A, but not to component B. The fixed partition is permeable to component B, but not to component A. The right-hand sliding wall is impermeable to either component.

The coupled sliding walls are then pushed quasi-statically to the right until $V_\beta = V_\delta = 0$ and $V_\alpha = V_\gamma = V_0$. Chamber α then contains pure A and chamber γ contains pure B. The initial mixture, of volume V_0, thereby is separated into two pure components, each of volume V_0. According to Gibbs's theorem the final entropy should be equal to the initial entropy, and we shall now see directly that this is, in fact, true.

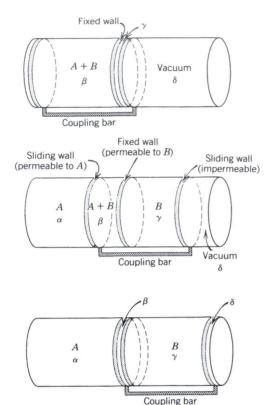

FIGURE 3.2
Separation of a mixture of ideal gases, demonstrating Gibbs's theorem.

We first note that the second of equations 3.39, stating that the energy is a function of only T and the mole number, ensures that the final energy is equal to the initial energy of the system. Thus $-T\Delta S$ is equal to the work done in moving the coupled walls.

The condition of equilibrium with respect to transfer of component A across the left-hand wall is $\mu_{A,\alpha} = \mu_{A,\beta}$. It is left to Problem 3.4-14 to show that the conditions $\mu_{A,\alpha} = \mu_{A,\beta}$ and $\mu_{B,\beta} = \mu_{B,\gamma}$ imply that

$$P_\alpha = P_\gamma \quad \text{and} \quad P_\beta = 2P_\alpha$$

That is, *the total force on the coupled moveable walls* $(P_\alpha - P_\beta + P_\gamma)$ *vanishes.* Thus no work is done in moving the walls, and consequently no entropy change accompanies the process. The entropy of the original mixture of A and B, in a common volume V_0, is precisely equal to the entropy of pure A and pure B, each in a separate volume V_0. This is Gibbs's theorem.

Finally, we note that the simple ideal gas considered in this section is a special case of the general ideal gas, which encompasses a very wide class of real gases at low or moderate pressures. The general ideal gas is again characterized by the mechanical equation of state $PV = NRT$ (equation 3.25), and by an energy that again is a function of the temperature only—but not simply a linear function. The general ideal gas will be discussed in detail in Chapter 13, and statistical mechanical derivations of the fundamental equations will emerge in Chapter 16.

PROBLEMS

Note that Problems 3.4-1, 3.4-2, 3.4-3, and 3.4-8 refer to "quasi-static processes"; such processes are to be interpreted not as real processes but merely as loci of equilibrium states. Thus we can apply thermodynamics to such quasi-static "processes"; the work done in a quasi-static change of volume (from V_1 to V_2) is $W = -\int P \, dV$ and the heat transfer is $Q = \int T \, dS$. The relationship of real processes to these idealized "quasi-static processes" will be discussed in Chapter 4.

3.4-1. A "constant volume ideal gas thermometer" is constructed as shown (schematically) in Fig. 3.3. The bulb containing the gas is constructed of a material with a negligibly small coefficient of thermal expansion. The point A is a reference point marked on the stem of the bulb. The bulb is connected by a flexible tube to a reservoir of liquid mercury, open to the atmosphere. The mercury reservoir is raised or lowered until the mercury miniscus coincides with the reference point A. The height h of the mercury column is then read.
a) Show that the pressure of the gas is the sum of the external (atmospheric) pressure plus the height h of the mercury column multiplied by the weight per unit volume of mercury (as measured at the temperature of interest).
b) Using the equation of state of the ideal gas, explain how the temperature of the gas is then evaluated.

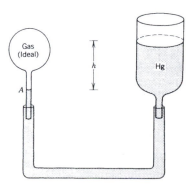

FIGURE 3.3
Constant-volume ideal gas thermometer.

c) Describe a "constant pressure ideal gas thermometer" (in which a changing volume is directly measured at constant pressure).

3.4-2. Show that the relation between the volume and the pressure of a monatomic ideal gas undergoing a quasi-static adiabatic compression ($dQ = T\,dS = 0$, $S = $ constant) is

$$Pv^{5/3} = \left(P_0 v_0^{5/3} e^{-2s_0/3R} \right) e^{2s/3R} = \text{constant}$$

Sketch a family of such "adiabats" in a graph of P versus V. Find the corresponding relation for a simple ideal gas.

3.4-3. Two moles of a monatomic ideal gas are at a temperature of 0°C and a volume of 45 liters. The gas is expanded adiabatically ($dQ = 0$) and quasi-statically until its temperature falls to $-50°C$. What are its initial and final pressures and its final volume?

Answer:

$P_i = 0.1$ MPa, $V_f = 61 \times 10^{-3}$ m^3

3.4-4. By carrying out the integral $\int P\,dV$, compute the work done by the gas in Problem 3.4-3. Also compute the initial and final energies, and corroborate that the difference in these energies is the work done.

3.4-5. In a particular engine a gas is compressed in the initial stroke of the piston. Measurements of the instantaneous temperature, carried out during the compression, reveal that the temperature increases according to

$$T = \left(\frac{V}{V_0} \right)^{\eta} T_0$$

where T_0 and V_0 are the initial temperature and volume, and η is a constant. The gas is compressed to the volume V_1 (where $V_1 < V_0$). Assume the gas to be monatomic ideal, and assume the process to be quasi-static.
a) Calculate the work W done on the gas.
b) Calculate the change in energy ΔU of the gas.
c) Calculate the heat transfer Q to the gas (through the cylinder walls) by using the results of (a) and (b).
d) Calculate the heat transfer directly by integrating $dQ = T\,dS$.
e) From the result of (c) or (d), for what value of η is $Q = 0$? Show that for this value of η the locus traversed coincides with an adiabat (as calculated in Problem 3.4-2).

3.4-6. Find the three equations of state of the "simple ideal gas" (equation 3.34). Show that these equations of state satisfy the Euler relation.

3.4-7. Find the four equations of state of a two-component mixture of simple ideal gases (equations 3.39). Show that these equations of state satisfy the Euler relation.

3.4-8. If a monatomic ideal gas is permitted to expand into an evacuated region, thereby increasing its volume from V to λV, and if the walls are rigid and adiabatic, what is the ratio of the initial and final pressures? What is the ratio of the initial and final temperatures? What is the difference of the initial and final entropies?

3.4-9. A tank has a volume of 0.1 m³ and is filled with He gas at a pressure of 5×10^6 Pa. A second tank has a volume of 0.15 m³ and is filled with He gas at a pressure of 6×10^6 Pa. A valve connecting the two tanks is opened. Assuming He to be a monatomic ideal gas and the walls of the tanks to be adiabatic and rigid, find the final pressure of the system.

Hint: Note that the internal energy is constant.

Answer:
$$P_f = 5.6 \times 10^6 \text{ Pa}$$

3.4-10.

a) If the temperatures within the two tanks of Problem 3.4-9, before opening the valve, had been $T = 300$ K and 350 K, respectively, what would the final temperature be?

b) If the first tank had contained He at an initial temperature of 300 K, and the second had contained a diatomic ideal gas with $c = 5/2$ and an initial temperature of 350 K, what would the final temperature be?

Answer:
a) $T_f = 330$ K
b) $T_f = 337$ K

3.4-11. Show that the pressure of a multicomponent simple ideal gas can be written as the sum of "partial pressures" P_j, where $P_j \equiv N_j RT/V$. These "partial pressures" are purely formal quantities not subject to experimental observation. (From the mechanistic viewpoint of kinetic theory the partial pressure P_i is the contribution to the total pressure that results from bombardment of the wall by molecules of species i—a distinction that can be made only when the molecules are noninteracting, as in an ideal gas.)

3.4-12. Show that μ_j, the electrochemical potential of the jth component in a multicomponent simple ideal gas, satisfies

$$\mu_j = RT \ln\left(\frac{N_j v_0}{V}\right) + (\text{function of } T)$$

and find the explicit form of the "function of T."

Show that μ_j can be expressed in terms of the "partial pressure" (Problem 3.4-11) and the temperature.

3.4-13. An impermeable, diathermal, and rigid partition divides a container into two subvolumes, each of volume V. The subvolumes contain, respectively, one mole of H_2 and three moles of Ne. The system is maintained at constant temperature T. The partition is suddenly made permeable to H_2, but not to Ne, and equilibrium is allowed to reestablish. Find the mole numbers and the pressures.

3.4-14. Use the results of Problems 3.4-11 and 3.4-12 to establish the results $P_\alpha = P_\gamma$ and $P_\beta = 2P_\alpha$ in the demonstration of Gibbs's theorem at the end of this section.

3.4-15. An impermeable, diathermal and rigid partition divides a container into two subvolumes, of volumes nV_0 and mV_0. The subvolumes contain, respectively, n moles of H_2 and m moles of Ne, each to be considered as a simple ideal gas. The system is maintained at constant temperature T. The partition is suddenly ruptured and equilibrium is allowed to re-establish. Find the initial pressure in each subvolume and the final pressure. Find the change in entropy of the system. How is this result related to the "entropy of mixing" (the last term in equation 3.40)?

3-5 THE "IDEAL VAN DER WAALS FLUID"

Real gases seldom satisfy the ideal gas equation of state except in the limit of low density. An improvement on the mechanical equation of state (3.28) was suggested by J. D. van der Waals in 1873.

$$P = \frac{RT}{v - b} - \frac{a}{v^2} \tag{3.41}$$

Here a and b are two empirical constants characteristic of the particular gas. In strictly quantitative terms the success of the equation has been modest, and for detailed practical applications it has been supplanted by more complicated empirical equations with five or more empirical constants. Nevertheless the van der Waals equation is remarkably successful in representing the qualitative features of real fluids, including the gas–liquid phase transition.

The heuristic reasoning that underlies the van der Waals equation is intuitively plausible and informative, although that reasoning lies outside the domain of thermodynamics. The ideal gas equation $P = RT/v$ is known to follow from a model of point molecules moving independently and colliding with the walls to exert the pressure P. Two simple corrections to this picture are plausible. The first correction recognizes that the molecules are not point particles, but that each has a nonzero volume b/N_A. Accordingly, the volume V in the ideal gas equation is replaced by $V - Nb$; the total volume diminished by the volume Nb occupied by the molecules themselves.

The second correction arises from the existence of forces between the molecules. A molecule in the interior of the vessel is acted upon by intermolecular forces in all directions, which thereby tend to cancel. But a molecule approaching the wall of the container experiences a net backward attraction to the remaining molecules, and this force in turn reduces the effective pressure that the molecule exerts on colliding with the container wall. This diminution of the pressure should be proportional to the number of interacting *pairs* of molecules, or upon the square of the number of molecules per unit volume $(1/v^2)$; hence the second term in the van der Waals equation.

Statistical mechanics provides a more quantitative and formal derivation of the van der Waals equation, but it also reveals that there are an infinite series of higher order corrections beyond those given in equation 3.41. The truncation of the higher order terms to give the simple van der Waals equation results in an equation with appropriate *qualitative* features and with reasonable (but not optimum) quantitative accuracy.

The van der Waals equation must be supplemented with a thermal equation of state in order to define the system fully. It is instructive not simply to appeal to experiment, but rather to inquire as to the simplest possible (and reasonable) thermal equation of state that can be paired with the van der Waals equation of state. Unfortunately we are not free simply to adopt the thermal equation of state of an ideal gas, for thermodynamic formalism imposes a consistency condition between the two equations of state. We shall be forced to alter the ideal gas equation slightly.

We write the van der Waals equation as

$$\frac{P}{T} = \frac{R}{v-b} - \frac{a}{v^2}\frac{1}{T} \tag{3.42}$$

and the sought for additional equation of state should be of the form

$$\frac{1}{T} = f(u,v) \tag{3.43}$$

These two equations would permit us to integrate the molar equation

$$ds = \frac{1}{T}\,du + \frac{P}{T}\,dv \tag{3.44}$$

to obtain the fundamental equation. However, if ds is to be a perfect differential, it is required that the mixed second-order partial derivatives should be equal

$$\frac{\partial^2 s}{\partial v\,\partial u} = \frac{\partial^2 s}{\partial u\,\partial v} \tag{3.45}$$

or

$$\frac{\partial}{\partial v}\left(\frac{1}{T}\right)_u = \frac{\partial}{\partial u}\left(\frac{P}{T}\right)_v \tag{3.46}$$

whence

$$\frac{\partial}{\partial v}\left(\frac{1}{T}\right)_u = \frac{\partial}{\partial u}\left(\frac{R}{v-b} - \frac{a}{v^2}\frac{1}{T}\right)_v$$

$$= -\frac{a}{v^2}\frac{\partial}{\partial u}\left(\frac{1}{T}\right)_v \tag{3.47}$$

This condition can be written as

$$\frac{\partial}{\partial(1/v)}\left(\frac{1}{T}\right)_u = \frac{\partial}{\partial(u/a)}\left(\frac{1}{T}\right)_v \tag{3.48}$$

That is, the function $1/T$ must depend on the two variables $1/v$ and u/a in such a way that the two derivatives are equal. One possible way of accomplishing this is to have $1/T$ depend only on the sum $(1/v + u/a)$. We first recall that for a simple ideal gas $1/T = cR/u$; this suggests that the simplest possible change consistent with the van der Waals equation is

$$\frac{1}{T} = \frac{cR}{u + a/v} \tag{3.49}$$

For purposes of illustration throughout this text we shall refer to the hypothetical system characterized by the van der Waals equation of state (3.41) and by equation 3.49 as the "ideal van der Waals fluid."

We should note that equation 3.41, although referred to as the "van der Waals equation of state," is not in the appropriate form of an equation of state. However, from equations 3.49 and 3.42 we obtain

$$\frac{P}{T} = \frac{R}{v-b} - \frac{acR}{uv^2 + av} \tag{3.50}$$

The two preceding equations are the proper equations of state in the entropy representation, expressing $1/T$ and P/T as functions of u and v.

With the two equations of state we are now able to obtain the fundamental relation. It is left to the reader to show that

$$S = NR\ln\left[(v-b)(u+a/v)^c\right] + Ns_0 \tag{3.51}$$

where s_0 is a constant. As in the case of the ideal gas the fundamental

TABLE 3.1
**Van der Waals Constants and Molar Heat
Capacities of Common Gases**[a]

Gas	a (Pa-m^6)	b ($10^{-6}m^3$)	c
He	0.00346	23.7	1.5
Ne	0.0215	17.1	1.5
H_2	0.0248	26.6	2.5
A	0.132	30.2	1.5
N_2	0.136	38.5	2.5
O_2	0.138	32.6	2.5
CO	0.151	39.9	2.5
CO_2	0.401	42.7	3.5
N_2O	0.384	44.2	3.5
H_2O	0.544	30.5	3.1
Cl_2	0.659	56.3	2.8
SO_2	0.680	56.4	3.5

[a] Adapted from Paul S. Epstein, *Textbook of Thermodynamics*,
Wiley, New York, 1937.

equation does not satisfy the Nernst theorem, and it cannot be valid at
very low temperatures.

We shall see later (in Chapter 9) that the ideal van der Waals fluid is
unstable in certain regions of temperature and pressure, and that it
spontaneously separates into two phases ("liquid" and "gas"). The fundamental equation (3.51) is a very rich one for the illustration of thermodynamic principles.

The van der Waals constants for various real gases are given in Table
3.1. The constants a and b are obtained by empirical curve fitting to the
van der Waals isotherms in the vicinity of 273 K; they represent more
distant isotherms less satisfactorily. The values of c are based on the
molar heat capacities at room temperatures.

PROBLEMS

3.5-1. Are each of the listed pairs of equations of state compatible (recall
equation 3.46)? If so, find the fundamental equation of the system.
a) $u = aPv$ and $Pv^2 = bT$
b) $u = aPv^2$ and $Pv^2 = bT$
c) $P = \dfrac{u}{v} \cdot \dfrac{c + buv}{a + buv}$ and $T = \dfrac{u}{a + buv}$

3.5-2. Find the relationship between the volume and the temperature of an ideal
van der Waals fluid in a quasi-static adiabatic expansion (i.e., in an isentropic
expansion, with $dQ = T\,dS = 0$, or $S = $ constant).

3.5-3. Repeat Problem 3.4-3 for CO_2, rather than for a monatomic ideal gas. Assume CO_2 can be represented by an ideal van der Waals fluid with constants as given in Table 3.1.

At what approximate pressure would the term $(-a/v^2)$ in the van der Waals equation of state make a 10% correction to the pressure at room temperature?

Answer:
$$V_f = 0.091 \text{ m}^3$$

3.5-4. Repeat parts (*a*), (*b*), and (*c*) of problem 3.4-5, assuming that $\eta = -\frac{1}{2}$ and that the gas is an ideal van der Waals fluid.

Show that your results for ΔU and for W (and hence for Q) reduce to the results of Problem 3.4-5 (for $\eta = -\frac{1}{2}$) as the van der Waals constants a and b go to zero, and $c = \frac{3}{2}$. Recall that $\ln(1 + x) \simeq x$, for small x.

3.5-5. Consider a van der Waals gas contained in the apparatus described in Problem 3.4-1 (i.e., in the "constant volume gas thermometer").

a) Assuming it to be known in advance that the gas obeys a van der Waals equation of state, show that knowledge of *two* reference temperatures enables one to evaluate the van der Waals constants a and b.

b) Knowing the constants a and b, show that the apparatus can then be used as a thermometer, to measure any other temperature.

c) Show that knowledge of *three* reference temperatures enables one to determine whether a gas satisfies the van der Waals equation of state, and if it does, enables one to measure any other temperature.

3.5-6. One mole of a monatomic ideal gas and one mole of Cl_2 are contained in a rigid cylinder and are separated by a moveable internal piston. If the gases are at a temperature of 300 K the piston is observed to be precisely in the center of the cylinder. Find the pressure of each gas. Treat Cl_2 as a van der Waals gas (see Table 3.1).

Answer:
$$P = 3.5 \times 10^7 \text{ Pa}$$

3-6   ELECTROMAGNETIC RADIATION

If the walls of any "empty" vessel are maintained at a temperature T it is found that the vessel is, in fact, the repository of electromagnetic energy. The quantum theorist might consider the vessel as containing photons, the engineer might view the vessel as a resonant cavity supporting electromagnetic modes, whereas the classical thermodynamicist might eschew any such mechanistic models. From any viewpoint, the empirical equations of state of such an electromagnetic cavity are the "Stefan–Boltzmann Law"

$$U = bVT^4 \tag{3.52}$$

and

$$P = \frac{U}{3V} \tag{3.53}$$

where b is a particular constant ($b = 7.56 \times 10^{-16}$ J/m^3 K^4) which will be evaluated from basic principles in Section 16.8. It will be noted that these empirical equations of state are functions of U and V, but not of N. This observation calls our attention to the fact that in the "empty" cavity there exist no conserved particles to be counted by a parameter N. The electromagnetic radiation within the cavity is governed by a fundamental equation of the form $S = S(U, V)$ in which there are only two rather than three independent extensive parameters!

For electromagnetic radiation the two known equations of state constitute a complete set, which need only be substituted in the truncated Euler relation

$$S = \frac{1}{T}U + \frac{P}{T}V \tag{3.54}$$

to provide a fundamental relation. For this purpose we rewrite equations 3.52 and 3.53 in the appropriate form of entropic equations of state

$$\frac{1}{T} = b^{1/4}\left(\frac{V}{U}\right)^{1/4} \tag{3.55}$$

and

$$\frac{P}{T} = \frac{1}{3}b^{1/4}\left(\frac{U}{V}\right)^{3/4} \tag{3.56}$$

so that the fundamental relation becomes, on substitution into 3.54

$$S = \tfrac{4}{3}b^{1/4}U^{3/4}V^{1/4} \tag{3.57}$$

PROBLEMS

3.6-1. The universe is considered by cosmologists to be an expanding electromagnetic cavity containing radiation that now is at a temperature of 2.7 K. What will be the temperature of the radiation when the volume of the universe is twice its present value? Assume the expansion to be isentropic (this being a nonobvious prediction of cosmological model calculations).

3-6.2. Assuming the electromagnetic radiation filling the universe to be in equilibrium at $T = 2.7$ K, what is the pressure associated with this radiation? Express the answer both in pascals and in atmospheres.

3.6-3. The density of matter (primarily hydrogen atoms) in intergalactic space is such that its contribution to the pressure is of the order of 10^{-23} Pa.

a) What is the approximate density of matter (in atoms/m^3) in intergalactic space?

b) What is the ratio of the kinetic energy of matter to the energy of radiation in intergalactic space? (Recall Problems 3.6-1 and 3.6-2.)

c) What is the ratio of the *total* matter energy (i.e., the sum of the kinetic energy plus the relativistic energy mc^2) to the energy of radiation in intergalactic space?

3-7 THE "RUBBER BAND"

A somewhat different utility of the thermodynamic formalism is illustrated by consideration of the physical properties of a rubber band; thermodynamics constrains and guides the construction of simple phenomenological models for physical systems.

Let us suppose that we are interested in building a descriptive model for the properties of a rubber band. The rubber band consists of a bundle of long-chain polymer molecules. The quantities of macroscopic interest are the length L, the tension $\mathcal{T}$, the temperature T, and the energy U of the rubber band. The length plays a role analogous to the volume and the tension plays a role analogous to the negative pressure ($\mathcal{T} \sim -P$). An analogue of the mole number might be associated with the number of monomer units in the rubber band (but that number is not generally variable and it can be taken here as constant and suppressed in the analysis).

A qualitative representation of experimental observations can be summarized in two properties. First, at constant length the tension *increases* with the temperature—a rather startling property which is in striking contrast to the behavior of a stretched metallic wire. Second, the energy is observed to be essentially independent of the length, at least for lengths shorter than the "elastic limit" of the rubber band (a length corresponding to the "unkinking" or straightening of the polymer chains).

The simplest representation of the latter observation would be the equation

$$U = cL_0 T \tag{3.58}$$

where c is a constant and L_0 (also constant) is the unstretched length of the rubber band. The linearity of the length with tension, between the unstretched length L_0 and the elastic limit length L_1, is represented by

$$\mathcal{T} = bT \frac{L - L_0}{L_1 - L_0}, \qquad L_0 < L < L_1 \tag{3.59}$$

where b is a constant. The insertion of the factor T in this equation (rather than T^2 or some other function of T) is dictated by the thermodynamic condition of consistency of the two equations of state. That is, as in equation 3.46

$$\frac{\partial}{\partial L}\left(\frac{1}{T}\right)_U = \frac{\partial}{\partial U}\left(-\frac{\mathscr{T}}{T}\right)_L \tag{3.60}$$

which dictates the linear factor T in equation (3.59). Then

$$dS = \frac{1}{T} dU - \frac{\mathscr{T}}{T} dL = cL_0 \frac{dU}{U} - b\frac{L - L_0}{L_1 - L_0} dL \tag{3.61}$$

and the fundamental equation correspondingly is

$$S = S_0 + cL_0 \ln\frac{U}{U_0} - \frac{b}{2(L_1 - L_0)}(L - L_0)^2 \tag{3.62}$$

Although this fundamental equation has been constructed on the basis only of the most qualitative of information, it does represent empirical properties reasonably and, most important, consistently. The model illustrates the manner in which thermodynamics guides the scientist in elementary model building.

A somewhat more sophisticated model of polymer elasticity will be derived by statistical mechanical methods in Chapter 15.

PROBLEMS

3.7-1. For the rubber band model, calculate the fractional change in $(L - L_0)$ that results from an increase δT in temperature, at constant tension. Express the result in terms of the length and the temperature.

3.7-2. A rubber band is stretched by an amount dL, at constant T. Calculate the heat transfer dQ to the rubber band. Also calculate the work done. How are these related and why?

3.7-3. If the energy of the unstretched rubber band were found to increase quadratically with T, so that equation 3.58 were to be replaced by $U = cL_0T^2$, would equation 3.59 require alteration? Again find the fundamental equation of the rubber band.

3-8 UNCONSTRAINABLE VARIABLES; MAGNETIC SYSTEMS

In the preceding sections we have seen examples of several specific systems, emphasizing the great diversity of types of systems to which thermodynamics applies and illustrating the constraints on analytic modeling of simple systems. In this section we give an example of a magnetic system. Here we have an additional purpose, for although the general structure of thermodynamics is represented by the examples already given, particular "idiosyncrasies" are associated with certain thermodynamic parameters. Magnetic systems are particularly prone to such individual peculiarities, and they well illustrate the special considerations that occasionally are required.

In order to ensure magnetic homogeneity we focus attention on ellipsoidal samples in homogeneous external fields, with one symmetry axis of the sample parallel to the external field. For simplicity we assume no magnetocrystalline anisotropy, or, if such exists, that the "easy axis" lies parallel to the external field. Furthermore we initially consider only paramagnetic or diamagnetic systems—that is, systems in which the magnetization vanishes in the absence of an externally imposed magnetic field. In our eventual consideration of phase transitions we shall include the transition to the ferromagnetic phase, in which the system develops a spontaneous magnetization.

As shown in Appendix B, the extensive parameter that characterizes the magnetic state is the magnetic dipole moment I of the system. The fundamental equation of the system is of the form $U = U(S, V, I, N)$. In the more general case of an ellipsoidal sample that is *not* coaxial with the external field, the single parameter I would be replaced by the three cartesian coordinates of the magnetic moment: $U(S, V, I_x, I_y, I_z, N)$. The thermodynamic structure of the problem is most conveniently illustrated in the one-parameter case.

The intensive parameter conjugate to the magnetic moment I is B_e, the *external magnetic field* that would exist in the absence of the system

$$B_e = \left(\frac{\partial U}{\partial I} \right)_{S, V, N} \tag{3.63}$$

The unit of B_e is the tesla (T), and the units of I are Joules/Tesla (J/T).

It is necessary to note a subtlety of definition implicit in these identifications of extensive and intensive parameters (see Appendix B). The energy U is here construed as the energy of the material system alone; in addition the "vacuum" occupied by the system must be assigned an energy $\frac{1}{2}\mu_0^{-1}B_e^2V$ (where μ_0, the permeability of free space, has the value $\mu_0 = 4\pi \times 10^{-7}$ tesla-meters/ampere). Thus the *total energy* within the spatial region occupied by a system is $U + \frac{1}{2}\mu_0^{-1}B_e^2V$. Whether the "vacuum term" in the energy is associated with the system or is treated separately (as we do) is a matter of arbitrary choice, but considerable confusion can arise if different conventions are not carefully distinguished. To repeat, the energy U is the *change* in energy within a particular region in the field when the material system is introduced; it excludes the energy $\frac{1}{2}\mu_0^{-1}B_e^2V$ of the region prior to the introduction of the system.

The Euler relation for a magnetic system is now

$$U = TS - PV + B_e I + \mu N \qquad (3.64)$$

and the Gibbs–Duhem relation is

$$S\,dT - V\,dP + I\,dB_e + N\,d\mu = 0 \qquad (3.65)$$

An "idiosyncrasy" of magnetic systems becomes evident if we attempt to consider problems analogous to those of Sections 2.7 and 2.8—namely, the condition of equilibrium of two subsystems following the removal of a constraint. We soon discover that we do not have the capability of constraining the magnetic moment; *in practice the magnetic moment is always unconstrained!* We *can* specify and control the magnetic field applied to a sample (just as we can control the pressure), and we thereby can bring about a desired value of the magnetic moment. We can even hold that value of the magnetic moment constant by monitoring its value and by continually adjusting the magnetic field—again, just as we might keep the volume of a system constant by a feedback mechanism that continually adjusts the external pressure. But that is very different from simply enclosing the system in a restrictive wall. *There exist no walls restrictive with respect to magnetic moment.*

Despite the fact that the magnetic moment is an unconstrainable variable, the over-all structure of thermodynamic theory still applies. The fundamental equation, the equations of state, the Gibbs–Duhem, and the Euler relations maintain their mutual relationships. The nonavailability of walls restrictive to magnetic moment can be viewed as a "mere experimental quirk," that does not significantly influence the applicability of thermodynamic theory.

Finally, to anchor the discussion of magnetic systems in an explicit example, the fundamental equation of a simple paramagnetic model system is

$$U = NRT_0 \exp\left[\frac{S}{NR} + \frac{I^2}{N^2 I_0^2} \right] \qquad (3.66)$$

where T_0 and I_0 are positive constants. This model does not describe any particular known system—it is devised to provide a simple, tractable model on which examples and problems can be based, and to illustrate characteristic thermomagnetic interactions. We shall leave it to the problems to explore some of these properties.

With the magnetic case always in mind as a prototype for generalizations, we return to explicit consideration of simple systems.

PROBLEMS

3.8-1. Calculate the three equations of state of the paramagnetic model of equation 3.66. That is, calculate $T(S, I, N)$, $B_e(S, I, N)$, and $\mu(S, I, N)$. (Note that the fundamental equation of this problem is independent of V, and that more generally there would be four equations of state.) Show that the three equations of state satisfy the Euler relation.

3.8-2. Repeat Problem 3.8-1 for a system with the fundamental equation

$$U = \frac{\mu_0}{2N\chi} I^2 + N\varepsilon \exp(2S/NR)$$

where χ and ε are positive constants.

3-9   MOLAR HEAT CAPACITY AND OTHER DERIVATIVES

The first derivatives of the fundamental equation have been seen to have important physical significance. The various second derivatives are descriptive of material properties, and these second derivatives often are the quantities of most direct physical interest. Accordingly we exhibit a few particularly useful second derivatives and illustrate their utility. In Chapter 7 we shall return to study the formal structure of such second derivatives, demonstrating that only a small number are independent and that all others can be related to these few by a systematic "reduction scheme." For simple nonmagnetic systems the basic set of derivatives (to which a wide set of others can be related) are just three.

The *coefficient of thermal expansion* is defined by

$$\alpha \equiv \frac{1}{v}\left(\frac{\partial v}{\partial T}\right)_P = \frac{1}{V}\left(\frac{\partial V}{\partial T}\right)_P \tag{3.67}$$

The coefficient of thermal expansion is the fractional increase in the volume per unit increase in the temperature of a system maintained at constant pressure (and constant mole numbers).

The *isothermal compressibility* is defined by

$$\kappa_T \equiv -\frac{1}{v}\left(\frac{\partial v}{\partial P}\right)_T = -\frac{1}{V}\left(\frac{\partial V}{\partial P}\right)_T \tag{3.68}$$

The isothermal compressibility is the fractional decrease in volume per unit increase in pressure at constant temperature.

The *molar heat capacity at constant pressure* is defined by

$$c_P \equiv T\left(\frac{\partial s}{\partial T}\right)_P = \frac{T}{N}\left(\frac{\partial S}{\partial T}\right)_P = \frac{1}{N}\left(\frac{dQ}{dT}\right)_P \tag{3.69}$$

The molar heat capacity at constant pressure is the quasi-static heat flux per mole required to produce unit increase in the temperature of a system maintained at constant pressure.

For systems of constant mole number all other second derivatives can be expressed in terms of these three, and these three are therefore normally tabulated as functions of temperature and pressure for a wide variety of materials.

The origin of the relationships among second derivatives can be understood in principle at this point, although we postpone a full exploration to Chapter 7. Perhaps the simplest such relationship is the identity

$$\left(\frac{\partial T}{\partial V}\right)_{S,N} = -\left(\frac{\partial P}{\partial S}\right)_{V,N} \tag{3.70}$$

which follows directly from the elementary theorem of calculus to the effect that the two mixed second partial derivatives of U with respect to V and S are equal

$$\frac{\partial}{\partial V}\left(\frac{\partial U}{\partial S}\right) = \frac{\partial}{\partial S}\left(\frac{\partial U}{\partial V}\right) \tag{3.71}$$

The two quantities appearing in equation (3.70) have direct physical interpretations and each can be measured. The quantity $(\partial T/\partial V)_{S,N}$ is the temperature change associated with adiabatic expansion of the volume; the quantity $(\partial P/\partial S)_{V,N}$, when written as $T(dP/dQ)_{V,N}$ is the product of the temperature and the change in pressure associated with an introduction of heat dQ into a system at constant volume. The prediction of equality of these apparently unrelated quantities is a nontrivial result; in effect, the first "triumph" of the theory. Needless to say, the prediction is corroborated by experiment.

The analogue of equation 3.70, in the entropy representation, is

$$\frac{\partial}{\partial V}\left(\frac{1}{T}\right)_{U,N} = \frac{\partial}{\partial U}\left(\frac{P}{T}\right)_{V,N} \tag{3.72}$$

and we recognize that this is precisely the identity that we invoked in equation 3.46 in our quest for a thermal equation of state to be paired with the van der Waals equation.

In Chapter 7 we show in considerable detail that these equalities are prototypes of a general class of analogous relationships referred to as Maxwell relations. Although the Maxwell relations have the simple form of equality of two derivatives, they, in turn, are degenerate cases of a more general theorem that asserts that there must exist a relation among any *four* derivatives. These general relations will permit any second derivative (at constant N) to be expressed in terms of the basic set c_p, α, and κ_T.

To illustrate such anticipated relationships we first introduce two additional second derivatives of practical interest; the adiabatic compressibility κ_S and the molar heat capacity at constant volume c_v.

The adiabatic compressibility is defined by

$$\kappa_s = -\frac{1}{v}\left(\frac{\partial v}{\partial P}\right)_s = -\frac{1}{V}\left(\frac{\partial V}{\partial P}\right)_S \tag{3.73}$$

This quantity characterizes the fractional decrease in volume associated with an isentropic increase in pressure (i.e., for a system that is adiabatically insulated).

The molar heat capacity at constant volume, defined by

$$c_v \equiv T\left(\frac{\partial s}{\partial T}\right)_v = \frac{T}{N}\left(\frac{\partial S}{\partial T}\right)_V = \frac{1}{N}\left(\frac{dQ}{dT}\right)_V \tag{3.74}$$

measures the quasi-static heat flux per mole required to produce unit increase in the temperature of a system maintained at constant volume.

In Chapter 7 we show that

$$c_P = c_v + \frac{TV\alpha^2}{N\kappa_T} \tag{3.75}$$

and

$$\kappa_T = \kappa_S + \frac{TV\alpha^2}{Nc_P} \tag{3.76}$$

Again, our purpose here is not to focus on the detailed relationships (3.75) and (3.76), but to introduce definitions of c_p, α, and κ_T, to call attention to the fact that c_p, α, and κ_T are normally tabulated as functions of T and P, and to stress that all other derivatives (such as c_v and κ_S) can be related to c_p, α, and κ_T. A systematic approach to all such equalities, and a mnemonic device for recalling them as needed, is presented in Chapter 7.

Problem 3.9-6 is particularly recommended to the student.

Example

For a particular material c_P, α, and κ_T are tabulated as functions of T and P. Find the molar volume v as a function of T and P.

Solution

We consider the "T–P plane." The quantities c_P, α, and κ_T are known at all points in the plane, and we seek to evaluate $v(T, P)$ at an arbitrary point in the plane. Then

$$dv = \left(\frac{\partial v}{\partial P}\right)_T dP + \left(\frac{\partial v}{\partial T}\right)_P dT$$

$$= -v\kappa_T\, dP + v\alpha\, dT$$

or

$$\frac{dv}{v} = -\kappa_T\, dP + \alpha\, dT$$

If (T_0, P_0) is a chosen reference point in the plane, and if (T', P') is a point of interest, we can integrate along the path shown (or any other convenient path). For the path that we have chosen the term in dP vanishes for the "horizontal" section of the path, and the term in dT vanishes for the "vertical" section of the path, so that

$$\int \frac{dv}{v} = \int_{T_0}^{T'} \alpha(T, P_0)\, dT - \int_{P_0}^{P'} \kappa_T(T', P)\, dP$$

or

$$\ln\frac{v'}{v_0} = \int_{T_0}^{T'} \alpha(T, P_0)\, dT - \int_{P_0}^{P'} \kappa_T(T', P)\, dP$$

The value of the molar volume at the reference point (v_0) must be specified; we are then able to relate all other volumes to this volume.

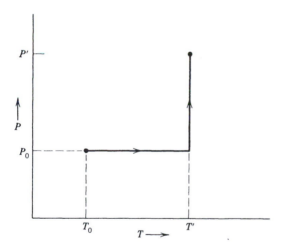

PROBLEMS

3.9-1.

a) Show that for the multicomponent simple ideal gas

$$c_v = \bar{c}R$$

$$\alpha = 1/T$$

$$\kappa_T = 1/P$$

and
$$\kappa_S = \frac{\bar{c}}{\bar{c}+1}\frac{1}{P}$$

$$c_P = (\bar{c}+1)R \qquad \text{where } \bar{c} = \sum_j c_j x_j = \frac{1}{N}\sum_j c_j N_j$$

b) What is the value of $\bar{c}$ for a monatomic ideal gas?

c) Using the values found in part (*a*), corroborate equations 3.75 and 3.76.

3.9-2. Corroborate equation 3.70 for a multicomponent simple ideal gas, showing that both the right- and left-hand members of the equation equal $-T/\bar{c}V$ (where $\bar{c}$ is defined in Problem 3.9-1).

3.9-3. Compute the coefficient of expansion α and the isothermal compressibility κ_T in terms of P and v for a system with the van der Waals equation of state (equation 3.41).

3.9-4. Compute c_P, c_v, κ_S, and κ_T for the system in Problem 1.10-1(*a*). With these values corroborate the validity of equations 3.75 and 3.76.

3.9-5. From equations 3.75 and 3.76 show that

$$c_P/c_v = \kappa_T/\kappa_S$$

3.9-6. A simple fundamental equation that exhibits some of the qualitative properties of typical crystaline solids is

$$u = Ae^{b(v-v_0)^2}s^{4/3}e^{s/3R}$$

where A, b, and v_0 are positive constants.

a) Show that the system satisfies the Nernst theorem.

b) Show that c_v is proportional to T^3 at low temperature. This is commonly observed (and was explained by P. Debye by a statistical mechanical analysis, which will be developed in Chapter 16).

c) Show that $c_v \to 3k_B$ at high temperatures. This is the "equipartition value," which is observed and which will be demonstrated by statistical mechanical analysis in Chapter 16.

d) Show that for zero pressure the coefficient of thermal expansion vanishes in this model—a result that is incorrect. *Hint*: Calculate the value of v at $P = 0$.

3.9-7. The density of mercury at various temperatures is given here in grams/cm^3.

13.6202 ($-10°$C)	13.5217 (30°C)	13.3283 (110°C)
13.5955 (0°C)	13.4973 (40°C)	13.1148 (200°C)
13.5708 (10°C)	13.4729 (50°C)	12.8806 (300°C)
13.5462 (20°C)	13.3522 (100°C)	12.8572 (310°C)

Calculate α at 0°C, at 45°C, at 105°C, and at 305°C.

Should the stem of a mercury-in-glass thermometer be marked off in equal divisions for equal temperature intervals if the coefficient of thermal expansion of glass is assumed to be strictly constant?

3.9-8. For a particular material c_P, α, and κ_T can be represented empirically by power series in the vicinity of T_0, P_0, as follows

$$c_P = c_P^0 + A_c\tau + B_c\tau^2 + D_c p + E_c p^2 + F_c \tau p$$

$$\alpha = \alpha^0 + A_\alpha\tau + B_\alpha\tau^2 + D_\alpha p + E_\alpha p^2 + F_\alpha\tau p$$

$$\kappa_T = \kappa^0 + A_\kappa\tau + B_\kappa\tau^2 + D_\kappa p + E_\kappa p^2 + F_\kappa\tau p \qquad \text{where } \tau \equiv T - T_0; \ p \equiv P - P_0$$

Find the molar volume explicitly as a function of T and P in the vicinity of (T_0, P_0).

3.9-9. Calculate the molar entropy $s(T, P_0)$ for fixed pressure P_0 and for temperature T in the vicinity of T_0. Assume that c_p, α, and κ_T are given in the vicinity of (T_0, P_0) as in the preceding problem, and assume that $s(T_0, P_0)$ is known.

3.9-10. By analogy with equations 3.70 and 3.71 show that for a paramagnetic system

$$\left(\frac{\partial B_e}{\partial S}\right)_{I,V,N} = \left(\frac{\partial T}{\partial I}\right)_{S,V,N}$$

or, inverting,

$$T\left(\frac{\partial S}{\partial B_e}\right)_{I,V,N} = T\left(\frac{\partial I}{\partial T}\right)_{S,V,N}$$

Interpret the physical meaning of this relationship.

3.9-11. By analogy with equations 3.70 and 3.71 show that for a paramagnetic system

$$\left(\frac{\partial B_e}{\partial V}\right)_{S,I,N} = -\left(\frac{\partial P}{\partial I}\right)_{S,V,N}$$

3.9-12. The magnetic analogues of the molar heat capacities c_P and c_v are c_B and c_I. Calculate $c_B(T, B_e, N)$ and $c_I(T, B_e, N)$ for the paramagnetic model of equation 3.66. (Note that no distinction need be made between $c_{I,V}$ and $c_{I,P}$ for this model, because of the absence of a dependence on volume in the fundamental relation (3.66). Generally all four heat capacities exist and are distinct.)

3.9-13. The (isothermal) molar magnetic susceptibility is defined by

$$\chi \equiv \frac{\mu_0}{N}\left(\frac{\partial I}{\partial B_e}\right)_T$$

Show that the susceptibility of the paramagnetic model of equation 3.66 varies inversely with the temperature, and evaluate χ_1, defined as the value of χ for $T = 1$ K.

3.9-14. Calculate the adiabatic molar susceptibility

$$\chi_s \equiv \frac{\mu_0}{N}\left(\frac{\partial I}{\partial B_e}\right)_S$$

as a function of T and B_e for the paramagnetic model of equation 3.66.

3.9-15. Calculate the isothermal and adiabatic molar susceptibilities (defined in Problems 3.9-13 and 3.9-14) for the system with fundamental equation

$$U = \frac{\mu_0}{2} \frac{I^2}{N\chi} + N\varepsilon \exp(2S/NR)$$

How are each of these related to the constant "χ" appearing in the fundamental relation?

3.9-16. Show that for the system of Problem 3.8-2

$$\left(\frac{\partial T}{\partial B_e}\right)_S = \left(\frac{\partial T}{\partial I}\right)_S = \left(\frac{\partial S}{\partial I}\right)_T = \left(\frac{\partial S}{\partial B_e}\right)_T = 0$$

and

$$\left(\frac{\partial B_e}{\partial T}\right)_I = \left(\frac{\partial B_e}{\partial S}\right)_I = \left(\frac{\partial I}{\partial T}\right)_{B_e} = \left(\frac{\partial I}{\partial S}\right)_{B_e} = 0$$

That is, there is no "coupling" between the thermal and magnetic properties. What is the (atypical) feature of the equation of state of this system that leads to these results?

3.9-17. Calculate the heat transfer to a particular system if 1 mole is taken from (T_0, P_0) to $(2T_0, 2P_0)$ along a straight line in the T–P plane. For this system it is known that:

$$\alpha(T, P) = \alpha^0 \cdot \left(\frac{T}{T_0}\right)^{\frac{1}{2}}, \text{ where } \alpha^0 \text{ is a constant}$$

$$c_P(T, P) = c_P^0, \text{ a constant}$$

$$\kappa_T(T, P) = \kappa_T^0, \text{ a constant}$$

Hint: Use the relation $(\partial s/\partial P)_T = -(\partial v/\partial T)_P$, analogous to equations 3.70 through 3.72 (and to be derived systematically in Chapter 7), to establish that $dQ = T\,ds = c_P\,dT - Tv\alpha\,dP$.

4

REVERSIBLE PROCESSES AND THE MAXIMUM WORK THEOREM

4-1 POSSIBLE AND IMPOSSIBLE PROCESSES

An engineer may confront the problem of designing a device to accomplish some specified task—perhaps to lift an elevator to the upper floors of a tall building. Accordingly the engineer contrives a linkage or "engine" that conditionally permits transfer of energy from a furnace to the elevator; *if* heat flows from the furnace then, by virtue of the interconnection of various pistons, levers, and cams, the elevator is required to rise. But "nature" (i.e., the laws of physics) exercises the crucial decision—will the proposition be accepted or will the device sit dormant and inactive, with no heat leaving the furnace and no rise in height of the elevator? The outcome is conditioned by two criteria. First, the engine must obey the laws of mechanics (including, of course, the conservation of energy). Second, the process must maximally increase the entropy.

Patent registration offices are replete with failed inventions of impeccable conditional logic (if *A* occurs then *B* must occur)—ingenious devices that conform to all the laws of mechanics but that nevertheless sit stubbornly inert, in mute refusal to decrease the entropy. Others operate, but with unintended results, increasing the entropy more effectively than envisaged by the inventor.

If, however, the net changes to be effected correspond to a maximal permissible increase in the total entropy, with no change in total energy, then no fundamental law precludes the existence of an appropriate process. It may require considerable ingenuity to devise the appropriate engine, but such an engine can be assumed to be permissible in principle.

Example 1

A particular system is constrained to constant mole number and volume, so that no work can be done on or by the system. Furthermore, the heat capacity of the system is C, a constant. The fundamental equation of the system, for constant volume, is $S = S_0 + C \ln(U/U_0)$, so $U = CT$.

Two such systems, with equal heat capacities, have initial temperatures T_{10} and T_{20}, with $T_{10} < T_{20}$. An engine is to be designed to lift an elevator (i.e., to deliver work to a purely mechanical system), drawing energy from the two thermodynamic systems. What is the maximum work that can be so delivered?

Solution

The two thermal systems will be left at some common temperature T_f. The change in energy of the two thermal systems accordingly will be

$$\Delta U = 2CT_f - C(T_{10} + T_{20})$$

and the work delivered to the mechanical system (the "elevator") will be $W = -\Delta U$, or

$$W = C(T_{10} + T_{20} - 2T_f)$$

The change in total entropy will occur entirely in the two thermal systems, for which

$$\Delta S = C \ln \frac{T_f}{T_{10}} + C \ln \frac{T_f}{T_{20}} = 2C \ln \frac{T_f}{\sqrt{T_{10}T_{20}}}$$

To maximize W we clearly wish to minimize T_f (*cf.* the second equation preceding), and by the third equation this dictates that we minimize ΔS. The minimum possible ΔS is zero, corresponding to a reversible process. Hence the optimum engine will be one for which

$$T_f = \sqrt{T_{10}T_{20}}$$

and

$$W = C\left(T_{10} + T_{20} - 2\sqrt{T_{10}T_{20}}\right)$$

As a postscript, we note that the assumption that the two thermal systems are left at a common temperature is not necessary; W can be minimized with respect to T_{1f} and T_{2f} separately, with the same result. The simplifying assumption of a common temperature follows from a self-consistent argument, for if the final temperature were different we could obtain additional work by the method described.

Example 2

An interesting variant of Example 1 is one in which three bodies (each of the type described in Example 1, with $U = CT$) have initial temperatures of 300 K, 350 K, and 400 K, respectively. It is desired to raise *one* body to as high a temperature as possible, independent of the final temperatures of the other two (and without changing the state of any external system). What is the maximum achievable temperature of the single body?

Solution

Designate the three initial temperatures, measured in units of 100 K, as T_1, T_2, and T_3 ($T_1 = 3$, $T_2 = 3.5$, and $T_3 = 4$). Similarly, designate the high temperature achieved by one of the bodies (in the same units) as T_h. It is evident that the two remaining bodies will be left at the *same* temperature T_c (for if they were to be left at different temperatures we could extract work, as in Example 1, and insert it as heat to further raise the temperature of the hot body). Then energy conservation requires

$$T_h + 2T_c = T_1 + T_2 + T_3 = 10.5$$

The total entropy change is

$$\Delta S = C \ln\left(\frac{T_c^2 T_h}{T_1 T_2 T_3} \right)$$

and the requirement that this be positive implies

$$T_c^2 T_h \geq T_1 T_2 T_3 \quad (=42)$$

Eliminating T_c by the energy conservation condition

$$\left(5.25 - \frac{T_h}{2}\right)^2 T_h \geq 42$$

A plot of the left-hand side of this equation is shown. The plot is restricted to values of T_h between 0 and 10.5, the latter bound following from the energy conservation condition and the requirement that T_c be positive. The plot indi-

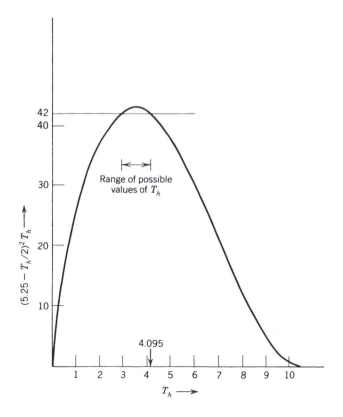

cates that the maximum value of T_h, for which the ordinate is greater than 42, is

$$T_h = 4.095 \quad (\text{or } T_h = 409.5 \text{ K})$$

and furthermore that this value satisfies the equality, and therefore corresponds to a reversible process.

Another solution to this problem will be developed in Problem 4.6-7.

PROBLEMS

4.1-1. One mole of a monatomic ideal gas and one mole of an ideal van der Waals fluid (Section 3.5) with $c = 3/2$ are contained separately in vessels of fixed volumes v_1 and v_2. The temperature of the ideal gas is T_1 and that of the van der Waals fluid is T_2. It is desired to bring the ideal gas to temperature T_2, maintaining the total energy constant. What is the final temperature of the van der Waals fluid? What restrictions apply among the parameters $(T_1, T_2, a, b, v_1, v_2)$ if it is to be possible to design an engine to accomplish this temperature inversion (assuming, as always, that no external system is to be altered in the process)?

4.1-2. A rubber band (Section 3.7) is initially at temperature T_B and length L_B. One mole of a monatomic ideal gas is initially at temperature T_G and volume V_G. The ideal gas, maintained at constant volume V_G, is to be heated to a final temperature T_G'. The energy required is to be supplied entirely by the rubber band. Need the length of the rubber band be changed, and, if so, by what amount?

Answer:

If $l \equiv L_B - L_0$,

$$l^2 - (l')^2 \geq 2b^{-1}cL_0(L_1 - L_0)\ln\left(1 - \frac{3R}{2RL_0}\frac{T_G' - T_G}{T_B}\right) + 3Rb^{-1}(L_1 - L_0)\ln(T_G'/T_G)$$

4.1-3. Suppose the two systems in Example 1 were to have heat capacities of the form $C(T) = DT^n$, with $n > 0$:

a) Show that for such systems $U = U_0 + DT^{n+1}/(n + 1)$ and $S = S_0 + DT^n/n$. What is the fundamental equation of such a system?

b) If the initial temperature of the two systems were T_{10} and T_{20} what would be the maximum delivered work (leaving the two systems at a common temperature)?

Answer:

b) for $n = 2$:

$$W = \frac{D}{3}\left[T_{10}^3 + T_{20}^3 - \frac{1}{\sqrt{2}}(T_{10}^2 + T_{20}^2)^{\frac{3}{2}}\right]$$

4-2 QUASI-STATIC AND REVERSIBLE PROCESSES

The central principle of entropy maximization spawns various theorems of more specific content when specialized to particular classes of processes. We shall turn our attention to such theorems after a preliminary refinement of the descriptions of states and of processes.

To describe and characterize thermodynamic states, and then to describe possible processes, it is useful to define a *thermodynamic configuration space*. The thermodynamic configuration space of a simple system is an abstract space spanned by coordinate axes that correspond to the entropy S and to the extensive parameters $U, V, N_1, \ldots, N_r$ of the system. The fundamental equation of the system $S = S(U, V, N_1, \ldots, N_r)$ defines a surface in the thermodynamic configuration space, as indicated schematically in Fig. 4.1. It should be noted that the surface of Fig. 4.1 conforms to the requirements that $(\partial S / \partial U) \ldots, x_j, \ldots$ ($\equiv 1/T$) be positive, and that U be a single valued function of $S, \ldots, x_j, \ldots$.

By definition, each point in the configuration space represents an equilibrium state. Representation of a nonequilibrium state would require a space of immensely greater dimension.

The fundamental equation of a composite system can be represented by a surface in a thermodynamic configuration space with coordinate axes

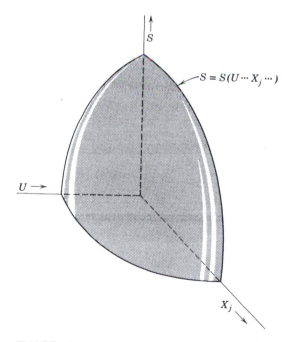

FIGURE 4.1
The hyper-surface $S = S(U, \ldots, X_j, \ldots)$ in the thermodynamic configuration space of a simple system.

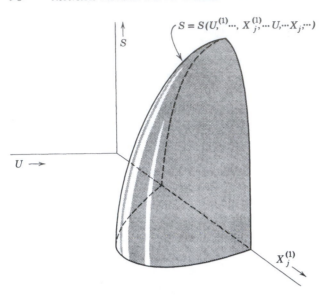

FIGURE 4.2
The hypersurface $S = S(U^{(1)}, \ldots, X_j^{(1)}, \ldots, U, \ldots, X_j, \ldots)$ in the thermodynamic configuration space of a composite system.

corresponding to the extensive parameters of all of the subsystems. For a composite system of two simple subsystems the coordinate axes can be associated with the total entropy S and the extensive parameters of the two subsystems. A more convenient choice is the total entropy S, the extensive parameters of the first subsystem $(U^{(1)}, V^{(1)}, N_1^{(1)}, N_2^{(1)}, \ldots)$, and the extensive parameters of the composite system $(U, V, N_1, N_2, \ldots)$. An appropriate section of the thermodynamic configuration space of a composite system is sketched in Fig. 4.2.

Consider an arbitrary curve drawn on the hypersurface of Fig. 4.3, from an initial state to a terminal state. Such a curve is known as a *quasi-static locus* or a *quasi-static process*. A quasi-static process is thus defined in terms of a dense succession of *equilibrium* states. It is to be stressed that a quasi-static process therefore is an idealized concept, quite distinct from a real physical process, for a real process always involves nonequilibrium intermediate states having no representation in the thermodynamic configuration space. Furthermore, a quasi-static process, in contrast to a real process, does not involve considerations of rates, velocities, or time. The quasi-static process simply is an ordered succession of equilibrium states, whereas a real process is a *temporal* succession of equilibrium and *nonequilibrium* states.

Although no real process is identical to a quasi-static process, it is possible to contrive real processes that have a close relationship to quasi-static processes. In particular, it is possible to lead a system through a succession of states that coincides at any desired number of points with

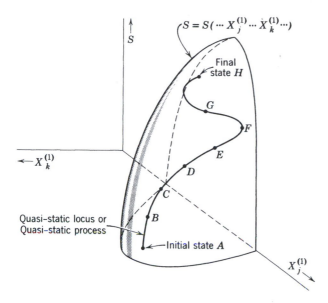

FIGURE 4.3

Representation of a quasi-static process in the thermodynamic configuration space.

a given quasi-static locus. Thus consider a system originally in the state A of Fig. 4.3, and consider the quasi-static locus passing through the points $A, B, C, \ldots, H$. We remove a constraint which permits the system to proceed from A to B but not to points further along the locus. The system "disappears" from the point A and subsequently appears at B, having passed en route through nonrepresentable nonequilibrium states. If the constraint is further relaxed, making the state C accessible, the system disappears from B and subsequently reappears at C. Repetition of the operation leads the system to states $D, E, \ldots, H$. By such a succession of real processes we construct a process that is an approximation to the abstract quasi-static process shown in the figure. By spacing the points $A, B, C, \ldots$ arbitrarily closely along the quasi-static locus we approximate the quasi-static locus arbitrarily closely.

The identification of $-P\,dV$ *as the mechanical work and of* $T\,dS$ *as the heat transfer is valid only for quasi-static processes.*

Consider a *closed* system that is to be led along the sequence of states $A, B, C, \ldots, H$ approximating a quasi-static locus. The system is induced to go from A to B by the removal of some internal constraint. The closed system proceeds to B if (and only if) the state B has maximum entropy among all newly accessible states. In particular the state B must have higher entropy than the state A. Accordingly, the physical process joining states A and B in a closed system has unique directionality. It proceeds from the state A, of lower entropy, to the state B, of higher entropy, but not inversely. Such processes are *irreversible*.

A quasi-static locus can be approximated by a real process in a closed system only if the entropy is monotonically nondecreasing along the quasi-static locus.

The limiting case of a quasi-static process in which the increase in the entropy becomes vanishingly small is called a reversible process (Fig. 4.4). For such a process the final entropy is equal to the initial entropy, and the process can be traversed in either direction.

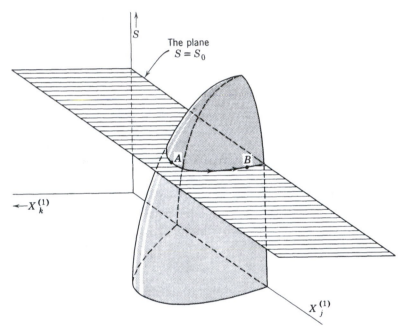

FIGURE 4.4
A reversible process, along a quasi-static isentropic locus.

PROBLEMS

4.2-1. Does every reversible process coincide with a quasi-static locus? Does every quasi-static locus coincide with a reversible process? For any real process starting in a state A and terminating in a state H, does there exist some quasi-static locus with the same two terminal states A and H? Does there exist some reversible process with the same two terminal states?

4.2-2. Consider a monatomic ideal gas in a cylinder fitted with a piston. The walls of the cylinder and the piston are adiabatic. The system is initially in equilibrium, but the external pressure is slowly decreased. The energy change of the gas in the resultant expansion dV is $dU = -P\,dV$. Show, from equation 3.34, that $dS = 0$, so that the quasi-static adiabatic expansion is isentropic and reversible.

4.2-3. A monatomic ideal gas is permitted to expand by a free expansion from V to $V + dV$ (recall Problem 3.4-8). Show that

$$dS = \frac{NR}{V}\, dV$$

In a series of such infinitesimal free expansions, leading from V_i to V_f, show that

$$\Delta S = NR \ln\!\left(\frac{V_f}{V_i}\right)$$

Whether this atypical (and infamous) "continuous free expansion" process should be considered as quasi-static is a delicate point. On the positive side is the observation that the terminal states of the infinitesimal expansions can be spaced as closely as one wishes along the locus. On the negative side is the realization that the system necessarily passes through nonequilibrium states during each expansion; the irreversibility of the microexpansions is essential and irreducible. The fact that $dS > 0$ whereas $dQ = 0$ is inconsistent with the presumptive applicability of the relation $dQ = T\,dS$ to all quasi-static processes. We *define* (by somewhat circular logic!) the continuous free expansion process as being "essentially irreversible" and *non-quasi-static*.

4.2-4. In the temperature range of interest a system obeys the equations

$$T = Av^2/s \qquad P = -2Av \ln(s/s_0)$$

where A is a positive constant. The system undergoes a free expansion from v_0 to v_f (with $v_f > v_0$). Find the final temperature T_f in terms of the initial temperature T_0, v_0, and v_f. Find the increase in molar entropy.

4-3   RELAXATION TIMES AND IRREVERSIBILITY

Consider a system that is to be led along the quasi-static locus of Fig. 4.3. The constraints are to be removed step by step, the system being permitted at each step to come to a new equilibrium state lying on the locus. After each slight relaxation of a constraint we must wait until the system fully achieves equilibrium, then we proceed with the next slight relaxation of the constraint and we wait again, and so forth. Although this is the theoretically prescribed procedure, the practical realization of the process seldom follows this prescription. In practice the constraints usually are relaxed continuously, at some "sufficiently slow" rate.

The rate at which constraints can be relaxed as a system approximates a quasi-static locus is characterized by the *relaxation time* τ of the system. For a given system, with a given relaxation time τ, processes that occur in times short compared to τ are not quasi-static, whereas processes that occur in times long compared to τ can be approximately quasi-static.

The physical considerations that determine the relaxation time can be illustrated by the adiabatic expansion of a gas (recall Problem 4.2-2). If the piston is permitted to move outward only extremely slowly the process is quasi-static (and reversible). If, however, the external pressure is decreased rapidly the resulting rapid motion of the piston is accompanied by turbulence and inhomogeneous flow within the cylinder (and by an entropy increase that "drives" these processes). The process is then neither quasi-static nor reversible. To estimate the relaxation time we first recognize that a slight outward motion of the piston reduces the density of the gas immediately adjacent to the piston. If the expansion is to be reversible this local "rarefaction" in the gas must be homogenized by hydrodynamic flow processes before the piston again moves appreciably. The rarefaction itself propagates through the gas with the velocity of sound, reflects from the walls of the cylinder, and gradually dissipates. The mechanism of dissipation involves both diffusive reflection from the walls and viscous damping within the gas. The simplest case would perhaps be that in which the cylinder walls are so rough that a single reflection would effectively dissipate the rarefaction pulse—admittedly not the common situation, but sufficient for our purely illustrative purposes. Then the relaxation time would be on the order of the time required for the rarefaction to propagate across the system, or $\tau \simeq V^{\frac{1}{3}}/c$, where the cube root of the volume is taken as a measure of the "length" of the system and c is the velocity of sound in the gas. If the adiabatic expansion of the gas in the cylinder is performed in times much longer than this relaxation time the expansion occurs reversibly and isentropically. If the expansion is performed in times comparable to or shorter than the relaxation time there is an irreversible increase in entropy within the system and the expansion, though adiabatic, is not isentropic.

PROBLEMS

4.3-1. A cylinder of length L and cross-sectional area A is divided into two equal-volume chambers by a piston, held at the midpoint of the cylinder by a setscrew. One chamber of the cylinder contains N moles of a monatomic ideal gas at temperature T_0. This same chamber contains a spring connected to the piston and to the end-wall of the cylinder; the unstretched length of the spring is $L/2$, so that it exerts no force on the piston when the piston is at its initial midpoint position. The force constant of the spring is K_{spring}. The other chamber of the cylinder is evacuated. The setscrew is suddenly removed. Find the volume and temperature of the gas when equilibrium is achieved. Assume the walls and the piston to be adiabatic and the heat capacities of the spring, piston, and walls to be negligible.

Discuss the nature of the processes that lead to the final equilibrium state. If there were gas in each chamber of the cylinder the problem as stated would be indeterminate! Why?

4-4 HEAT FLOW: COUPLED SYSTEMS
AND REVERSAL OF PROCESSES

Perhaps the most characteristic of all thermodynamic processes is the quasi-static transfer of heat between two systems, and it is instructive to examine this process with some care.

In the simplest case we consider the transfer of heat dQ from one system at temperature T to another at the *same* temperature. Such a process is reversible, the increase in entropy of the recipient subsystem dQ/T being exactly counterbalanced by the decrease in entropy $-dQ/T$ of the donor subsystem.

In contrast, suppose that the two subsystems have different initial temperatures T_{10} and T_{20}, with $T_{10} < T_{20}$. Further, let the heat capacities (at constant volume) be $C_1(T)$ and $C_2(T)$. Then if a quantity of heat dQ_1 is quasi-statically inserted into system 1 (at constant volume) the entropy increase is

$$dS_1 = \frac{dQ_1}{T_1} = C_1(T_1)\frac{dT_1}{T_1} \tag{4.1}$$

and similarly for subsystem 2. If such infinitesimal transfers of heat from the hotter to the colder body continue until the two temperatures become equal, then energy conservation requires

$$\Delta U = \int_{T_{10}}^{T_f} C_1(T_1)\, dT_1 + \int_{T_{20}}^{T_f} C_2(T_2)\, dT_2 = 0 \tag{4.2}$$

which determines T_f. The resultant change in entropy is

$$\Delta S = \int_{T_{10}}^{T_f} \frac{C_1(T_1)}{T_1}\, dT_1 + \int_{T_{20}}^{T_f} \frac{C_2(T_2)}{T_2}\, dT_2 \tag{4.3}$$

In the particular case in which C_1 and C_2 are independent of T the energy conservation condition gives

$$T_f = \frac{C_1 T_{10} + C_2 T_{20}}{C_1 + C_2} \tag{4.4}$$

and the entropy increase is

$$\Delta S = C_1 \ln\left(\frac{T_f}{T_{10}}\right) + C_2 \ln\left(\frac{T_f}{T_{20}}\right) \tag{4.5}$$

It is left to Problem 4.4-3 to demonstrate that this expression for ΔS is intrinsically positive.

Several aspects of the heat transfer process deserve reflection.

First, we note that the process, though quasi-static, is irreversible; it is represented in thermodynamic configuration space by a quasi-static locus of monotonically increasing S.

Second, the process can be associated with the *spontaneous* flow of heat from a hot to a cold system providing (*a*) that the intermediate wall through which the heat flow occurs is thin enough that its mass (and hence its contribution to the thermodynamic properties of the system) is negligible, and (*b*) that the rate of heat flow is sufficiently slow (i.e., the thermal resistivity of the wall is sufficiently high) that the temperature remains spatially homogeneous within each subsystem.

Third, we note that the entropy of one of the subsystems is *decreased*, whereas that of the other subsystem is increased. *It is possible to decrease the entropy of any particular system, providing that this decrease is linked to an even greater entropy increase in some other system.* In this sense an irreversible process within a given system *can* be "reversed"—*with the hidden cost paid elsewhere.*

PROBLEMS

4.4-1. Each of two bodies has a heat capacity given, in the temperature range of interest, by

$$C = A + BT$$

where $A = 8$ J/K and $B = 2 \times 10^{-2}$ J/K^2. If the two bodies are initially at temperatures $T_{10} = 400$ K and $T_{20} = 200$ K, and if they are brought into thermal contact, what is the final temperature and what is the change in entropy?

4.4-2. Consider again the system of Problem 4.4-1. Let a third body be available, with heat capacity

$$C_3 = BT$$

and with an initial temperature of T_{30}. Bodies 1 and 2 are separated, and body 3 is put into thermal contact with body 2. What must the initial temperature T_{30} be in order thereby to restore body 2 to its initial state? By how much is the entropy of body 2 decreased in this second process?

4.4-3. Prove that the entropy change in a heat flow process, as given in equation 4.5, is intrinsically positive.

4.4-4. Show that if two bodies have equal heat capacities, each of which is constant (independent of temperature), the equilibrium temperature achieved by direct thermal contact is the arithmetic average of the initial temperatures.

4.4-5. Over a limited temperature range the heat capacity at constant volume of a particular type of system is inversely proportional to the temperature.

a) What is the temperature dependence of the energy, at constant volume, for this type of system?

b) If two such systems, at initial temperatures T_{10} and T_{20}, are put into thermal contact what is the equilibrium temperature of the pair?

4.4-6. A series of $N + 1$ large vats of water have temperatures $T_0, T_1, T_2, \ldots, T_N$ (with $T_n > T_{n-1}$). A small body with heat capacity C (and with a constant volume, independent of temperature) is initially in thermal equilibrium with the vat of temperature T_0. The body is removed from this vat and immersed in the vat of temperature T_1. The process is repeated until, after N steps, the body is in equilibrium with the vat of temperature T_N. The sequence is then reversed, until the body is once again in the initial vat, at temperature T_0. Assuming the ratio of temperatures of successive vats to be a constant, or

$$T_n/T_{n-1} = (T_N/T_0)^{1/N}$$

and neglecting the (small) change in temperature of any vat, calculate the change in total entropy as
a) the body is successively taken "up the sequence" (from T_0 to T_N), and
b) the body is brought back "down the sequence" (from T_N to T_0).
What is the total change in entropy in the sum of the two sequences above?

Calculate the leading nontrivial limit of these results as $N \to \infty$, keeping T_0 and T_N constant. Note that for large N

$$N(x^{1/N} - 1) \simeq \ln x + (\ln x)^2/2N + \cdots$$

4-5   THE MAXIMUM WORK THEOREM

The propensity of physical systems to increase their entropy can be channeled to deliver useful work. All such applications are governed by the *maximum work theorem.*

Consider a system that is to be taken from a specified initial state to a specified final state. Also available are two auxiliary systems, into one of which work can be transferred, and into the other of which heat can be transferred. Then the maximum work theorem states that *for all processes leading from the specified initial state to the specified final state of the primary system, the delivery of work is maximum (and the delivery of heat is minimum) for a reversible process.* Furthermore *the delivery of work (and of heat) is identical for every reversible process.*

The repository system into which work is delivered is called a "reversible work source." *Reversible work sources are defined as systems enclosed by adiabatic impermeable walls and characterized by relaxation times sufficiently short that all processes within them are essentially quasi-static.* From the thermodynamic point of view the "conservative" (nonfrictional) systems considered in the theory of mechanics are reversible work sources.

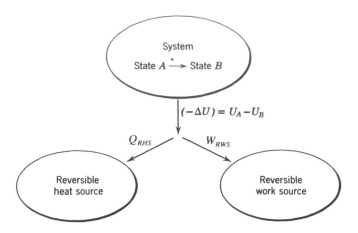

FIGURE 4.5

Maximum work process. The delivered work W_{RWS} is maximum and the delivered heat Q_{RHS} is minimum if the entire process is reversible ($\Delta S_{Total} = 0$).

The repository system into which heat is delivered is called a "reversible heat source"[1]. *Reversible heat sources are defined as systems enclosed by rigid impermeable walls and characterized by relaxation times sufficiently short that all processes of interest within them are essentially quasi-static.* If the temperature of the reversible heat source is T the transfer of heat dQ to the reversible heat source increases its entropy according to the quasi-static relationship $dQ = T\, dS$. The external interactions of a reversible heat source accordingly are fully described by its heat capacity $C(T)$ (the definition of the reversible heat source implies that this heat capacity is at constant volume, but we shall not so indicate by an explicit subscript). The energy change of the reversible heat source is $dU = dQ = C(T)\, dT$ and the entropy change is $dS = [C(T)/T]\, dT$. The various transfers envisaged in the maximum work theorem are indicated schematically in Fig. 4.5.

The proof of the maximum work theorem is almost immediate. Consider two processes. Each leads to the same energy change ΔU and the same entropy change ΔS within the primary subsystem, for these are determined by the specified initial and final states. The two processes differ only in the apportionment of the energy difference $(-\Delta U)$ between the reversible work source and the reversible heat source ($-\Delta U = W_{RWS} + Q_{RHS}$). But the process that delivers the maximum possible work to the reversible work source correspondingly delivers the least possible heat to the reversible heat source, *and therefore leads to the least possible entropy increase of the reversible heat source (and thence of the entire system).*

[1] The use of the term *source* might be construed as biasing the terminology in favor of *extraction* of heat, as contrasted with *injection*; such a bias is not intended.

The *absolute* minimum of ΔS_{total}, for *all* possible processes, is attained by *any reversible process* (for all of which $\Delta S_{\text{total}} = 0$).

To recapitulate, *energy conservation requires* $\Delta U + W_{\text{RWS}} + Q_{\text{RHS}} = 0$. *With* ΔU *fixed, to maximize* W_{RWS} *is to minimize* Q_{RHS}. *This is achieved by minimizing* $S_{\text{RHS}}^{\text{final}}$ (since S_{RHS} increases monotonically with positive heat input Q_{RHS}). *The minimum* $S_{\text{RHS}}^{\text{final}}$ *therefore is achieved by minimum* ΔS_{total}, *or by* $\Delta S_{\text{total}} = 0$.

The foregoing "descriptive" proof can be cast into more formal language, and this is particularly revealing in the case in which the initial and final states of the subsystem are so close that all differences can be expressed as differentials. Then energy conservation requires

$$dU + dQ_{\text{RHS}} + dW_{\text{RWS}} = 0 \tag{4.6}$$

whereas the entropy maximum principle requires

$$dS_{\text{tot}} = dS + \frac{dQ_{\text{RHS}}}{T_{\text{RHS}}} \geq 0 \tag{4.7}$$

It follows that

$$dW_{\text{RWS}} \leq T_{\text{RHS}}dS - dU \tag{4.8}$$

The quantities on the right-hand side are all specified. In particular dS and dU are the entropy and energy differences of the primary subsystem in the specified final and initial states. The maximum work transfer dW_{RWS} corresponds to the equality sign in equation 4.8, and therefore in equation 4.7 ($dS_{\text{tot}} = 0$).

It is useful to calculate the maximum delivered work which, from equation 4.8 and from the identity $dU = dQ + dW$, becomes

$$dW_{\text{RWS}} \text{ (maximum)} = \left(\frac{T_{\text{RHS}}}{T}\right)dQ - dU$$

$$= [1 - (T_{\text{RHS}}/T)](-dQ) + (-dW) \tag{4.9}$$

That is, *in an infinitesimal process, the maximum work that can be delivered to the reversible work source is the sum of*:

(*a*)   *the work* $(-dW)$ *directly extracted from the subsystem,*
(*b*)   *a fraction* $(1 - T_{\text{RHS}}/T)$ *of the heat* $(-dQ)$ *directly extracted from the subsystem.*

The fraction $(1 - T_{\text{RHS}}/T)$ of the extracted heat that can be "converted" to work in an infinitesimal process is called the *thermodynamic engine efficiency*, and we shall return to a discussion of that quantity in Section 4.5. However, *it generally is preferable to solve maximum work problems in terms of an overall accounting of energy and entropy changes* (rather than to integrate over the thermodynamic engine efficiency).

Returning to the total (noninfinitesimal) process, the energy conservation condition becomes

$$\Delta U_{\text{subsystem}} + Q_{\text{RHS}} + W_{\text{RWS}} = 0 \qquad (4.10)$$

whereas the reversibility condition is

$$\Delta S_{\text{total}} = \Delta S_{\text{subsystem}} + \int dQ_{\text{RHS}}/T_{\text{RHS}} = 0 \qquad (4.11)$$

In order to evaluate the latter integral it is necessary to know the heat capacity $C_{\text{RHS}}(T) = dQ_{\text{RHS}}/dT_{\text{RHS}}$ of the reversible heat source. Given $C_{\text{RHS}}(T)$ the integral can be evaluated, and one can then also infer the net heat transfer Q_{RHS}. Equation 4.10 in turn evaluates W_{RWS}. *Equations 4.10 and 4.11, evaluated as described, provide the solution of all problems based on the maximum work theorem.*

The problem is further simplified if the reversible heat source is a *thermal reservoir. A thermal reservoir is defined as a reversible heat source that is so large that any heat transfer of interest does not alter the temperature of the thermal reservoir.* Equivalently, a thermal reservoir is a reversible heat source characterized by a *fixed* and definite temperature. For such a system equation 4.11 reduces simply to

$$\Delta S_{\text{total}} = \Delta S_{\text{subsystem}} + \frac{Q_{\text{res}}}{T_{\text{res}}} = 0 \qquad (4.12)$$

and Q_{res} ($= Q_{\text{RHS}}$) can be eliminated between equations 4.10 and 4.12, giving

$$W_{\text{RWS}} = T_{\text{res}}\Delta S_{\text{subsystem}} - \Delta U_{\text{subsystem}} \qquad (4.13)$$

Finally, it should be recognized that the specified final state of the subsystem may have a larger energy than the initial state. In that case the theorem remains formally true but the "delivered work" may be negative. This work which must be supplied to the subsystem will then be *least* (the *delivered* work remains algebraically maximum) for a reversible process.

Example 1
One mole of an ideal van der Waals fluid is to be taken by an unspecified process from the state T_0, v_0 to the state T_f, v_f. A second system is constrained to have a

fixed volume and its initial temperature is T_{20}; its heat capacity is linear in the temperature

$$C_2(T) = DT \qquad (D = \text{constant})$$

What is the maximum work that can be delivered to a reversible work source?

Solution

The solution parallels those of the problems in Section 4.1 despite the slightly different formulations. The second system is a reversible heat source; for it the dependence of energy on temperature is

$$U_2(T) = \int C_2(T)\, dT = \tfrac{1}{2} DT^2 + \text{constant}$$

and the dependence of entropy on temperature is

$$S_2(T) = \int \frac{C_2(T)}{T}\, dT = DT + \text{constant}$$

For the primary fluid system the dependence of energy and entropy on T and v is given in equations 3.49 and 3.51 from which we find

$$\Delta U_1 = cR(T_f - T_0) - \frac{a}{v_f} + \frac{a}{v_0}$$

$$\Delta S_1 = R\ln\left(\frac{v_f - b}{v_0 - b}\right) + cR\ln\frac{T_f}{T_0}$$

The second system (the reversible heat source) changes temperature from T_{20} to some as yet unknown temperature T_{2f}, so that

$$\Delta U_2 = \tfrac{1}{2} D\left(T_{2f}^2 - T_{20}^2\right)$$

and

$$\Delta S_2 = D\left(T_{2f} - T_{20}\right)$$

The value of T_{2f} is determined by the reversibility condition

$$\Delta S_1 + \Delta S_2 = R\ln\left(\frac{v_f - b}{v_0 - b}\right) + cR\ln\frac{T_f}{T_0} + D\left(T_{2f} - T_{20}\right) = 0$$

or

$$T_{2f} = T_{20} - RD^{-1}\ln\left(\frac{v_f - b}{v_0 - b}\right) - cRD^{-1}\ln\frac{T_f}{T_0}$$

The conservation of energy then determines the work W_3 delivered to the reversible work source

$$W_3 + \Delta U_2 + \Delta U_1 = 0$$

whence

$$W_3 = -\left[\frac{1}{2}D\left(T_{2f}^2 - T_{20}^2\right)\right] - \left[cR(T_f - T_0) - \frac{a}{v_f} + \frac{a}{v_0}\right]$$

where we recall that T_f is given, whereas T_{2f} has been found.

An equivalent problem, but with a somewhat simpler system (a monatomic ideal gas and a thermal reservoir) is formulated in Problem 4.5-1. In each of these problems we do not commit ourselves to any specific process by which the result might be realized, but such a specific process is developed in Problem 4.5-2 (which, with 4.5-1, is strongly recommended to the reader).

Example 2   Isotope Separation
In the separation of U^{235} and U^{238} to produce enriched fuels for atomic power plants the naturally occurring uranium is reacted with fluorine to form uranium hexafluoride (UF_6). The uranium hexafluoride is a gas at room temperature and atmospheric pressure. The naturally occurring mole fraction of U^{235} is 0.0072, or 0.72%. It is desired to process 10 moles of natural UF_6 to produce 1 mole of 2% enriched material, leaving 9 moles of partially depleted material. The UF_6 gas can be represented approximately as a polyatomic, multicomponent simple ideal gas with $c = 7/2$ (equation 3.40). Assuming the separation process to be carried out at a temperature of 300 K and a pressure of 1 atm, and assuming the ambient atmosphere (at 300 K) to act as a thermal reservoir, what is the minimum amount of work required to carry out the enrichment process? Where does this work (energy) ultimately reside?

Solution
The problem is an example of the maximum work theorem in which the minimum work *required* corresponds to the maximum work "delivered." The initial state of the system is 10 moles of natural UF_6 at $T = 300$ K and $P = 1$ atm. The final state of the system is 1 mole of enriched gas and 9 moles of depleted gas at the same temperature and pressure. The cold reservoir is also at the same temperature.

We find the changes of entropy and of energy of the system. From the fundamental equation (3.40) we find the equations of state to be the familiar forms

$$U = 7/2\, NRT \qquad PV = NRT$$

These enable us to write the entropy as a function of T and P.

$$S = \sum_{j=1}^{2} N_j s_{0j} + \left(\frac{7}{2}\right) NR \ln\left(\frac{T}{T_0}\right) - NR \ln\left(\frac{P}{P_0}\right) - NR \sum_{j=1}^{2} x_j \ln x_j$$

This last term—the "entropy of mixing" as defined following equation 3.40—is the significant term in the isotope separation process.

We first calculate the mole fraction of $U^{235}F_6$ in the 9 moles of depleted material; this is found to be 0.578%. Accordingly the change in entropy is

$$\Delta S = -R[0.02 \ln 0.02 + 0.98 \ln 0.98] - 9R[0.00578 \ln 0.00578$$
$$+ 0.994 \ln 0.994] + 10R[0.0072 \ln 0.0072 + 0.9928 \ln 0.9928]$$
$$= -0.0081R = -0.067 \text{ J/K}$$

The gas *ejects* heat.

There is no change in the energy of the gas, and all the energy supplied as work is transferred to the ambient atmosphere as heat. That work, or heat, is

$$-W_{RWS} = Q_{res} = -T\Delta S = 300 \times 0.067 = 20 \text{ J}$$

If there existed a semipermeable membrane, permeable to $U^{235}F_6$ but not to $U^{238}F_6$, the separation could be accomplished simply. Unfortunately no such membrane exists. The methods employed in practice are all dynamic (non-quasi-static) processes that exploit the small mass difference of the two isotopes—in ultracentrifuges, in mass spectrometers, or in gaseous diffusion.

PROBLEMS

4.5-1. One mole of a monatomic ideal gas is contained in a cylinder of volume 10^{-3} m³ at a temperature of 400 K. The gas is to be brought to a final state of volume 2×10^{-3} m³ and temperature 400 K. A thermal reservoir of temperature 300 K is available, as is a reversible work source. What is the maximum work that can be delivered to the reversible work source?

Answer:
$W_{RWS} = 300\ R \ln 2$

4.5-2. Consider the following process for the system of Problem 4.5-1. The ideal gas is first expanded adiabatically (and isentropically) until its temperature falls to 300 K; the gas does work on the reversible work source in this expansion. The gas is then expanded while in thermal contact with the thermal reservoir. And finally the gas is compressed adiabatically until its volume and temperature reach the specified values (2×10^{-3} m³ and 400 K).
a) Draw the three steps of this process on a $T - V$ diagram, giving the equation of each curve and labelling the numerical coordinates of the vertices.
b) To what volume must the gas be expanded in the second step so that the third (adiabatic) compression leads to the desired final state?
c) Calculate the work and heat transfers in each step of the process and show that the overall results are identical to those obtained by the general approach of Example 1.

4.5-3. Describe how the gas of the preceding two problems could be brought to the desired final state by a free expansion. What are the work and heat transfers in this case? Are these results consistent with the maximum work theorem?

4.5-4. The gaseous system of Problem 4.5-1 is to be restored to its initial state. Both states have temperature 400 K, and the energies of the two states are equal ($U = 600\ R$). Need any work be supplied, and if so, what is the *minimum* supplied work? Note that the thermal reservoir of temperature 300 K remains accessible.

4.5-5. If the thermal reservoir of Problem 4.5-1 were to be replaced by a reversible heat source having a heat capacity of the form

$$C(T) = \left(2 + \frac{T}{150}\right)R$$

and an initial temperature of $T_{RHS,0} = 300$ K, again calculate the maximum delivered work.

Before doing the calculation, would you expect the delivered work to be greater, equal to, or smaller than that calculated in Prob. 4.5-1? Why?

4.5-6. A system can be taken from state A to state B (where $S_B = S_A$) either (a) directly along the adiabat $S =$ constant, or (b) along the isochore AC and the isobar CB. The difference in the work done by the system is the area enclosed between the two paths in a $P-V$ diagram. Does this contravene the statement that the work delivered to a reversible work source is the same for every reversible process? Explain!

4.5-7. Consider the maximum work theorem in the case in which the specified final state of the subsystem has lower energy than the initial state. Then the essential logic of the theorem can be summarized as follows: "Extraction of heat from the subsystem decreases its entropy. Consequently a portion of the extracted heat must be sacrificed to a reversible heat source to effect a net increase in entropy; otherwise the process will not proceed. The remainder of the extracted heat is available as work."

Similarly summarize the essential logic of the theorem in the case in which the final state of the subsystem has larger energy and larger entropy than the initial state.

4.5-8. If $S_B < S_A$ and $U_B > U_A$ does this imply that the delivered work is negative? Prove your assertion assuming the reversible heat source to be a thermal reservoir.

Does postulate III, which states that S is a monotonically increasing function of U, disbar the conditions assumed here? Explain.

4.5-9. Two identical bodies each have constant and equal heat capacities ($C_1 = C_2 = C$, a constant). In addition a reversible work source is available. The initial temperatures of the two bodies are T_{10} and T_{20}. What is the maximum work that can be delivered to the reversible work source, leaving the two bodies in thermal equilibrium? What is the corresponding equilibrium temperature? Is this the minimum attainable equilibrium temperature, and if so, why? What is the maximum attainable equilibrium temperature?

For $C = 8$ J/K, $T_{10} = 100°C$ and $T_{20} = 0°C$ calculate the maximum delivered work and the possible range of final equilibrium temperature.

Answer:
$$T_f^{min} = 46°C \qquad T_f^{max} = 50°C$$
$$W^{max} = C[\sqrt{T_{10}} - \sqrt{T_{20}}]^2 = 62.2 \text{ J}$$

4.5-10. Two identical bodies each have heat capacities (at constant volume) of
$$C(T) = a/T$$
The initial temperatures are T_{10} and T_{20}, with $T_{20} > T_{10}$. The two bodies are to be brought to thermal equilibrium with each other (maintaining both volumes constant) while delivering as much work as possible to a reversible work source. What is the final equilibrium temperature and what is the maximum work delivered to the reversible work source?

Evaluate your answer for $T_{20} = T_{10}$ and for $T_{20} = 2T_{10}$.

Answer:
$$W = a \ln(9/8) \text{ if } T_{20} = 2T_{10}$$

4.5-11. Two bodies have heat capacities (at constant volume) of
$$C_1 = aT$$
$$C_2 = 2bT$$
The initial temperatures are T_{10} and T_{20}, with $T_{20} > T_{10}$. The two bodies are to be brought to thermal equilibrium (maintaining both volumes constant) while delivering as much work as possible to a reversible work source. What is the final equilibrium temperature and what is the (maximum) work delivered to the reversible work source?

4.5-12. One mole of an ideal van der Waals fluid is contained in a cylinder fitted with a piston. The initial temperature of the gas is T_i and the initial volume is v_i. A reversible heat source with a constant heat capacity C and with an initial temperature T_0 is available. The gas is to be compressed to a volume of v_f and brought into thermal equilibrium with the reversible heat source. What is the maximum work that can be delivered to the reversible work source and what is the final temperature?

Answer:
$$T_f = \left[\left(\frac{v_i - b}{v_f - b} \right)^R T_i^{cR} T_0 \right]^{1/(cR + C)}$$

4.5-13. A system has a temperature-independent heat capacity C. The system is initially at temperature T_i and a heat reservoir is available, at temperature T_c (with $T_c < T_i$). Find the maximum work recoverable as the system is cooled to the temperature of the reservoir.

4.5-14. If the temperature of the atmosphere is 5°C on a winter day and if 1 kg of water at 90°C is available, how much work can be obtained as the water is cooled to the ambient temperature? Assume that the volume of the water is constant, and assume that the molar heat capacity at constant volume is 75 J/mole K and is independent of temperature.

Answer:
$$45 \times 10^3 \text{J}$$

4.5-15. A rigid cylinder contains an internal adiabatic piston separating it into two chambers, of volumes V_{10} and V_{20}. The first chamber contains one mole of a monatomic ideal gas at temperature T_{10}. The second chamber contains one mole of a simple diatomic ideal gas ($c = 5/2$) at temperature T_{20}. In addition a thermal reservoir at temperature T_c is available. What is the maximum work that can be delivered to a reversible work source, and what are the corresponding volumes and temperatures of the two subsystems?

4.5-16. Each of three identical bodies has a temperature-independent heat capacity C. The three bodies have initial temperatures $T_3 > T_2 > T_1$. What is the maximum amount of work that can be extracted leaving the three bodies at a common final temperature?

4.5-17. Each of two bodies has a heat capacity given by

$$C = A + 2BT$$

where $A = 8$ J/K and $B = 2 \times 10^{-2}$ J/K^2. If the bodies are initially at temperatures of 200 K and 400 K, and if a reversible work source is available, what is the minimum final common temperature to which the two bodies can be brought? If no work can be extracted from the reversible work source what is the maximum final common temperature to which the two bodies can be brought? What is the maximum amount of work that can be transferred to the reversible work source?

Answer:
$$T_{min} = 293 \text{ K}$$

4.5-18. A particular system has the equations of state

$$T = As/v^{1/2} \quad \text{and} \quad P = T^2/4Av^{1/2}$$

where A is a constant. One mole of this system is initially at a temperature T_1 and volume V_1. It is desired to cool the system to a temperature T_2 while compressing it to volume V_2 ($T_2 < T_1$; $V_2 < V_1$). A second system is available. It is initially at a temperature T_c ($T_c < T_2$). Its volume is kept constant throughout, and its heat capacity is

$$C_V = BT^{1/2} \quad (B = \text{constant})$$

What is the minimum amount of work that must be supplied by an external agent to accomplish this goal?

4.5-19. A particular type of system obeys the equations

$$T = \frac{u}{b} \quad \text{and} \quad P = avT$$

where a and b are constants. Two such systems, each of 1 mole, are initially at temperatures T_1 and T_2 (with $T_2 > T_1$) and each has a volume v_0. The systems are to be brought to a common temperature T_f, with each at the same final volume v_f. The process is to be such as to deliver maximum work to a reversible work source.

a) What is the final temperature T_f?

b) How much work can be delivered? Express the result in terms of T_1, T_2, v_0, v_f, and the constants a and b.

4.5-20. Suppose that we have a system in some initial state (we may think of a tank of hot, compressed gas as an example) and we wish to use it as a source of work. Practical considerations require that the system be left finally at atmospheric temperature and pressure, in equilibrium with the ambient atmosphere. Show, first, that the system does work on the atmosphere, and that the work actually available for useful purposes is therefore less than that calculated by a straightforward application of the maximum work theorem. In engineering parlance this net available work is called the "availability".

a) Show that the availability is given by

$$\text{Availability} = \left(U_0 + P_{atm}V_0 - T_{atm}S_0\right) - \left(U_f + P_{atm}V_f - T_{atm}S_f\right)$$

where the subscript *f* denotes the final state, in which the pressure is P_{atm} and the temperature is T_{atm}.

b) If the original system were to undergo an internal chemical reaction during the process considered, would that invalidate this formula for the availability?

4.5-21. An antarctic meteorological station suddenly loses all of its fuel. It has N moles of an inert "ideal van der Waals fluid" at a high temperature T_h and a high pressure P_h. The (constant) temperature of the environment is T_0 and the atmospheric pressure is P_0. If operation of the station requires a continuous power $\mathscr{P}$, what is the longest conceivable time, t_{max}, that the station can operate? Calculate t_{max} in terms of $T_h, T_0, P_h, P_0, \mathscr{P}, N$ and the van der Waals constants a, b, and c.

Note that this is a problem in availability, as defined and discussed in Problem 4.5-20. In giving the solution it is not required that the molar volume v_h be solved explicitly in terms of T_h and P_h; it is sufficient simply to designate it as $v_h(T_h, P_h)$ and similarly for $v_0(T_0, P_0)$.

4.5-22. A "geothermal" power source is available to drive an oxygen production plant. The geothermal source is simply a well containing 10^3 m^3 of water, initially at 100°C; nearby there is a huge ("infinite") lake at 5°C. The oxygen is to be separated from air, the separation being carried out at 1 atm of pressure and at 20°C. Assume air to be $\frac{1}{5}$ oxygen and $\frac{4}{5}$ nitrogen (in mole fractions), and assume that it can be treated as a mixture of ideal gases. How many moles of O_2 can be produced in principle (i.e., assuming perfect thermodynamic efficiency) before exhausting the power source?

4-6 COEFFICIENTS OF ENGINE, REFRIGERATOR, AND HEAT PUMP PERFORMANCE

As we saw in equations 4.6 and 4.7, in an infinitesimal reversible process involving a "hot" subsystem, a "cold" reversible heat source, and a reversible work source

$$\left(dQ_h + dW_h\right) + dQ_c + dW_{RWS} = 0 \tag{4.14}$$

and

$$dS_h + \frac{dQ_c}{T_c} = 0 \qquad (4.15)$$

where we now indicate the "hot" system by the subscript h and the "cold" reversible heat source by the subscript c. In such a process the delivered work dW_{RWS} is algebraically maximum. This fact leads to criteria for the operation of various types of useful devices.

The most immediately evident system of interest is a *thermodynamic engine*. Here the "hot subsystem" may be a furnace or a steam boiler, whereas the "cold" reversible heat source may be the ambient atmosphere or, for a large power plant, a river or lake. The measure of performance is the fraction of the heat $(-dQ_h)$ withdrawn[2] from the hot system that is converted to work dW_{RWS}. Taking $dW_h = 0$ in equation 4.14 (it is simply additive to the delivered work in equation 4.9) we find the *thermodynamic engine efficiency ε_e.*

$$\varepsilon_e = \frac{dW_{RWS}}{(-dQ_h)} = 1 - \frac{T_c}{T_h} \qquad (4.16)$$

The relationship of the various energy exchanges is indicated in Fig. 4.6a.

For a subsystem of given temperature T_h, the thermodynamic engine efficiency increases as T_c decreases. That is, the lower the temperature of the cold system (to which heat is delivered), the higher the engine efficiency. The maximum possible efficiency, $\varepsilon_e = 1$, occurs if the temperature of the cold heat source is equal to zero. If a reservoir at zero temperature were available as a heat repository, heat could be freely and fully converted into work (and the world "energy shortage" would not exist![3]).

A *refrigerator* is simply a thermodynamic engine operated in reverse (Fig. 4.7b). The purpose of the device is to extract heat from the cold system and, with the input of the minimum amount of work, to eject that heat into the comparatively hot ambient atmosphere. Equations 4.14 and

---

[2] The problem of signs may be confusing. Throughout this book the symbols W and Q, or dW and dQ, indicate work and heat *inputs*. Heat withdrawn from a system is $(-Q)$ or $(-dQ)$. Thus if 5 J are withdrawn from the hot subsystem we would write that the heat *withdrawn* is $(-Q_h) = 5$ J, whereas Q_h, the heat *input*, would be -5 J. For clarity in this chapter we use the parentheses to serve as a reminder that $(-Q_h)$ is to be considered as a positive quantity in the particular example being discussed.

[3] The energy shortage is, in any case, a misnomer. Energy is conserved! The shortage is one of "entropy sinks"—of systems of low entropy. Given such systems we could bargain with nature, offering to allow the entropy of such a system to increase (as by allowing a hydrocarbon to oxidize, or heat to flow to a low temperature sink, or a gas to expand) if useful tasks were simultaneously done. There is only a "neg-entropy" shortage!

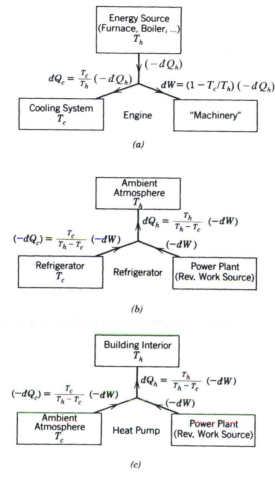

FIGURE 4.6
Engine, refrigerator, and heat pump. In this diagram $dW \equiv dW_{RWS}$

4.15 remain true, but the *coefficient of refrigerator performance* represents the appropriate criterion for this device—the ratio of the heat removed from the refrigerator (the cold system) to the work that must be purchased from the power company. That is

$$\varepsilon_r \equiv \frac{(-dQ_c)}{(-dW_{RWS})} = \frac{T_c}{T_h - T_c} \tag{4.17}$$

If the temperatures T_h and T_c are equal, the coefficient of refrigerator performance becomes infinite: no work is then required to transfer heat from one system to the other. The coefficient of performance becomes progressively smaller as the temperature T_c decreases relative to T_h. And if the temperature T_c approaches zero, the coefficient of performance also approaches zero (assuming T_h fixed). It therefore requires huge amounts of work to extract even trivially small quantities of heat from a system near $T_c = 0$.

We now turn our attention to the *heat pump*. In this case we are interested in heating a warm system, extracting some heat from a cold system, and extracting some work from a reversible work source. In a practical case the warm system may be the interior of a home in winter, the cold system is the outdoors, and the reversible work source is again the power company. In effect, we heat the home by removing the door of a refrigerator and pushing it up to an open window. The inside of the refrigerator is exposed to the outdoors, and the refrigerator attempts (with negligible success) further to cool the outdoors. The heat extracted from this huge reservoir, together with the energy purchased from the power company, is ejected directly into the room from the cooling coils in the back of the refrigerator.

The *coefficient of heat pump performance* ε_p is the ratio of the heat delivered to the hot system to the work extracted from the reversible work source.

$$\varepsilon_p = \frac{dQ}{(-dW_{\text{RWS}})} = \frac{T_h}{T_h - T_c} \tag{4.18}$$

PROBLEMS

4.6-1. A temperature of 0.001 K is accessible in low temperature laboratories with moderate effort. If the price of energy purchased from the electric utility company is 15¢/kW h what would be the minimum cost of extraction of one watt-hour of heat from a system at 0.001 K? The "warm reservoir" is the ambient atmosphere at 300 K.

Answer:
$45

4.6-2. A home is to be maintained at 70°F, and the external temperature is 50°F. One method of heating the home is to purchase work from the power company and to convert it directly into heat: This is the method used in common electric room heaters. Alternatively, the purchased work can be used to operate a heat pump. What is the ratio of the costs if the heat pump attains the ideal thermodynamic coefficient of performance?

4.6-3. A household refrigerator is maintained at a temperature of 35°F. Every time the door is opened, warm material is placed inside, introducing an average of 50 kcal, but making only a small change in the temperature of the refrigerator.

The door is opened 15 times a day, and the refrigerator operates at 15% of the ideal coefficient of performance. The cost of work is 15¢/kW h. What is the monthly bill for operating this refrigerator?

4.6-4. Heat is extracted from a bath of liquid helium at a temperature of 4.2 K. The high-temperature reservoir is a bath of liquid nitrogen at a temperature of 77.3 K. How many Joules of heat are introduced into the nitrogen bath for each Joule extracted from the helium bath?

4.6-5. Assume that a particular body has the equation of state $U = NCT$ with $NC = 10$ J/K and assume that this equation of state is valid throughout the temperature range from 0.5 K to room temperature. How much work must be expended to cool this body from room temperature (300 K) to 0.5 K, using the ambient atmosphere as the hot reservoir?

Answer:
16.2 kJ.

4.6-6. One mole of a monatomic ideal gas is allowed to expand isothermally from an initial volume of 10 liters to a final volume of 15 liters, the temperature being maintained at 400 K. The work delivered is used to drive a thermodynamic refrigerator operating between reservoirs of temperatures 200 and 300 K. What is the maximum amount of heat withdrawn from the low-temperature reservoir?

4.6-7. Give a "constructive solution" of Example 2 of Section 4.1. Your solution may be based on the following procedure for achieving maximum temperature of the hot body. A thermodynamic engine is operated between the two cooler bodies, extracting work until the two cooler bodies reach a common temperature. This work is then used as the input to a heat pump, extracting heat from the cooler pair and heating the hot body. Show that this procedure leads to the same result as was obtained in the example.

4.6-8. Assume that 1 mole of an ideal van der Waals fluid is expanded isothermally, at temperature T_h, from an initial volume V_i to a final volume V_f. A thermal reservoir at temperature T_c is available. Apply equation 4.9 to a differential process and integrate to calculate the work delivered to a reversible work source. Corroborate by overall energy and entropy conservation.
Hint: Remember to add the *direct* work transfer $P\,dV$ to obtain the total work delivered to the reversible work source (as in equation 4.9).

4.6-9. Two moles of a monatomic ideal gas are to be taken from an initial state (P_i, V_i) to a final state $(P_f = B^2 P_i, V_f = V_i/B)$, where B is a constant. A reversible work source and a thermal reservoir of temperature T_c are available. Find the maximum work that can be delivered to the reversible work source.

Given values of B, P_i and T_c, for what values of V_i is the maximum delivered work positive?

4.6-10. Assume the process in Problem 4.6-9 to occur along the locus $P = B/V^2$, where $B = P_i V_i^2$. Apply the thermodynamic engine efficiency to a differential process and integrate to corroborate the result obtained in Problem 4.6-9. Recall the hint given in Problem 4.6-8.

4.6-11. Assume the process in Problem 4.6-9 to occur along a straight-line locus in the $T-V$ plane. Integrate along this locus and again corroborate the results of Problems 4.6-9 and 4.6-10.

4-7 THE CARNOT CYCLE

Throughout this chapter we have given little attention to *specific* processes, purposefully stressing that the delivery of maximum work is a general attribute of *all* reversible processes. It is useful nevertheless to consider briefly one particular type of process—the "Carnot cycle"—both because it elucidates certain general features and because this process has played a critically important role in the historical development of thermo-dynamic theory.

A system is to be taken from a particular initial state to a given final state while exchanging heat and work with reversible heat and work sources. To describe a particular process it is not sufficient merely to describe the path of the system in its thermodynamic configuration space. *The critical features of the process concern the manner in which the extracted heat and work are conveyed to the reversible heat and work sources.* For that purpose auxiliary systems may be employed. The aux-iliary systems are the "tool" or "devices" used to accomplish the task at hand, or, in a common terminology, they constitute the *physical engines* by which the process is effected.

Any thermodynamic system—a gas in a cylinder and piston, a magnetic substance in a controllable magnetic field, or certain chemical systems—can be employed as the auxiliary system. It is only required that the auxiliary system be restored, at the end of the process, to its initial state; the *auxiliary system must not enter into the overall energy or entropy accounting.* It is this cyclic nature of the process within the auxiliary system that is reflected in the name of the Carnot "cycle."

For clarity we temporarily assume that the primary system and the reversible heat source are each thermal reservoirs, the primary system being a "hot reservoir" and the reversible heat source being a "cold reservoir"; this restriction merely permits us to consider finite heat and work transfers rather than infinitesimal transfers.

The Carnot cycle is accomplished in four steps, and the changes of the temperature and the entropy of the *auxiliary* system are plotted for each of these steps in Fig. 4.7.

1. The auxiliary system, originally at the same temperature as the primary system (the hot reservoir), is placed in contact with that reservoir and with the reversible work source. The auxiliary system is then caused to undergo an isothermal process by changing some convenient extensive

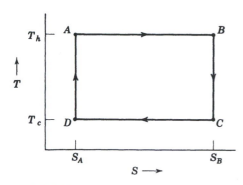 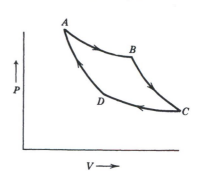

FIGURE 4.7
The T–S and P–V diagrams for the auxiliary system in the Carnot cycle.

parameter; if the auxiliary system is a gas it may be caused to expand isothermally, if it is a magnetic system its magnetic moment may be decreased isothermally, and so forth. In this process a flux of heat occurs from the hot reservoir to the auxiliary system, and a transfer of work ($\int P\,dV$ or its magnetic or other analogue) occurs from the auxiliary system to the reversible work source. This is the isothermal step $A \to B$ in Fig. 4.7.

2. The auxiliary system, now in contact only with the reversible work source, is adiabatically expanded (or adiabatically demagnetized, etc.) until its temperature falls to that of the cold reservoir. A further transfer of work occurs from the auxiliary system to the reversible work source. The quasi-static adiabatic process occurs at constant entropy of the auxiliary system, as in $B \to C$ of Fig. 4.7.

3. The auxiliary system is isothermally compressed while in contact with the cold reservoir and the reversible work source. This compression is continued until the entropy of the auxiliary system attains its initial value. During this process there is a transfer of work from the reversible work source to the auxiliary system, and a transfer of heat from the auxiliary system to the cold reservoir. This is the step $C \to D$ in Fig. 4.7.

4. The auxiliary system is adiabatically compressed and receives work from the reversible work source. The compression brings the auxiliary system to its initial state and completes the cycle. Again the entropy of the auxiliary system is constant, from D to A in Fig. 4.7.

The heat withdrawn from the primary system (the hot reservoir) in process 1 is $T_h \Delta S$, and the heat transferred to the cold reservoir in process 3 is $T_c \Delta S$. The difference $(T_h - T_c) \Delta S$ is the net work transferred to the reversible work source in the complete cycle. On the T–S diagram of Fig. 4.7 the heat $T_h \Delta S$ withdrawn from the primary system is represented by the area bounded by the four points labeled ABS_BS_A, the heat ejected to the cold reservoir is represented by the area CDS_AS_B, and the net work delivered is represented by the area $ABCD$. The coefficient of performance is the ratio of the area $ABCD$ to the area ABS_BS_A or $(T_h - T_c)/T_h$.

The Carnot cycle can be represented on any of a number of other diagrams, such as a $P-V$ diagram or a $T-V$ diagram. The representation on a $P-V$ diagram is indicated in Fig. 4.7. The precise form of the curve BC, representing the dependence of P on V in an adiabatic (isentropic) process, would follow from the equation of state $P = P(S, V, N)$ of the auxiliary system.

If the hot and cold systems are merely reversible heat sources, rather than reservoirs, the Carnot cycle must be carried out in infinitesimal steps. The heat withdrawn from the primary (hot) system in process 1 is then $T_h dS$ rather than $T_h \Delta S$, and similarly for the other steps. There is clearly no difference in the essential results, although T_h and T_c are continually changing variables and the net evaluation of the process requires an integration over the differential steps.

It should be noted that real engines never attain ideal thermodynamic efficiency. Because of mechanical friction, and because they cannot be operated so slowly as to be truly quasi-static, they seldom attain more than 30 or 40% thermodynamic efficiency. Nevertheless, the upper limit on the efficiency, set by basic thermodynamic principles, is an important factor in engineering design. There are other factors as well, to which we shall return in Section 4.9.

Example

N moles of a monatomic ideal gas are to be employed as the auxiliary system in a Carnot cycle. The ideal gas is initially in contact with the hot reservoir, and in the first stage of the cycle it is expanded from volume V_A to volume V_B.[4] Calculate the work and heat transfers in each of the four steps of the cycle, in terms of T_h, T_c, V_A, V_B, and N. Directly corroborate that the efficiency of the cycle is the Carnot efficiency.

Solution

The data are given in terms of T and V; we therefore express the entropy and energy as functions of T, V, and N.

$$S = Ns_0 + NR \ln \left(\frac{T^{3/2} V N_0}{T_0^{3/2} V_0 N} \right)$$

and

$$U = \tfrac{3}{2} NRT$$

Then in the isothermal expansion at temperature T_h

$$\Delta S_{AB} = S_B - S_A = NR \ln \left(\frac{V_B}{V_A} \right) \quad \text{and} \quad \Delta U_{AB} = 0$$

---

[4]Note that in this example quantities such as U, S, V, Q refer to the *auxiliary* system rather than to the "primary system" (the hot reservoir).

whence

$$Q_{AB} = T_h \Delta S_{AB} = NRT_h \ln\left(\frac{V_B}{V_A}\right)$$

and

$$W_{AB} = -NRT_h \ln\left(\frac{V_B}{V_A}\right)$$

In the second step of the cycle the gas is expanded adiabatically until the temperature falls to T_c, the volume meanwhile increasing to V_C. From the equation for S, we see that $T^{\frac{3}{2}} V = $ constant, and

$$V_C = V_B \left(\frac{T_h}{T_c}\right)^{3/2}$$

and

$$Q_{BC} = 0 \qquad W_{BC} = \Delta U = \tfrac{3}{2} NR(T_c - T_h)$$

In the third step the gas is isothermally compressed to a volume V_D. This volume must be such that it lies on the same adiabat as V_A (see Fig. 4.7), so that

$$V_D = V_A \left(\frac{T_h}{T_c}\right)^{3/2}$$

Then, as in step 1,

$$Q_{CD} = NRT_c \ln\left(\frac{V_D}{V_C}\right) = NRT_c \ln\left(\frac{V_A}{V_B}\right)$$

and

$$W_{CD} = -NRT_c \ln\left(\frac{V_A}{V_B}\right)$$

Finally, in the adiabatic compression

$$Q_{DA} = 0$$

and

$$W_{DA} = U_{DA} = \tfrac{3}{2} NR(T_h - T_c)$$

From these results we obtain

$$W = W_{AB} + W_{BC} + W_{CD} + W_{DA} = -NR(T_h - T_c)\ln\left(\frac{V_B}{V_A}\right)$$

and

$$-W/Q_{AB} = (T_h - T_c)/T_h$$

which is the expected Carnot efficiency.

PROBLEMS

4.7-1. Repeat the calculation of Example 5 assuming the "working substance" of the auxiliary system to be 1 mole of an ideal van der Waals fluid rather than of a monatomic ideal gas (recall Section 3.5).

4.7-2. Calculate the work and heat transfers in each stage of the Carnot cycle for the auxiliary system being an "empty" cylinder (containing only electromagnetic radiation). The first step of the cycle is again specified to be an expansion from V_A to V_B. All results are to be expressed in terms of V_A, V_B, T_h, and T_c. Show that the ratio of the total work transfer to the first-stage heat transfer agrees with the Carnot efficiency.

4.7-3. A "primary subsystem" in the initial state A is to be brought reversibly to a specified final state B. A reversible work source and a thermal reservoir at temperature T_r are available, but no "auxiliary system" is to be employed. Is it possible to devise such a process? Prove your answer. Discuss Problem 4.5-2 in this context.

4.7-4. The fundamental equation of a particular fluid is $UN^{\frac{1}{2}}V^{\frac{3}{2}} = A(S - R)^3$ where $A = 2 \times 10^{-2}$ $(K^3m^{\frac{9}{2}}/J^3)$. Two moles of this fluid are used as the auxiliary system in a Carnot cycle, operating between two thermal reservoirs at temperature 100°C and 0°C. In the first isothermal expansion 10^6 J is extracted from the high-temperature reservoir. Find the heat transfer and the work transfer for each of the four processes in the Carnot cycle.

Calculate the efficiency of the cycle directly from the work and heat transfers just computed. Does this efficiency agree with the theoretical Carnot efficiency? *Hint*: Carnot cycle problems generally are best discussed in terms of a T–S diagram for the auxiliary system.

4.7-5. One mole of the "simple paramagnetic model system" of equation 3.66 is to be used as the auxiliary system of a Carnot cycle operating between reservoirs of temperature T_h and T_c. The auxiliary system initially has a magnetic moment I_i and is at temperature T_h. By decreasing the external field while the system is in contact with the high temperature reservoir, a quantity of heat Q_1 is absorbed from the reservoir; the system meanwhile does work $(-W_1)$ on the reversible work source (i.e., on the external system that creates the magnetic field and thereby induces the magnetic moment). Describe each step in the Carnot cycle and calculate the work and heat transfer in each step, expressing each in terms of T_h, T_c, Q_1, and the parameters T_0 and I_0 appearing in the fundamental equation.

4.7-6. Repeat Problem 4.7-4 using the "rubber band" model of Section 3.7 as the auxiliary system.

4-8   MEASURABILITY OF THE TEMPERATURE AND OF THE ENTROPY

The Carnot cycle not only illustrates the general principle of reversible processes as maximum work processes, but it provides us with an operational method for measurements of temperature. We recall that the entropy was introduced merely as an abstract function, the maxima of which determine the equilibrium states. The temperature was then defined in terms of a partial derivative of this function. It is clear that such a definition does not provide a direct recipe for an operational measurement of the temperature and that it is necessary therefore for such a procedure to be formulated explicitly.

In our discussion of the efficiency of thermodynamic engines we have seen that the efficiency of an engine working by reversible processes between two systems, of temperatures T_h and T_c, is

$$\varepsilon_e = 1 - T_c/T_h \tag{4.19}$$

The thermodynamic engine efficiency is defined in terms of fluxes of heat and work and is consequently operationally measurable. Thus a Carnot cycle provides us with an operational method of measuring the ratio of two temperatures.

Unfortunately, real processes are never truly quasi-static, so that real engines never quite exhibit the theoretical engine efficiency. Therefore, the ratio of two given temperatures must actually be determined in terms of the limiting maximum efficiency of all real engines, but this is a difficulty of practice rather than of principle.

The statement that the ratio of temperatures is a measurable quantity is tantamount to the statement that the scale of temperature is determined within an arbitrary multiplicative constant. The temperature of some arbitrarily chosen standard system may be assigned at will, and the temperatures of all other systems are then uniquely determined, with values directly proportional to the chosen temperature of the fiducial system.

The choice of a standard system, and the arbitrary assignment of some definite temperature to it, has been discussed in Section 2.6. We recall that the assignment of the number 273.16 to a system of ice, water, and vapor in mutual equilibrium leads to the absolute Kelvin scale of temperature. A Carnot cycle operating between this system and another system determines the ratio of the second temperature to 273.16 K and consequently determines the second temperature on the absolute Kelvin scale.

Having demonstrated that the temperature is operationally measurable we are able almost trivially to corroborate that the entropy too is measurable. The ability to measure the entropy underlies the utility of the entire thermodynamic formalism. It is also of particular interest because of the somewhat abstract nature of the entropy concept.

The method of measurement to be described yields only entropy differences, or relative entropies—these differences are then converted to absolute entropies by Postulate IV—the "Nernst postulate" (Section 1.10).

Consider a reversible process in a composite system, of which the system of interest is a subsystem. The subsystem is taken from some reference state (T_0, P_0) to the state of interest (T_1, P_1) by some path in the T–P plane. The change in entropy is

$$S_1 - S_0 = \int_{(T_0, P_0)}^{(T_1, P_1)} \left[\left(\frac{\partial S}{\partial T} \right)_P dT + \left(\frac{\partial S}{\partial P} \right)_T dP \right] \qquad (4.20)$$

$$= \int_{(T_0, P_0)}^{(T_1, P_1)} \left(\frac{\partial S}{\partial P} \right)_T \left[-\left(\frac{\partial P}{\partial T} \right)_S dT + dP \right] \qquad (4.21)$$

$$= \int_{(T_0, P_0)}^{(T_1, P_1)} (-V\alpha) \left[-\left(\frac{\partial P}{\partial T} \right)_S dT + dP \right] \qquad (4.22)$$

Equation 4.21 follows from the elementary identity A.22 of Appendix A. Equation 4.22 is less obvious, though the general methods to be developed in Chapter 7 will reduce such transformations to a straightforward procedure; an elementary but relatively cumbersome procedure is suggested in Problem 4.8-1.

Now each of the factors in the integrand is directly measurable; the factor $(\partial P/\partial T)_S$ requires only a measurement of pressure and temperature changes for a system enclosed by an adiabatic wall. Thus, the entropy difference of the two arbitrary states (T_0, P_0) and (T_1, P_1) is obtainable by numerical integration of measurable data.

PROBLEMS

4.8-1. To corroborate equation 4.22 show that

$$\left(\frac{\partial P}{\partial s} \right)_T = -\left(\frac{\partial T}{\partial v} \right)_P$$

First consider the right-hand side, and write generally that

$$dT = u_{ss} \, ds + u_{vs} \, dv$$

so that

$$\left(\frac{\partial T}{\partial v} \right)_P = u_{ss} \left(\frac{\partial s}{\partial v} \right)_P + u_{vs} = -u_{ss} \frac{u_{vv}}{u_{sv}} + u_{vs}$$

Similarly show that $\left(\frac{\partial P}{\partial s}\right)_T = u_{ss}u_{vv}/u_{sv} - u_{sv}$, establishing the required identity.

4-9 OTHER CRITERIA OF ENGINE PERFORMANCE; POWER OUTPUT AND "ENDOREVERSIBLE ENGINES"

As we have remarked earlier, maximum efficiency is not necessarily the primary concern in design of a real engine. Power output, simplicity, low initial cost, and various other considerations are also of importance, and, of course, these are generally in conflict. An informative perspective on the criteria of real engine performance is afforded by the "endoreversible engine problem."[5]

Let us suppose once again that two thermal reservoirs exist, at temperatures T_h and T_c, and that we wish to remove heat from the high temperature reservoir, delivering work to a reversible work source. We now know that the maximum possible efficiency is obtained by any reversible engine. However, considerations of the operation of such an engine immediately reveals that its power output (work delivered per unit time) is atrocious. Consider the very first stage of the process, in which heat is transferred to the system from the hot reservoir. If the working fluid of the engine is at the same temperature as the reservoir no heat will flow; whereas if it is at a lower temperature the heat flow process (and hence the entire cycle) becomes irreversible. In the Carnot engine the temperature difference is made "infinitely small," resulting in an "infinitely slow" process and an "infinitely small" power output.

To obtain a nonzero power output the extraction of heat from the high temperature reservoir and the insertion of heat into the low temperature reservoir must each be done irreversibly.

An *endoreversible engine* is defined as one in which the two processes of heat transfer (from and to the heat reservoirs) are the only irreversible processes in the cycle.

To analyze such an engine we assume, as usual, a high temperature thermal reservoir at temperature T_h, a low temperature thermal reservoir at temperature T_c, and a reversible work source. We assume the isothermal strokes of the engine cycle to be at T_w (w designating "warm") and T_t (t designating "tepid"), with $T_h > T_w > T_t > T_c$. Thus heat flows from the high temperature reservoir to the working fluid across a temperature difference of $T_h - T_w$, as indicated schematically in Fig. 4.8. Similarly, in the heat rejection stroke of the cycle the heat flows across the temperature difference $T_t - T_c$.

---

[5]F. L. Curzon and B. Ahlborn, *Amer. J. Phys.* **43**, 22 (1975). See also M. H. Rubin, *Phys. Rev.* **A19**, 1272 and 1279 (1979) (and references therein) for a sophisticated analysis and for further generalization of the theorem.

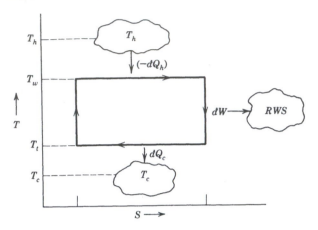

FIGURE 4.8
The endoreversible engine cycle.

Let us now suppose that the rate of heat flow from the high temperature reservoir to the system is proportional to the temperature difference $T_h - T_w$. If t_h is the time required to transfer an amount Q_h of energy, then

$$\frac{(-Q_h)}{t_h} = \sigma_h \cdot (T_h - T_w) \tag{4.23}$$

where σ_h is the conductance (the product of the thermal conductivity times the area divided by the thickness of the wall between the hot reservoir and the working fluid). A similar law holds for the rate of heat flow to the cold reservoir. Therefore the time required for the two isothermal strokes of the engine is

$$t = t_h + t_c = \frac{1}{\sigma_h} \frac{(-Q_h)}{T_h - T_w} + \frac{1}{\sigma_c} \frac{Q_c}{T_t - T_c} \tag{4.24}$$

We assume the time required for the two adiabatic strokes of the engine to be negligible relative to $(t_h + t_c)$, as these times are limited by relatively rapid relaxation times within the working fluid itself. Furthermore the relaxation times within the working fluid can be shortened by appropriate design of the piston and cylinder dimensions, internal baffles, and the like.

Now Q_h, Q_c, and the delivered work W are related by the Carnot efficiency of an engine working between the temperatures T_w and T_t, so that equation 4.24 becomes

$$t = \left[\frac{1}{\sigma_h} \frac{1}{T_h - T_w} \frac{T_w}{T_w - T_t} + \frac{1}{\sigma_c} \frac{1}{T_t - T_c} \frac{T_t}{T_w - T_t} \right] W \tag{4.25}$$

The power output of the engine is W/t, and this quantity is to be maximized with respect to the two as yet undetermined temperatures T_w and T_t. The optimum intermediate temperatures are then found to be

$$T_w = c(T_h)^{1/2} \qquad T_t = c(T_c)^{1/2} \tag{4.26}$$

where

$$c = \frac{\left[(\sigma_h T_h)^{1/2} + (\sigma_c T_c)^{1/2}\right]}{\left[\sigma_h^{1/2} + \sigma_c^{1/2}\right]} \tag{4.27}$$

and the optimum power delivered by the engine is

$$\text{power} = \left(\frac{W}{t}\right)_{\max} = \sigma_h \sigma_c \left[\frac{T_h^{1/2} - T_c^{1/2}}{\sigma_h^{1/2} + \sigma_c^{1/2}}\right]^2 \tag{4.28}$$

Let ε_{erp} denote the efficiency of such an "endoreversible engine maximized for *power*"; for which we find

$$\varepsilon_{erp} = 1 - (T_c/T_h)^{1/2} \tag{4.29}$$

Remarkably, the engine efficiency is not dependent on the conductances σ_h and σ_c!

Large power plants are evidently operated close to the criterion for maximum power output, as Curzon and Ahlborn demonstrate by data on three power plants, as shown in Table 4.1.

TABLE 4.1
Efficiencies of Power Plants as Compared with the Carnot Efficiency and with the Efficiency of an Endoreversible Engine Maximized for Power Output (ε_{erp}).[a]

Power Plant	T_c (°C)	T_h (°C)	ε (Carnot)	ε_{erp}	ε (observed)
West Thurrock (U.K.) coal fired steam plant	~ 25	565	0.64	0.40	0.36
CANDU (Canada) PHW nuclear reactor	~ 25	300	0.48	0.28	0.30
Larderello (Italy) geothermal steam plant	80	250	0.32	0.175	0.16

[a] From Curzon and Ahlborn.

PROBLEMS

4.9-1. Show that the efficiency of an endoreversible engine, maximized for power output, is always less than $\varepsilon_{\text{Carnot}}$. Plot the former efficiency as a function of the Carnot efficiency.

4.9-2. Suppose the conductance σ_h $(= \sigma_c)$ to be such that 1 kW is transferred to the system (as heat flux) if its temperature is 50 K below that of the high temperature reservoir. Assuming $T_h = 800$ K and $T_c = 300$ K, calculate the maximum power obtainable from an endoreversible engine, and find the temperatures T_w and T_t for which such an engine should be designed.

4.9-3. Consider an endoreversible engine for which the high temperature reservoir is boiling water (100°C) and the cold reservoir is at room temperature (taken as 20°C). Assuming the engine is operated at maximum power, what is the ratio of the amount of heat withdrawn from the high temperature reservoir (per kilowatt hour of delivered work) to that withdrawn by a Carnot engine? How much heat is withdrawn by each engine per kilowatt hour of delivered work?

Answer:
Ratio = 1.9

4.9-4. Assume that one cycle of the engine of Problem 4.9-3 takes 20 s and that the conductance $\sigma_h = \sigma_c = 100$ W/K. How much work is delivered per cycle? Assuming the "control volume" (i.e., the auxiliary system) is a gas, driven through a Carnot cycle, plot a T–S diagram of the gas during the cycle. Indicate numerical values for each vertex of the diagram (note that one value of the entropy can be assigned arbitrarily).

4-10   OTHER CYCLIC PROCESSES

In addition to Carnot and endoreversible engines, various other engines are of interest as they conform more or less closely to the actual operation of commonplace practical engines.

The *Otto cycle* (or, more precisely, the "air-standard Otto cycle") is a rough approximation to the operation of a gasoline engine. The cycle is shown in Fig. 4.9 in a V–S diagram. The working fluid (a mixture of air and gasoline vapor in the gasoline engine) is first compressed adiabatically

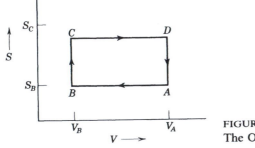

FIGURE 4.9
The Otto cycle.

$(A \rightarrow B)$. It is then heated at constant volume ($B \rightarrow C$); this step crudely describes the combustion of the gasoline in the gasoline engine. In the third step of the cycle the working fluid is expanded adiabatically in the "power stroke" ($C \rightarrow D$). Finally the working fluid is cooled isochorically to its initial state A.

In a real gasoline engine the working fluid chemically reacts ("burns") during the process $B \rightarrow C$; so that its mole number changes—an effect not represented in the Otto cycle. Furthermore the initial adiabatic compression is not quasi-static and therefore is certainly not isentropic. Nevertheless the idealized air-standard Otto cycle does provide a rough perspective for the analysis of gasoline engines.

In contrast to the Carnot cycle, the absorption of heat in step $B \rightarrow C$ of the idealized Otto cycle does not occur at constant temperature. Therefore the ideal engine efficiency is different for each infinitesimal step, and the over-all efficiency of the cycle must be computed by integration of the Carnot efficiency over the changing temperature. It follows that the efficiency of the Otto cycle depends upon the particular properties of the working fluid. It is left to the reader to corroborate that for an ideal gas with temperature independent heat capacities, the Otto cycle efficiency is

$$
\varepsilon_{\text{Otto}} = 1 - \left(\frac{V_B}{V_A} \right)^{\frac{(c_P - c_v)}{c_v}} \tag{4.30}
$$

The ratio V_A/V_B is called the compression ratio of the engine.

The *Brayton or Joule cycle* consists of two isentropic and two isobaric steps. It is shown on a P–S diagram in Fig. 4.10. In a working engine air (and fuel) is compressed adiabatically ($A \rightarrow B$), heated by fuel combustion at constant pressure ($B \rightarrow C$), expanded ($C \rightarrow D$), and rejected to the atmosphere. The process $D \rightarrow A$ occurs outside the engine, and a fresh charge of air is taken in to repeat the cycle. If the working gas is an ideal gas, with temperature independent heat capacities, the efficiency of a

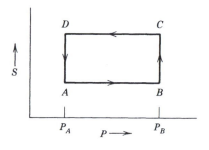

FIGURE 4.10
The Brayton or Joule cycle.

Brayton cycle is

$$\varepsilon_e = 1 - \left(\frac{P_A}{P_B}\right)^{\frac{(c_P - c_v)}{c_P}}$$

(4.31)

The *air-standard diesel cycle* consists of two isentropic processes, alternating with isochoric and isobaric steps. The cycle is represented in Fig. 4.11. After compression of the air and fuel mixture ($A \rightarrow B$), the fuel combustion occurs at constant pressure ($B \rightarrow C$). The gas is adiabatically expanded ($C \rightarrow D$) and then cooled at constant volume ($D \rightarrow A$).

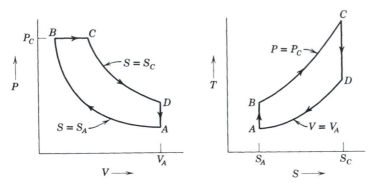

FIGURE 4.11
The air-standard diesel cycle.

PROBLEMS

4.10-1. Assuming that the working gas is a monatomic ideal gas, plot a T–S diagram for the Otto cycle.

4.10-2. Assuming that the working gas is a simple ideal gas (with temperature independent heat capacities), show that the engine efficiency of the Otto cycle is given by equation 4.30.

4.10-3. Assuming that the working gas is a simple ideal gas (with temperature independent heat capacities), show that the engine efficiency of the Brayton cycle is given by equation 4.31.

4.10-4. Assuming that the working gas is a monatomic ideal gas, plot a T–S diagram of the Brayton cycle.

4.10-5. Assuming that the working gas is a monatomic ideal gas, plot a T–S diagram of the air-standard diesel cycle.

5

ALTERNATIVE FORMULATIONS
AND
LEGENDRE TRANSFORMATIONS

5-1 THE ENERGY MINIMUM PRINCIPLE

In the preceding chapters we have inferred some of the most evident and immediate consequences of the principle of maximum entropy. Further consequences will lead to a wide range of other useful and fundamental results. But to facilitate those developments it proves to be useful now to reconsider the formal aspects of the theory and to note that the same content can be reformulated in several equivalent mathematical forms. Each of these alternative formulations is particularly convenient in particular types of problems, and the art of thermodynamic calculations lies largely in the selection of the particular theoretical formulation that most incisively "fits" the given problem. In the appropriate formulation thermodynamic problems tend to be remarkably simple; the converse is that they tend to be remarkably complicated in an inappropriate formalism!

Multiple equivalent formulations also appear in mechanics—Newtonian, Lagrangian, and Hamiltonian formalisms are tautologically equivalent. Again certain problems are much more tractable in a Lagrangian formalism than in a Newtonian formalism, or vice versa. But the difference in convenience of different formalisms is enormously greater in thermodynamics. It is for this reason that *the general theory of transformation among equivalent representations is here incorporated as a fundamental aspect of thermostatistical theory.*

In fact we have already considered two equivalent representations—the energy representation and the entropy representation. But the basic extremum principle has been formulated only in the entropy representation. If these two representations are to play parallel roles in the theory we must find an extremum principle in the energy representation, analogous to the entropy maximum principle. There is, indeed, such an extremum principle; the principle of maximum entropy is equivalent to, and can be

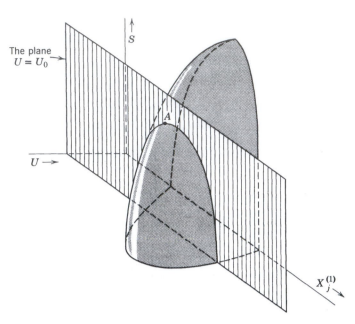

FIGURE 5.1
The equilibrium state A as a point of maximum S for constant U.

replaced by, a principle of minimum energy. Whereas the entropy maximum principle characterizes the equilibrium state as having maximum entropy for given total energy, the energy minimum principle characterizes the equilibrium state as having minimum energy for given total entropy.

Figure 5.1 shows a section of the thermodynamic configuration space for a composite system, as discussed in Section 4.1. The axes labeled S and U correspond to the total entropy and energy of the composite system, and the axis labeled $X_j^{(1)}$ corresponds to a particular extensive parameter of the first subsystem. Other axes, not shown explicitly in the figure, are $U^{(1)}$, X_j, and other pairs $X_k^{(1)}$, X_k.

The total energy of the composite system is a constant determined by the closure condition. The geometrical representation of this closure condition is the requirement that the state of the system lie on the plane $U = U_0$ in Fig. 5.1. The fundamental equation of the system is represented by the surface shown, and the representative point of the system therefore must be on the curve of intersection of the plane and the surface. If the parameter $X_j^{(1)}$ is unconstrained, the equilibrium state is the particular state that maximizes the entropy along the permitted curve; the state labeled A in Fig. 5.1.

The alternative representation of the equilibrium state A as a state of minimum energy for given entropy is illustrated in Fig. 5.2. Through the

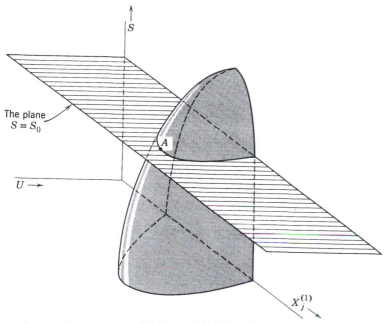

FIGURE 5.2
The equilibrium state A as a point of minimum U for constant S.

equilibrium point A is passed the plane $S = S_0$, which determines a curve of intersection with the fundamental surface. This curve consists of a family of states of constant entropy, and *the equilibrium state A is the state that minimizes the energy along this curve.*

The equivalence of the entropy maximum and the energy minimum principles clearly depends upon the fact that the geometrical form of the fundamental surface is generally as shown in Fig. 5.1 and 5.2. As discussed in Section 4.1, the form of the surface shown in the figures is determined by the postulates that $\partial S / \partial U > 0$ and that U is a single-valued continuous function of S; these analytic postulates accordingly are the underlying conditions for the equivalence of the two principles.

To recapitulate, we have made plausible, though we have not yet proved, that the following two principles are equivalent:

Entropy Maximum Principle. *The equilibrium value of any unconstrained internal parameter is such as to maximize the entropy for the given value of the total internal energy.*

Energy Minimum Principle. *The equilibrium value of any unconstrained internal parameter is such as to minimize the energy for the given value of the total entropy.*

The proof of the equivalence of the two extremum criteria can be formulated either as a physical argument or as a mathematical exercise. We turn first to the physical argument, to demonstrate that if the energy were *not* minimum the entropy could not be maximum in equilibrium, and inversely.

Assume, then, that the system is in equilibrium but that the energy does *not* have its smallest possible value consistent with the given entropy. We could then withdraw energy from the system (in the form of work) maintaining the entropy constant, and we could thereafter return this energy to the system in the form of heat. The entropy of the system would increase ($dQ = T dS$), and the system would be restored to its original energy but with an increased entropy. This is inconsistent with the principle that the initial equilibrium state is the state of maximum entropy! Hence we are forced to conclude that the original equilibrium state must have had minimum energy consistent with the prescribed entropy.

The inverse argument, that minimum energy implies maximum entropy, is similarly constructed (see Problem 5.1-1).

In a more formal demonstration we assume the entropy maximum principle

$$\left(\frac{\partial S}{\partial X}\right)_U = 0 \quad \text{and} \quad \left(\frac{\partial^2 S}{\partial X^2}\right)_U < 0 \tag{5.1}$$

where, for clarity, we have written X for $X_j^{(1)}$, and where it is implicit that all other X's are held constant throughout. Also, for clarity, we temporarily denote the first derivative $(\partial U/\partial X)_S$ by P. Then (by equation A.22 of Appendix A)

$$P \equiv \left(\frac{\partial U}{\partial X}\right)_S = -\frac{\left(\frac{\partial S}{\partial X}\right)_U}{\left(\frac{\partial S}{\partial U}\right)_X} = -T\left(\frac{\partial S}{\partial X}\right)_U = 0 \tag{5.2}$$

We conclude that U has an extremum. To classify that extremum as a maximum, a minimum, or a point of inflection we must study the sign of the second derivative $(\partial^2 U/\partial X^2)_S \equiv (\partial P/\partial X)_S$. But considering P as a function of U and X we have

$$\left(\frac{\partial^2 U}{\partial X^2}\right)_S = \left(\frac{\partial P}{\partial X}\right)_S = \left(\frac{\partial P}{\partial U}\right)_X\left(\frac{\partial U}{\partial X}\right)_S + \left(\frac{\partial P}{\partial X}\right)_U = \left(\frac{\partial P}{\partial U}\right)_X P + \left(\frac{\partial P}{\partial X}\right)_U \tag{5.3}$$

$$= \left(\frac{\partial P}{\partial X}\right)_U \quad \text{at } P = 0 \tag{5.4}$$

$$= \frac{\partial}{\partial X}\left[-\frac{\left(\frac{\partial S}{\partial X}\right)_U}{\left(\frac{\partial S}{\partial U}\right)_X} \right]_U \tag{5.5}$$

$$= -\frac{\frac{\partial^2 S}{\partial X^2}}{\frac{\partial S}{\partial U}} + \frac{\partial S}{\partial X}\frac{\frac{\partial^2 S}{\partial X \partial U}}{\left(\frac{\partial S}{\partial U}\right)^2} \tag{5.6}$$

$$= -T\frac{\partial^2 S}{\partial X^2} > 0 \quad \text{at} \quad \frac{\partial S}{\partial X} = 0 \tag{5.7}$$

so that U is a minimum. The inverse argument is identical in form.

As already indicated, the fact that precisely the same situation is described by the two extremal criteria is analogous to the isoperimetric problem in geometry. Thus a circle may be characterized either as the two dimensional figure of maximum area for given perimeter or, alternatively, as the two dimensional figure of minimum perimeter for given area.

The two alternative extremal criteria that characterize a circle are completely equivalent, and each applies to every circle. Yet they suggest two different ways of generating a circle. Suppose we are given a square and we wish to distort it continuously to generate a circle. We may keep its area constant and allow its bounding curve to contract as if it were a rubber band. We thereby generate a circle as the figure of minimum perimeter for the given area. Alternatively we might keep the perimeter of the given square constant and allow the area to increase, thereby obtaining a (different) circle, as the figure of maximum area for the given perimeter. However, after each of these circles is obtained *each satisfies both extremal conditions for its final values of area and perimeter*.

The physical situation pertaining to a thermodynamic system is very closely analogous to the geometrical situation described. Again, any equilibrium state can be characterized either as a state of maximum entropy for given energy or as a state of minimum energy for given entropy. But these two criteria nevertheless suggest two different ways of attaining equilibrium. As a specific illustration of these two approaches to equilibrium, consider a piston originally fixed at some point in a closed cylinder. We are interested in bringing the system to equilibrium without the constraint on the position of the piston. We can simply remove the constraint and allow the equilibrium to establish itself spontaneously; the entropy increases and the energy is maintained constant by the closure condition. This is the process suggested by the entropy maximum principle. Alternatively, we can permit the piston to move very slowly, reversibly doing work on an external agent until it has moved to the position that equalizes the pressure on the two sides. During this process energy is withdrawn from the system, but its entropy remains constant (the process is reversible and no heat flows). This is the process suggested by the energy minimum principle. The vital fact we wish to stress, however, is that *independent of whether the equilibrium is brought about by either of these two processes, or by any other process, the final equilibrium state in each case satisfies both extremal conditions.*

Finally, we illustrate the energy minimum principle by using it in place of the entropy maximum principle to solve the problem of thermal equilibrium, as treated in Section 2.4. We consider a closed composite system with an internal wall that is rigid, impermeable, and diathermal. Heat is free to flow between the two subsystems, and we wish to find the equilibrium state. The fundamental equation in the energy representation is

$$U = U^{(1)}\left(S^{(1)}, V^{(1)}, N_1^{(1)}, \ldots\right) + U^{(2)}\left(S^{(2)}, V^{(2)}, N_1^{(2)}, \ldots\right) \quad (5.8)$$

All volume and mole number parameters are constant and known. The variables that must be computed are $S^{(1)}$ and $S^{(2)}$. Now, despite the fact that the system is actually closed and that the total energy is fixed, the equilibrium state can be characterized as the state that would minimize the energy *if* energy changes were permitted. The virtual change in total energy associated with virtual heat fluxes in the two systems is

$$dU = T^{(1)} dS^{(1)} + T^{(2)} dS^{(2)} \quad (5.9)$$

The energy minimum condition states that $dU = 0$, subject to the condition of fixed total entropy:

$$S^{(1)} + S^{(2)} = \text{constant} \quad (5.10)$$

whence

$$dU = \left(T^{(1)} - T^{(2)}\right) dS^{(1)} = 0 \ . \quad (5.11)$$

and we conclude that

$$T^{(1)} = T^{(2)} \quad (5.12)$$

The energy minimum principle thus provides us with the same condition of thermal equilibrium as we previously found by using the entropy maximum principle.

Equation 5.12 is one equation in $S^{(1)}$ and $S^{(2)}$. The second equation is most conveniently taken as equation 5.8, in which the total energy U is known and which consequently involves only the two unknown quantities $S^{(1)}$ and $S^{(2)}$. Equations 5.8 and 5.12, in principle, permit a fully explicit solution of the problem.

In a precisely analogous fashion the equilibrium condition for a closed composite system with an internal moveable adiabatic wall is found to be equality of the pressure. This conclusion is straightforward in the energy representation but, as was observed in the last paragraph of Section 2.7, it is relatively delicate in the entropy representation.

PROBLEMS

5.1-1. Formulate a proof that the energy minimum principle implies the entropy maximum principle—the "inverse argument" referred to after equation 5.7. That is, show that if the entropy were not maximum at constant energy then the energy could not be minimum at constant entropy.

Hint: First show that the permissible increase in entropy in the system can be exploited to extract heat from a reversible heat source (initially at the same temperature as the system) and to deposit it in a reversible work source. The reversible heat source is thereby cooled. Continue the argument.

5.1-2. An adiabatic, impermeable and fixed piston separates a cylinder into two chambers of volumes $V_0/4$ and $3V_0/4$. Each chamber contains 1 mole of a monatomic ideal gas. The temperatures are T_s and T_l, the subscripts s and l referring to the small and large chambers, respectively.

a) The piston is made thermally conductive and moveable, and the system relaxes to a new equilibrium state, *maximizing its entropy while conserving its total energy*. Find this new equilibrium state.

b) Consider a small virtual change in the energy of the system, maintaining the entropy at the value attained in part (*a*). To accomplish this physically we can reimpose the adiabatic constraint and quasistatically displace the piston by imposition of an external force. Show that the external source of this force must do work *on* the system in order to displace the piston in *either* direction. Hence *the state attained in part (a) is a state of minimum energy at constant entropy.*

c) Reconsider the initial state and specify how equilibrium can be established by decreasing the energy at constant entropy. Find this equilibrium state.

d) Describe an operation that demonstrates that the equilibrium state attained in (*c*) is a state of maximum entropy at constant energy.

5-2   LEGENDRE TRANSFORMATIONS

In both the energy and entropy representations the extensive parameters play the roles of mathematically independent variables, whereas the intensive parameters arise as derived concepts. This situation is in direct contrast to the practical situation dictated by convenience in the laboratory. The experimenter frequently finds that the intensive parameters are the more easily measured and controlled and therefore is likely to think of the intensive parameters as operationally independent variables and of the extensive parameters as operationally derived quantities. The extreme instance of this situation is provided by the conjugate variables entropy and temperature. No practical instruments exist for the measurement and control of entropy, whereas thermometers and thermostats, for the measurement and control of the temperature, are common laboratory equipment. The question therefore arises as to the possibility of recasting the mathematical formalism in such a way that intensive parameters will replace extensive parameters as mathematically independent variables. We shall see that such a reformulation is, in fact, possible and that it leads to various other thermodynamic representations.

It is, perhaps, superfluous at this point to stress again that thermodynamics is logically complete and self-contained within either the entropy or the energy representations and that the introduction of the transformed representations is purely a matter of convenience. This is, admittedly, a convenience without which thermodynamics would be almost unusably awkward, but in principle it is still only a luxury rather than a logical necessity.

The purely formal aspects of the problem are as follows. We are given an equation (the fundamental relation) of the form

$$Y = Y(X_0, X_1, \ldots, X_t) \tag{5.13}$$

and it is desired to find a method whereby the derivatives

$$P_k \equiv \frac{\partial Y}{\partial X_k} \tag{5.14}$$

can be considered as independent variables without sacrificing any of the informational content of the given fundamental relation(5.13). This formal problem has its counterpart in geometry and in several other fields of physics. The solution of the problem, employing the mathematical technique of Legendre transformations, is most intuitive when given its geometrical interpretation; and it is this geometrical interpretation that we shall develop in this Section.

For simplicity, we first consider the mathematical case in which the fundamental relation is a function of only a single independent variable X.

$$Y = Y(X) \tag{5.15}$$

Geometrically, the fundamental relation is represented by a curve in a

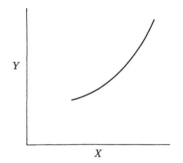

FIGURE 5.3

space (Fig. 5.3) with cartesian coordinates X and Y, and the derivative

$$P \equiv \frac{\partial Y}{\partial X} \tag{5.16}$$

is the slope of this curve. Now, if we desire to consider P as an independent variable in place of X, our first impulse might be simply to eliminate X between equations 5.15 and 5.16, thereby obtaining Y as a function of P

$$Y = Y(P) \tag{5.17}$$

A moment's reflection indicates, however, that we would sacrifice some of the mathematical content of the given fundamental relation (5.15) for, from the geometrical point of view, it is clear that knowledge of Y as a function of the slope dY/dX would not permit us to reconstruct the curve $Y = Y(X)$. In fact, each of the displaced curves shown in Fig. 5.4 corresponds equally well to the relation $Y = Y(P)$. From the analytical point of view the relation $Y = Y(P)$ is a first-order differential equation, and its integration gives $Y = Y(X)$ only to within an undetermined integration constant. Therefore we see that acceptance of $Y = Y(P)$ as a basic equation in place of $Y = Y(X)$ would involve the sacrifice of some information originally contained in the fundamental relation. Despite the

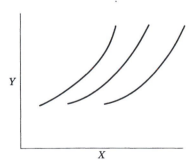

FIGURE 5.4

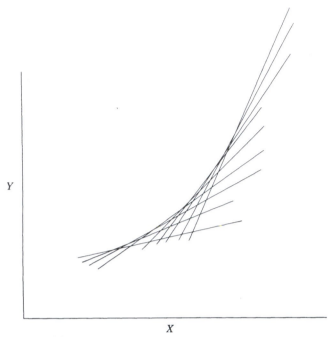

X

FIGURE 5.5

desirability of having P as a mathematically independent variable, this sacrifice of the informational content of the formalism would be completely unacceptable.

The practicable solution to the problem is supplied by the duality between conventional *point geometry* and the Pluecker *line geometry*. The essential concept in line geometry is that a given curve can be represented equally well either (*a*) as the envelope of a family of tangent lines (Fig. 5.5), or (*b*) as the locus of points satisfying the relation $Y = Y(X)$. Any equation that enables us to construct the family of tangent lines therefore determines the curve equally as well as the relation $Y = Y(X)$.

Just as every point in the plane is described by the two numbers X and Y, so every straight line in the plane can be described by the two numbers P and ψ, where P is the slope of the line and ψ is its intercept along the Y-axis. Then just as a relation $Y = Y(X)$ selects a subset of all possible points (X, Y), a relation $\psi = \psi(P)$ selects a subset of all possible lines (P, ψ). A knowledge of the intercepts ψ of the tangent lines as a function of the slopes P enables us to construct the family of tangent lines and thence the curve of which they are the envelope. Thus the relation

$$\psi = \psi(P) \qquad (5.18)$$

is completely equivalent to the fundamental relation $Y = Y(X)$. In this relation the independent variable is P, so that equation 5.18 provides a complete and satisfactory solution to the problem. As the relation $\psi = \psi(P)$ is mathematically equivalent to the relation $Y = Y(X)$, it can also be considered a fundamental relation; $Y = Y(X)$ is a fundamental relation in the "Y-representation"; whereas $\psi = \psi(P)$ is a fundamental relation in the "ψ-representation."

The reader is urged at this point actually to draw a reasonable number of straight lines, of various slopes P and of various Y-intercepts $\psi = -P^2$. The relation $\psi = -P^2$ thereby will be seen to characterize a parabola (which is more conventionally described as $Y = \frac{1}{4}X^2$). In ψ-representation the fundamental equation of the parabola is $\psi = -P^2$, whereas in Y-representation the fundamental equation of this same parabola is $Y = \frac{1}{4}X^2$.

The question now arises as to how we can compute the relation $\psi = \psi(P)$ if we are given the relation $Y = Y(X)$. The appropriate mathematical operation is known as a Legendre transformation. We consider a tangent line that goes through the point (X, Y) and has a slope P. If the intercept is ψ, we have (see Fig. 5.6)

$$P = \frac{Y - \psi}{X - 0} \tag{5.19}$$

or

$$\psi = Y - PX \tag{5.20}$$

Let us now suppose that we are given the equation

$$Y = Y(X) \tag{5.21}$$

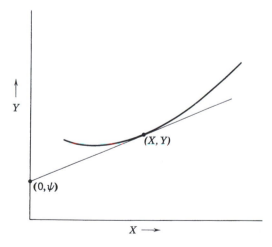

(X, Y)

$(0, \psi)$

Y

$X \longrightarrow$ FIGURE 5.6

and by differentiation we find

$$P = P(X) \tag{5.22}$$

Then by elimination[1] of X and Y among equations 5.20, 5.21, and 5.22 we obtain the desired relation between ψ and P. The basic identity of the Legendre transformation is equation 5.20, and this equation can be taken as the analytic definition of the function ψ. The function ψ is referred to as a *Legendre transform* of Y.

The inverse problem is that of recovering the relation $Y = Y(X)$ if the relation $\psi = \psi(P)$ is given. We shall see here that the relationship between (X, Y) and (P, ψ) is symmetrical with its inverse, except for a sign in the equation of the Legendre transformation. Taking the differential of equation 5.20 and recalling that $dY = P\,dX$, we find

$$d\psi = dY - P\,dX - X\,dP$$

$$= -X\,dP \tag{5.23}$$

or

$$-X = \frac{d\psi}{dP} \tag{5.24}$$

If the two variables ψ and P are eliminated[2] from the given equation $\psi = \psi(P)$ and from equations 5.24 and 5.20, we recover the relation $Y = Y(X)$. The symmetry between the Legendre transformation and its inverse is indicated by the following schematic comparison:

$Y = Y(X)$	$\psi = \psi(P)$
$P = \dfrac{dY}{dX}$	$-X = \dfrac{d\psi}{dP}$
$\psi = -PX + Y$	$Y = XP + \psi$
Elimination of X and Y yields	Elimination of P and ψ yields
$\psi = \psi(P)$	$Y = Y(X)$

The generalization of the Legendre transformation to functions of more than a single independent variable is simple and straightforward. In three dimensions Y is a function of X_0 and X_1, and the fundamental equation represents a surface. This surface can be considered as the locus of points

---

[1] This elimination is possible if P is not independent of X; that is, if $d^2Y/dX^2 \neq 0$. In the thermodynamic application this criterion will turn out to be identical to the criterion of stability. The criterion fails only at the "critical points," which are discussed in detail in Chapter 10.

[2] The condition that this be possible is that $d^2\psi/\partial P^2 \neq 0$, which will, in the thermodynamic application, be guaranteed by the stability of the system under consideration.

satisfying the fundamental equation $Y = Y(X_0, X_1)$, or it can be considered as the envelope of tangent planes. A plane can be characterized by its intercept ψ on the Y-axis and by the slopes P_0 and P_1 of its traces on the $Y - X_0$ and $Y - X_1$ planes. The fundamental equation then selects from all possible planes a subset described by $\psi = \psi(P_0, P_1)$.

In general the given fundamental relation

$$Y = Y(X_0, X_1, \ldots, X_t) \tag{5.25}$$

represents a hypersurface in a $(t + 2)$-dimensional space with cartesian coordinates $Y, X_0, X_1, \ldots, X_t$. The derivative

$$P_k = \frac{\partial Y}{\partial X_k} \tag{5.26}$$

is the partial slope of this hypersurface. The hypersurface may be equally well represented as the locus of points satisfying equation 5.25 or as the envelope of the tangent hyperplanes. The family of tangent hyperplanes can be characterized by giving the intercept of a hyperplane, ψ, as a function of the slopes $P_0, P_1, \ldots, P_t$. Then

$$\psi = Y - \sum_k P_k X_k \tag{5.27}$$

Taking the differential of this equation, we find

$$d\psi = -\sum_k X_k \, dP_k \tag{5.28}$$

whence

$$-X_k = \frac{\partial \psi}{\partial P_k} \tag{5.29}$$

A Legendre transformation is effected by eliminating Y and the X_k from $Y = Y(X_0, X_1, \ldots, X_t)$, the set of equations 5.26, and equation 5.27. The inverse transformation is effected by eliminating ψ and the P_k from $\psi = \psi(P_1, P_2, \ldots, P_r)$, the set of equations 5.29, and equation 5.27.

Finally, a Legendre transformation may be made only in some $(n + 2)$-dimensional subspace of the full $(t + 2)$-dimensional space of the relation $Y = Y(X_0, X_1, \ldots, X_t)$. Of course the subspace must contain the Y-coordinate but may involve any choice of $n + 1$ coordinates from the set $X_0, X_1, \ldots, X_t$. For convenience of notation, we order the coordinates so that the Legendre transformation is made in the subspace of the first $n + 1$ coordinates (and of Y); the coordinates $X_{n+1}, X_{n+2}, \ldots, X_t$ are left untransformed. Such a partial Legendre transformation is effected merely by considering the variables $X_{n+1}, X_{n+2}, \ldots, X_t$ as constants in the transformation. The resulting Legendre transform must be denoted by some explicit notation that indicates which of the independent variables have participated in the transformation. We employ the notation $Y[P_0, P_1, \ldots, P_n]$ to denote the function obtained by making a Legendre transformation with respect to $X_0, X_1, \ldots, X_n$ on the function $Y(X_0, X_1, \ldots, X_t)$. Thus $Y[P_0, P_1, \ldots, P_n]$ is a function of the independent variables $P_0, P_1, \ldots, P_n, X_{n+1}, \ldots, X_t$. The various relations involved in a partial Legendre transformation and its inverse are indicated in the following table.

$Y = Y(X_0, X_1, \ldots, X_t)$	$Y[P_0, P_1, \ldots, P_n] =$ function of $P_0, P_1, \ldots, P_n, X_{n+1}, \ldots, X_t$ (5.30)
$P_k = \dfrac{\partial Y}{\partial X_k}$	$-X_k = \dfrac{\partial Y[P_0, \ldots, P_n]}{\partial P_k} \qquad k \leq n$ (5.31)
	$P_k = \dfrac{\partial Y[P_0, \ldots, P_n]}{\partial X_k} \qquad k > n$
The partial differentiation denotes constancy of all the natural variables of Y other than X_k (i.e., of all X_j with $j \neq k$)	The partial differentiation denotes constancy of all the natural variables of $Y(P_0, \ldots, P_n)$ other than that with respect to which the differentiation is being carried out.
$dY = \displaystyle\sum_0^t P_k \, dX_k$	$dY[P_0, \ldots, P_n]$ $= -\displaystyle\sum_0^n X_k \, dP_k + \sum_{n+1}^t P_k \, dX_k$ (5.32)
$Y[P_0, \ldots, P_n] = Y - \displaystyle\sum_0^n P_k X_k$	$Y = Y[P_0, \ldots, P_n] + \displaystyle\sum_0^n X_k P_k$ (5.33)
Elimination of Y and $X_0, X_1, \ldots, X_n$ from equations 5.30, 5.33, and the first $n + 1$ equations of 5.31 yields the transformed fundamental relation.	Elimination of $Y[P_0, \ldots, P_n]$ and $P_0, P_1, \ldots, P_n$ from equations 5.30, 5.33, and the first $n + 1$ equations of 5.31 yields the original fundamental relation.

In this section we have divorced the mathematical aspects of Legendre transformations from the physical applications. Before proceeding to the thermodynamic applications in the succeeding sections of this chapter, it may be of interest to indicate very briefly the application of the formalism to Lagrangian and Hamiltonian mechanics, which perhaps may be a more familiar field of physics than thermodynamics. The Lagrangian principle guarantees that a particular function, the Lagrangian, completely characterizes the dynamics of a mechanical system. The Lagrangian is a function of $2r$ variables, r of which are *generalized coordinates* and r of which are *generalized velocities*. Thus the equation

$$L = L(v_1, v_2, \ldots, v_r, q_1, q_2, \ldots, q_r) \tag{5.34}$$

plays the role of a fundamental relation. The *generalized momenta* are defined as derivatives of the Lagrangian function

$$P_k \equiv \frac{\partial L}{\partial v_k} \tag{5.35}$$

If it is desired to replace the velocities by the momenta as independent variables, we must make a partial Legendre transformation with respect to the velocities. We thereby introduce a new function, called the Hamiltonian, defined by [3]

$$(-H) = L - \sum_1^r P_k v_k \tag{5.36}$$

A complete dynamical formalism can then be based on the new fundamental relation

$$H = H(P_1, P_2, \ldots, P_r, q_1, q_2, \ldots, q_r) \tag{5.37}$$

Furthermore, by equation 5.31 the derivative of H with respect to P_k is the velocity v_k, which is one of the Hamiltonian dynamical equations. Thus, if an equation of the form 5.34 is considered as a dynamical fundamental equation in the Lagrangian representation, the Hamiltonian equation (5.37) is the equivalent fundamental equation expressed in the Hamiltonian representation.

PROBLEMS

5.2-1. The equation $y = x^2/10$ describes a parabola.
a) Find the equation of this parabola in the "line geometry representation" $\psi = \psi(P)$.
b) On a sheet of graph paper (covering the range roughly from $x \simeq -15$ to $x \simeq +15$ and from $y \simeq -25$ to $y \simeq +25$) draw straight lines with slopes $P = 0$,

---

[3] In our usage the Legendre transform of the Lagrangian is the *negative* Hamiltonian. Actually, the accepted mathematical convention agrees with the usage in mechanics, and the function $-\psi$ would be called the Legendre transform of Y.

$\pm 0.5, \pm 1, \pm 2, \pm 3$ and with intercepts ψ satisfying the relationship $\psi = \psi(P)$ as found in part (a). (Drawing each straight line is facilitated by calculating its intercepts on the x-axis and on the y-axis.)

5.2-2. Let $y = Ae^{Bx}$.

a) Find $\psi(P)$.

b) Calculate the inverse Legendre transform of $\psi(P)$ and corroborate that this result is $y(x)$.

c) Taking $A = 2$ and $B = 0.5$, draw a family of tangent lines in accordance with the result found in (a), and check that the tangent curve goes through the expected points at $x = 0, 1$, and 2.

5-3 THERMODYNAMIC POTENTIALS

The application of the preceding formalism to thermodynamics is self-evident. The fundamental relation $Y = Y(X_0, X_1, \ldots)$ can be interpreted as the energy-language fundamental relation $U = U(S, X_1, X_2, \ldots, X_t)$ or $U = U(S, V, N_1, N_2, \ldots)$. The derivatives $P_0, P_1, \ldots$ correspond to the intensive parameters $T, -P, \mu_1, \mu_2, \ldots$. The Legendre transformed functions are called *thermodynamic potentials*, and we now specifically define several of the most common of them. In Chapter 6 we continue the discussion of these functions by deriving extremum principles for each potential, indicating the intuitive significance of each, and discussing its particular role in thermodynamic theory. But for the moment we concern ourselves merely with the formal aspects of the definitions of the several particular functions.

The *Helmholtz potential* or the *Helmholtz free energy*, is the partial Legendre transform of U that replaces the entropy by the temperature as the independent variable. The internationally adopted symbol for the Helmholtz potential is F. The natural variables of the Helmholtz potential are $T, V, N_1, N_2, \ldots$. That is, the functional relation $F = F(T, V, N_1, N_2, \ldots)$ constitutes a fundamental relation. In the systematic notation introduced in Section 5.2

$$F \equiv U[T] \tag{5.38}$$

The full relationship between the energy representation and the Helmholtz representation, is summarized in the following schematic comparison:

$U = U(S, V, N_1, N_2, \ldots)$	$F = F(T, V, N_1, N_2, \ldots)$ $\qquad$ (5.39)
$T = \partial U/\partial S$	$-S = \partial F/\partial T$ $\qquad$ (5.40)
$F = U - TS$	$U = F + TS$ $\qquad$ (5.41)
Elimination of U and S yields	Elimination of F and T yields
$F = F(T, V, N_1, N_2, \ldots)$	$U = U(S, V, N_1, N_2, \ldots)$

The complete differential dF is

$$dF = -S\,dT - P\,dV + \mu_1\,dN_1 + \mu_2\,dN_2 + \cdots \qquad (5.42)$$

The *enthalpy* is that partial Legendre transform of U that replaces the volume by the pressure as an independent variable. Following the recommendations of the International Unions of Physics and of Chemistry, and in agreement with almost universal usage, we adopt the symbol H for the enthalpy. The natural variables of this potential are $S, P, N_1, N_2, \ldots$ and

$$H \equiv U[P] \qquad (5.43)$$

The schematic representation of the relationship of the energy and enthalpy representations is as follows:

$U = U(S, V, N_1, N_2, \ldots)$	$H = H(S, P, N_1, N_2, \ldots)$ (5.44)
$-P = \partial U/\partial V$	$V = \partial H/\partial P$ (5.45)
$H = U + PV$	$U = H - PV$ (5.46)
Elimination of U and V yields	Elimination of H and P yields
$H = H(S, P, N_1, N_2, \ldots)$	$U = U(S, V, N_1, N_2, \ldots)$

Particular attention is called to the inversion of the signs in equations 5.45 and 5.46, resulting from the fact that $-P$ is the intensive parameter associated with V. The complete differential dH is

$$dH = T\,dS + V\,dP + \mu_1\,dN_1 + \mu_2\,dN_2 + \cdots \qquad (5.47)$$

The third of the common Legendre transforms of the energy is the *Gibbs potential*, or *Gibbs free energy*. This potential is the Legendre transform that simultaneously replaces the entropy by the temperature and the volume by the pressure as independent variables. The standard notation is G, and the natural variables are $T, P, N_1, N_2, \ldots$. We thus have

$$G \equiv U[T, P] \qquad (5.48)$$

and

$U = U(S, V, N_1, N_2, \ldots)$	$G = G(T, P, N_1, N_2, \ldots)$ (5.49)
$T = \partial U/\partial S$	$-S = \partial G/\partial T$ (5.50)
$-P = \partial U/\partial V$	$V = \partial G/\partial P$ (5.51)
$G = U - TS + PV$	$U = G + TS - PV$ (5.52)
Elimination of U, S, and V yields	Elimination of G, T, and P yields
$G = G(T, P, N_1, N_2, \ldots)$	$U = U(S, V, N_1, N_2, \ldots)$

The complete differential dG is

$$dG = -S\,dT + V\,dP + \mu_1\,dN_1 + \mu_2\,dN_2 + \cdots \qquad (5.53)$$

A thermodynamic potential which arises naturally in statistical mechanics is the *grand canonical potential*, $U[T, \mu]$. For this potential we have

$U = U(S, V, N)$	$U[T, \mu] = $ function of T, V, and μ (5.54)
$T = \partial U / \partial S$	$-S = \partial U[T, \mu] / \partial T$ (5.55)
$\mu = \partial U / \partial N$	$-N = \partial U[T, \mu] / \partial \mu$ (5.56)
$U[T, \mu] = U - TS - \mu N$	$U = U[T, \mu] + TS + \mu N$ (5.57)
Elimination of	Elimination of
U, S, and N yields	$U[T, \mu]$, T, and μ yields
$U[T, \mu]$ as a function of T, V, μ	$U = U(S, V, N)$

and

$$dU[T, \mu] = -S\,dT - P\,dV - N\,d\mu \qquad (5.58)$$

Other possible transforms of the energy for a simple system, which are used only infrequently and which consequently are unnamed, are $U[\mu_1]$, $U[P, \mu_1]$, $U[T, \mu_1, \mu_2]$, and so forth. The complete Legendre transform is $U[T, P, \mu_1, \mu_2, \ldots, \mu_r]$. The fact that $U(S, V, N_1, N_2, \ldots, N_t)$ is a homogeneous first-order function of its arguments causes this latter function to vanish identically. For

$$U[T, P, \mu_1, \ldots, \mu_r] = U - TS + PV - \mu_1 N_1 - \mu_2 N_2 - \cdots - \mu_r N_r$$

$$(5.59)$$

which, by the Euler relation (3.6), is identically zero

$$U[T, P, \mu_1, \ldots, \mu_r] \equiv 0 \qquad (5.60)$$

PROBLEMS

5.3-1. Find the fundamental equation of a monatomic ideal gas in the Helmholtz representation, in the enthalpy representation, and in the Gibbs representation. Assume the fundamental equation computed in Section 3.4. In each case find the equations of state by differentiation of the fundamental equation.

5.3-2. Find the fundamental equation of the ideal van der Waals fluid (Section 3.5) in the Helmholtz representation.

Perform an inverse Legendre transform on the Helmholtz potential and show that the fundamental equation in the energy representation is recovered.

5.3-3. Find the fundamental equation of electromagnetic radiation in the Helmholtz representation. Calculate the "thermal" and "mechanical" equations of state and corroborate that they agree with those given in Section 3.6.

5.3-4[4]. Justify the following recipe for obtaining a plot of $F(V)$ from a plot of $G(P)$ (the common dependent variables T and N being notationally suppressed for convenience).

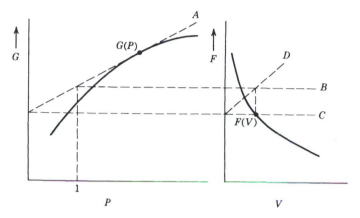

(1) At a chosen value of P draw the tangent line A.
(2) Draw horizontal lines B and C through the intersections of A with $P = 1$ and $P = 0$.
(3) Draw the 45° line D as shown and project the intersection of B and D onto the line C to obtain the point $F(V)$.
Hint: Identify the magnitude of the two vertical distances indicated in the G versus P diagram, and also the vertical separation of lines B and C.
   Note that the units of F and V are determined by the chosen units of G and P. Explain.
   Give the analogous construction for at least one other pair of potentials.
   Note that $G(P)$ is drawn as a concave function (i.e., negative curvature) and show that this is equivalent to the statement that $\kappa_T > 0$.

5.3-5. From the first acceptable fundamental equation in Problem 1.10-1 calculate the fundamental equation in Gibbs representation. Calculate $\alpha(T, P)$, $\kappa_T(T, P)$, and $c_p(T, P)$ by differentiation of G.

5.3-6. From the second acceptable fundamental equation in Problem 1.10-1 calculate the fundamental equation in enthalpy representation. Calculate $V(S, P, N)$ by differentiation.

5.3-7. The enthalpy of a particular system is

$$H = AS^2N^{-1}\ln\left(\frac{P}{P_0}\right)$$

[4]Adapted from H. E. Stanley, *Introduction to Phase Transitions and Critical Phenomena* (Oxford University Press, 1971).

where A is a positive constant. Calculate the molar heat capacity at constant volume c_v as a function of T and P.

5.3-8. In Chapter 15 it is shown by a statistical mechanical calculation that the fundamental equation of a system of $\tilde{N}$ "atoms" each of which can exist in an atomic state with energy ε_u or in an atomic state with energy ε_d (and in no other state) is

$$F = -\tilde{N}k_B T \ln\left(e^{-\beta\varepsilon_u} + e^{-\beta\varepsilon_d}\right)$$

Here k_B is Boltzmann's constant and $\beta = 1/k_B T$. Show that the fundamental equation of this system, in entropy representation, is

$$S = NR \ln\left(\frac{1 + Y^{\varepsilon_d/\varepsilon_u}}{Y^Y}\right)$$

where

$$Y \equiv \frac{U - \tilde{N}\varepsilon_u}{\tilde{N}\varepsilon_d - U}$$

Hint: Introduce $\beta = (k_B T)^{-1}$, and show first that $U = F + \beta \partial F/\partial\beta = \partial(\beta F)/\partial\beta$. Also, for definiteness, assume $\varepsilon_u < \varepsilon_d$, and note that $\tilde{N}k_B = NR$ where $\tilde{N}$ is the number of atoms and N is the number of moles.

5.3-9. Show, for the two-level system of Problem 5.3-8, that as the temperature increases from zero to infinity the energy increases from $\tilde{N}\varepsilon_u$ to $\tilde{N}(\varepsilon_u + \varepsilon_d)/2$. Thus, at zero temperature all atoms are in their "ground state" (with energy ε_u), and at infinite temperature the atoms are equally likely to be in either state. Energies higher than $N(\varepsilon_u + \varepsilon_d)/2$ are inaccessible in thermal equilibrium! (This upper bound on the energy is a consequence of the unphysical oversimplification of the model; it will be discussed again in Section 15.3.)

Show that the Helmholtz potential of a mixture of simple ideal gases is the sum of the Helmholtz potentials of each individual gas:

5.3-10.

a) Show that the Helmholtz potential of a mixture of simple ideal gases is the sum of the Helmholtz potentials of each individual gas:

$$F(T, V, N_1, \ldots, N_r) = F(T, V, N_1) + \cdots + F(T, V, N_r)$$

Recall the fundamental equation of the mixture, as given in equation 3.40.
An analogous additivity does not hold for any other potential expressed in terms of its natural variables.

5.3-11. A mixture of two monatomic ideal gases is contained in a volume V at temperature T. The mole numbers are N_1 and N_2. Calculate the chemical potentials μ_1 and μ_2. Recall Problems 5.3-1 and 5.3-10.

Assuming the system to be in contact with a reservoir of given T and μ_1, through a diathermal wall permeable to the first component but not to the second, calculate the pressure in the system.

5.3-12. A system obeys the fundamental relation

$$(s - s_0)^4 = Avu^2$$

Calculate the Gibbs potential $G(T, P, N)$.

5.3-13. For a particular system it is found that

$$u = \left(\tfrac{3}{2}\right) Pv$$

and

$$P = AvT^4$$

Find a fundamental equation, the molar Gibbs potential, and the Helmholtz potential for this system.

5.3-14. For a particular system (of 1 mole) the quantity $(v + a)f$ is known to be a function of the temperature only $(= Y(T))$. Here v is the molar volume, f is the molar Helmholtz potential, a is a constant, and $Y(T)$ denotes an unspecified function of temperature. It is also known that the molar heat capacity c_v is

$$c_v = b(v)T^{\frac{1}{2}}$$

where $b(v)$ is an unspecified function of v.

a) Evaluate $Y(T)$ and $b(v)$.

b) The system is to be taken from an initial state (T_0, v_0) to a final state (T_f, v_f). A thermal reservoir of temperature T_r is available, as is a reversible work source. What is the maximum work that can be delivered to the reversible work source? (Note that the answer may involve constants unevaluated by the stated conditions, but that the answer should be fully explicit otherwise.)

5-4 GENERALIZED MASSIEU FUNCTIONS

Whereas the most common functions definable in terms of Legendre transformations are those mentioned in Section 5.3, another set can be defined by performing the Legendre transformation on the entropy rather than on the energy. That is, the fundamental relation in the form $S = S(U, V, N_1, N_2, \ldots)$ can be taken as the relation on which the transformation is performed. Such Legendre transforms of the entropy were invented by Massieu in 1869 and actually predated the transforms of the energy introduced by Gibbs in 1875. We refer to the transforms of the entropy as *Massieu functions*, as distinguished from the *thermodynamic potentials* transformed from the energy. The Massieu functions will turn out to be particularly useful in the theory of irreversible thermodynamics, and they also arise naturally in statistical mechanics and in the theory of thermal fluctuations. Three representative Massieu functions are $S[1/T]$, in which the internal energy is replaced by the reciprocal temperature as independent variable; $S[P/T]$, in which the volume is replaced by P/T as independent variable; and $S[1/T, P/T]$, in which both replacements are made simultaneously. Clearly

$$S\left[\frac{1}{T}\right] \equiv S - \frac{1}{T}U = -\frac{F}{T} \tag{5.61}$$

$$S\left[\frac{P}{T}\right] \equiv S - \frac{P}{T} \cdot V \tag{5.62}$$

and

$$S\left[\frac{1}{T}, \frac{P}{T}\right] = S - \frac{1}{T}U - \frac{P}{T} \cdot V = -\frac{G}{T} \tag{5.63}$$

Thus, of the three, only $S[P/T]$ is not trivially related to one of the previously introduced thermodynamic potentials. For this function

$S = S(U, V, N_1, N_2, \ldots)$	$S[P/T]$ = function of
	$U, P/T, N_1, N_2, \ldots$, (5.64)
$P/T = \partial S/\partial V$	$-V = \partial S[P/T]/\partial(P/T)$ (5.65)
$S[P/T] = S - (P/T)V$	$S = S[P/T] + (P/T)V$ (5.66)
Elimination of	Elimination of
S and V yields $S[P/T]$	$S[P/T]$ and P/T yields
as a function of $U, P/T, N_1, N_2, \ldots$	$S = S(U, V, N_1, N_2, \ldots)$

and

$$dS[P/T] = (1/T)\,dU - V\,d(P/T) - (\mu_1/T)\,dN_1 - \frac{\mu_2}{T}\,dN_2 \ldots$$

$$\tag{5.67}$$

Other Massieu functions may be invented and analyzed by the reader as a particular need for them arises.

PROBLEMS

5.4-1. Find the fundamental equation of a monatomic ideal gas in the representation

$$S\left[\frac{P}{T}, \frac{\mu}{T}\right]$$

Find the equations of state by differentiation of this fundamental equation.

5.4-2. Find the fundamental equation of electromagnetic radiation (Section 3.6)
a) in the representation $S[1/T]$
b) in the representation $S[P/T]$

5.4-3. Find the fundamental equation of the ideal van der Waals fluid in the representation $S[1/T]$. Show that $S[1/T]$ is equal to $-F/T$ (recall that F was computed in Problem 5.3-2).

6

THE EXTREMUM PRINCIPLE
IN THE LEGENDRE
TRANSFORMED REPRESENTATIONS

6-1 THE MINIMUM PRINCIPLES FOR THE POTENTIALS

We have seen that the Legendre transformation permits expression of the fundamental equation in terms of a set of independent variables chosen to be particularly convenient for a given problem. Clearly, however, the advantage of being able to write the fundamental equation in various representations would be lost if the extremum principle were not itself expressible in those representations. We are concerned, therefore, with the reformulation of the basic extremum principle in forms appropriate to the Legendre transformed representations.

For definiteness consider a composite system in contact with a thermal reservoir. Suppose further that some internal constraint has been removed. We seek the mathematical condition that will permit us to predict the equilibrium state. For this purpose we first review the solution of the problem by the energy minimum principle.

In the equilibrium state the total energy of the composite system-plus-reservoir is minimum:

$$d(U + U^r) = 0 \tag{6.1}$$

and

$$d^2(U + U^r) = d^2U > 0 \tag{6.2}$$

subject to the isentropic condition

$$d(S + S^r) = 0 \tag{6.3}$$

The quantity d^2U^r has been put equal to zero in equation 6.2 because d^2U^r is a sum of products of the form

$$\frac{\partial^2 U^r}{\partial X_j^r \partial X_k^r} \, dX_j^r \, dX_k^r$$

which vanish for a reservoir (the coefficient varying as the reciprocal of the mole number of the reservoir).

The other closure conditions depend upon the particular form of the internal constraints in the composite system. If the internal wall is movable and impermeable, we have

$$dN_j^{(1)} = dN_j^{(2)} = d(V^{(1)} + V^{(2)}) = 0 \qquad \text{(for all } j) \tag{6.4}$$

whereas, if the internal wall is rigid and permeable to the kth component, we have

$$d(N_k^{(1)} + N_k^{(2)}) = dN_j^{(1)} = dN_j^{(2)} = dV^{(1)} = dV^{(2)} = 0 \qquad (j \neq k)$$

$$\tag{6.5}$$

These equations suffice to determine the equilibrium state.

The differential dU in equation 6.1 involves the terms $T^{(1)} dS^{(1)} + T^{(2)} dS^{(2)}$, which arise from heat flux among the subsystems and the reservoir, and terms such as $-P^{(1)} dV^{(1)} - P^{(2)} dV^{(2)}$ and $\mu_k^{(1)} dN_k^{(1)} + \mu_k^{(2)} dN_k^{(2)}$, which arise from processes within the composite system. The terms $T^{(1)} dS^{(1)} + T^{(2)} dS^{(2)}$ combine with the term $dU^r = T^r dS^r$ in equation 6.1 to yield

$$T^{(1)} dS^{(1)} + T^{(2)} dS^{(2)} + T^r dS^r = T^{(1)} dS^{(1)} + T^{(2)} dS^{(2)} - T^r d(S^{(1)} + S^{(2)})$$

$$= 0 \tag{6.6}$$

whence

$$T^{(1)} = T^{(2)} = T^r \tag{6.7}$$

Thus one evident aspect of the final equilibrium state is the fact that the reservoir maintains a constancy of temperature throughout the system. The remaining conditions of equilibrium naturally depend upon the specific form of the internal constraints in the composite system.

To this point we have merely reviewed the application of the energy minimum principle to the composite system (the subsystem plus the reservoir). We are finally ready to recast equations 6.1 and 6.2 into the language of another representation. We rewrite equation 6.1

$$d(U + U^r) = dU + T^r dS^r = 0 \tag{6.8}$$

or, by equation 6.3

$$dU - T^r dS = 0 \tag{6.9}$$

or, further, since T^r is a constant

$$d(U - T^r S) = 0 \tag{6.10}$$

Similarly, since T^r is a constant and S is an independent variable, equation 6.2 implies[1]

$$d^2 U = d^2(U - T^r S) > 0 \tag{6.11}$$

Thus the quantity $(U - T^r S)$ is minimum in the equilibrium state. Now the quantity $U - T^r S$ is suggestive by its form of the Helmholtz potential $U - TS$. We are therefore led to examine further the extremum properties of the quantity $(U - T^r S)$ and to ask how these may be related to the extremum properties of the Helmholtz potential. We have seen that an evident feature of the equilibrium is that the temperature of the composite system (i.e., of each of its subsystems) is equal to T^r. If we accept that part of the solution, we can immediately restrict our search for the equilibrium state among the manifold of states for which $T = T^r$. But over this manifold of states $U - TS$ is identical to $U - T^r S$. Then we can write equation 6.10 as

$$dF = d(U - TS) = 0 \tag{6.12}$$

subject to the auxiliary condition that

$$T = T^r \tag{6.13}$$

That is, the equilibrium state minimizes the Helmholtz potential, not absolutely, but over the manifold of states for which $T = T^r$. We thus arrive at the equilibrium condition in the Helmholtz potential representation.

Helmholtz Potential Minimum Principle. *The equilibrium value of any unconstrained internal parameter in a system in diathermal contact with a heat reservoir minimizes the Helmholtz potential over the manifold of states for which $T = T^r$.*

---

[1] $d^2 U$ represents the second-order terms in the expansion of U in powers of dS; the linear term $-T^r S$ in equation 6.11 contributes to the expansion only in first order (see equation A.9 of Appendix A).

The intuitive significance of this principle is clearly evident in equations 6.8 through 6.10. The energy of the system plus the reservoir is, of course, minimum. But the statement that the Helmholtz potential of the system alone is minimum is just another way of saying this, for $dF = d(U - TS)$, and the term $d(-TS)$ actually represents the change in energy of the reservoir (since $T = T'$ and $-dS = dS'$). It is now a simple matter to extend the foregoing considerations to the other common representations.

Consider a composite system in which all subsystems are in contact with a common pressure reservoir through walls nonrestrictive with respect to volume. We further assume that some internal constraint within the composite system has been removed. The first condition of equilibrium can be written

$$d(U + U') = dU - P'dV' = dU + P'dV = 0 \qquad (6.14)$$

or

$$d(U + P'V) = 0 \qquad (6.15)$$

Accepting the evident condition that $P = P'$, we can write

$$dH = d(U + PV) = 0 \qquad (6.16)$$

subject to the auxiliary restriction

$$P = P' \qquad (6.17)$$

Furthermore, since P' is a constant and V is an independent variable

$$d^2H = d^2(U + P'V) = d^2U > 0 \qquad (6.18)$$

so that the extremum is a minimum.

Enthalpy Minimum Principle. *The equilibrium value of any unconstrained internal parameter in a system in contact with a pressure reservoir minimizes the enthalpy over the manifold of states of constant pressure (equal to that of the pressure reservoir).*

Finally, consider a system in simultaneous contact with a thermal and a pressure reservoir. Again

$$d(U + U') = dU - T'dS + P'dV = 0 \qquad (6.19)$$

Accepting the evident conditions that $T = T'$ and $P = P'$, we can write

$$dG = d(U - TS + PV) = 0 \qquad (6.20)$$

subject to the auxiliary restrictions

$$T = T^r \qquad P = P^r \tag{6.21}$$

Again

$$d^2G = d^2(U - T^rS + P^rV) = d^2U > 0 \tag{6.22}$$

We thus obtain the equilibrium condition in the Gibbs representation.

Gibbs Potential Minimum Principle. *The equilibrium value of any unconstrained internal parameter in a system in contact with a thermal and a pressure reservoir minimizes the Gibbs potential at constant temperature and pressure (equal to those of the respective reservoirs).*

If the system is characterized by other extensive parameters in addition to the volume and the mole numbers the analysis is identical in form and the general result is now clear:

The General Minimum Principle for Legendre Transforms of the Energy. *The equilibrium value of any unconstrained internal parameter in a system in contact with a set of reservoirs (with intensive parameters $P_1^r, P_2^r, \dots$) minimizes the thermodynamic potential $U[P_1, P_2, \dots]$ at constant $P_1, P_2, \dots$ (equal to $P_1^r, P_2^r, \dots$).*

6-2 THE HELMHOLTZ POTENTIAL

For a composite system in thermal contact with a thermal reservoir the equilibrium state minimizes the Helmholtz potential over the manifold of states of constant temperature (equal to that of the reservoir). In practice many processes are carried out in rigid vessels with diathermal walls, so that the ambient atmosphere acts as a thermal reservoir; for these the Helmholtz potential representation is admirably suited.

The Helmholtz potential is a natural function of the variables $T, V, N_1, N_2, \dots$. The condition that T is constant reduces the number of variables in the problem, and F effectively becomes a function only of the variables V and $N_1, N_2, \dots$. This is in marked contrast to the manner in which constancy of T would have to be handled in the energy representation: there U would be a function of $S, V, N_1, N_2, \dots$ but the auxiliary condition $T = T^r$ would imply a relation among these variables. Particularly in the absence of explicit knowledge of the equation of state $T = T(S, V, N)$ this auxiliary restriction would lead to considerable awkwardness in the analytic procedures in the energy representation.

As an illustration of the use of the Helmholtz potential we first consider a composite system composed of two simple systems separated by a

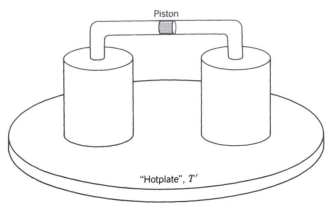

Piston

"Hotplate", T^r

FIGURE 6.1

movable, adiabatic, impermeable wall (such as a solid insulating piston). The subsystems are each in thermal contact with a thermal reservoir of temperature T^r (Fig. 6.1). The problem, then, is to predict the volumes $V^{(1)}$ and $V^{(2)}$ of the two subsystems. We write

$$P^{(1)}\left(T^r, V^{(1)}, N_1^{(1)}, N_2^{(1)}, \ldots\right) = P^{(2)}\left(T^r, V^{(2)}, N_1^{(2)}, N_2^{(2)}, \ldots\right) \quad (6.23)$$

This is one equation involving the two variables $V^{(1)}$ and $V^{(2)}$; all other arguments are constant. The closure condition

$$V^{(1)} + V^{(2)} = V, \text{ a constant} \quad (6.24)$$

provides the other required equation, permitting explicit solution for $V^{(1)}$ and $V^{(2)}$.

In the energy representation we would also have found equality of the pressures, as in equation 6.23, but the pressures would be functions of the entropies, volumes, and mole numbers. We would then require the equations of state to relate the entropies to the temperature and the volumes; the two simultaneous equations, 6.23 and 6.24, would be replaced by four.

Although this reduction of four equations to two may seem to be a modest achievement, such a reduction is a very great convenience in more complex situations. Perhaps of even greater conceptual value is the fact that the Helmholtz representation permits us to focus our thought processes exclusively on the subsystem of interest, relegating the reservoir only to an implicit role. And finally, for technical mathematical reasons to be elaborated in Chapter 16, statistical mechanical calculations are *enormously* simpler in Helmholtz representations, permitting calculations that would otherwise be totally intractable.

For a system in contact with a thermal reservoir the Helmholtz potential can be interpreted as the *available work at constant temperature*.

Consider a system that interacts with a reversible work source while being in thermal contact with a thermal reservoir. In a reversible process the work input to the reversible work source is equal to the decrease in energy of the system and the reservoir

$$dW_{RWS} = -dU - dU' = -dU - T'dS' \qquad (6.25)$$

$$= -dU + T'dS = -d(U - T'S) \qquad (6.26)$$

$$= -dF \qquad (6.27)$$

Thus *the work delivered in a reversible process, by a system in contact with a thermal reservoir, is equal to the decrease in the Helmholtz potential of the system.* The Helmholtz potential is often referred to as the Helmholtz "free energy," though the term *available work at constant temperature* would be less subject to misinterpretation.

Example 1

A cylinder contains an internal piston on each side of which is one mole of a monatomic ideal gas. The walls of the cylinder are diathermal, and the system is immersed in a large bath of liquid (a heat reservoir) at temperature 0°C. The initial volumes of the two gaseous subsystems (on either side of the piston) are 10 liters and 1 liter, respectively. The piston is now moved reversibly, so that the final volumes are 6 liters and 5 liters, respectively. How much work is delivered?

Solution

As the reader has shown in Problem 5.3-1, the fundamental equation of a monatomic ideal gas in the Helmholtz potential representation is

$$F = NRT \left\{ \frac{F_0}{N_0 RT_0} - \ln \left[\left(\frac{T}{T_0} \right)^{3/2} \frac{V}{V_0} \left(\frac{N}{N_0} \right)^{-1} \right] \right\}$$

At constant T and N this is simply

$$F = \text{constant} - NRT \ln V$$

The change in Helmholtz potential is

$$\Delta F = -NRT[\ln 6 + \ln 5 - \ln 10 - \ln 1] = -NRT \ln 3 = -2.5 \text{ kJ}$$

Thus 2.5 kJ of work are delivered in this process.

It is interesting to note that all of the energy comes from the thermal reservoir. The energy of a monatomic ideal gas is simply $\frac{3}{2}NRT$ and therefore it is constant at constant temperature. The fact that we withdraw heat from the temperature reservoir and deliver it *entirely* as work to the reversible work source does not, however, violate the Carnot efficiency principle because the gaseous subsystems are not left in their initial state. Despite the fact that the energy of these subsystems remains constant, their *entropy* increases.

PROBLEMS

6.2-1. Calculate the pressure on each side of the internal piston in Example 1, for arbitrary position of the piston. By integration then calculate the work done in Example 1 and corroborate the result there obtained.

6.2-2. Two ideal van der Waals fluids are contained in a cylinder, separated by an internal moveable piston. There is one mole of each fluid, and the two fluids have the same values of the van der Waals constants b and c; the respective values of the van der Waals constant "a" are a_1 and a_2. The entire system is in contact with a thermal reservoir of temperature T. Calculate the Helmholtz potential of the composite system as a function of T and of the total volume V. If the total volume is doubled (while allowing the internal piston to adjust), what is the work done by the system? Recall Problem 5.3-2.

6.2-3. Two subsystems are contained within a cylinder and are separated by an internal piston. Each subsystem is a mixture of one mole of helium gas and one mole of neon gas (each to be considered as a monatomic ideal gas). The piston is in the center of the cylinder, each subsystem occupying a volume of 10 liters. The walls of the cylinder are diathermal, and the system is in contact with a thermal reservoir at a temperature of 100°C. The piston is permeable to helium but impermeable to neon.

Recalling (from Problem 5.3-10) that the Helmholtz potential of a mixture of simple ideal gases is the sum of the individual Helmholtz potentials (each expressed as a function of temperature and volume), show that in the present case

$$F = N\frac{T}{T_0}f_0 - \frac{3}{2}NRT\ln\frac{T}{T_0} - N_1RT\ln\left(\frac{V}{V_0}\frac{N_0}{N_1}\right)$$
$$- N_2^{(1)}RT\ln\frac{V^{(1)}N_0}{V_0N_2^{(1)}} - N_2^{(2)}RT\ln\frac{V^{(2)}N_0}{V_0N_2^{(2)}}$$

where T_0, f_0, V_0, and N_0 are attributes of a standard state (recall Problem 5.3-1), N is the total mole number, $N_2^{(1)}$ is the mole number of neon (component 2) in subsystem 1, and $V^{(1)}$ and $V^{(2)}$ are the volumes of subsystems 1 and 2, respectively.

How much work is required to push the piston to such a position that the volumes of the subsystems are 5 liters and 15 liters? Carry out the calculation both by calculating the change in F and by a direct integration (as in Problem 6.2-1).

Answer:

work $= RT\ln(\frac{4}{3}) = 893$ J

6-3   THE ENTHALPY: THE JOULE–THOMSON OR "THROTTLING" PROCESS

For a composite system in interaction with a pressure reservoir the equilibrium state minimizes the enthalpy over the manifold of states of constant pressure. The enthalpy representation would be appropriate to processes carried out in adiabatically insulated cylinders fitted with adiabatically insulated pistons subject externally to atmospheric pressure, but this is not a very common experimental design. In processes carried out in open vessels, such as in the exercises commonly performed in an elementary chemistry laboratory, the ambient atmosphere acts as a pressure reservoir, but it also acts as a thermal reservoir: for the analysis of such processes only the Gibbs representation invokes the full power of Legendre transformations. Nevertheless, there are particular situations uniquely adapted to the enthalpy representation, as we shall see shortly.

More immediately evident is the interpretation of the enthalpy as a "potential for heat." From the differential form

$$dH = T\,dS + V\,dP + \mu_1\,dN_1 + \mu_2\,dN_2 + \cdots \qquad (6.28)$$

it is evident that for a system in contact with a pressure reservoir and enclosed by impermeable walls

$$dH = dQ \qquad \text{(where } P, N_1, N_2,\ldots \text{ are constant)} \qquad (6.29)$$

That is, *heat added to a system at constant pressure and at constant values of all the remaining extensive parameters* (other than S and V) *appears as an increase in the enthalpy.*

This statement may be compared to an analogous relation for the energy

$$dU = dQ \qquad \text{(where } V, N_1, N_2,\ldots \text{ are constant)} \qquad (6.30)$$

and similar results for any Legendre transform in which the entropy is not among the transformed variables.

Because heating of a system is so frequently done while the system is maintained at constant pressure by the ambient atmosphere, the enthalpy is generally useful in discussion of heat transfers. The enthalpy accordingly is sometimes referred to as the "heat content" of the system (but it should be stressed again that "heat" refers to a mode of energy *flux* rather than to an attribute of a state of a thermodynamic system).

To illustrate the significance of the enthalpy as a "potential for heat," suppose that a system is to be maintained at constant pressure and its volume is to be changed from V_i to V_f. We desire to compute the heat absorbed by the system. As the pressure is constant, the heat flux is equal to the change in the enthalpy

$$Q_{i \to f} \equiv \int dQ = H_f - H_i \qquad (6.31)$$

If we were to know the fundamental equation

$$H = H(S, P, N) \qquad (6.32)$$

then, by differentiation

$$V = \frac{\partial H}{\partial P} = V(S, P, N) \tag{6.33}$$

and we could eliminate the entropy to find H as a function of V, P, and N. Then

$$Q_{i \to f} = H(V_f, P, N) - H(V_i, P, N) \tag{6.34}$$

A process of great practical importance, for which an enthalpy representation is extremely convenient, is the Joule–Thomson or "throttling" process. This process is commonly used to cool and liquify gases and as a second-stage refrigerator in "cryogenic" (low-temperature) laboratories.

In the Joule–Thomson process or "Joule–Kelvin" process (William Thomson was only later granted peerage as Lord Kelvin) a gas is allowed to seep through a porous barrier from a region of high pressure to a region of low pressure (Fig. 6.2). The porous barrier or "throttling valve" was originally a wad of cotton tamped into a pipe; in a laboratory demonstration it is now more apt to be glass fibers, and in industrial practice it is generally a porous ceramic termination to a pipe (Fig. 6.3). The process can be made continuous by using a mechanical pump to return the gas from the region of low pressure to the region of high pressure. Depending on certain conditions, to be developed in a moment, the gas is either heated or cooled in passing through the throttling valve.

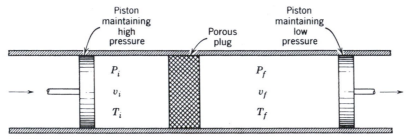

FIGURE 6.2
Schematic representation of the Joule-Thomson process.

For real gases and for given initial and final pressures, the change in temperature is generally positive down to a particular temperature, and it is negative below that temperature. The temperature at which the process changes from a heating to a cooling process is called the *inversion temperature*; it depends upon the particular gas and upon both the initial and final pressures. In order that the throttling process operate as an effective cooling process the gas must first be precooled below its inversion temperature.

To show that the Joule–Thomson process occurs at constant enthalpy consider one mole of the gas undergoing a throttling process. The piston

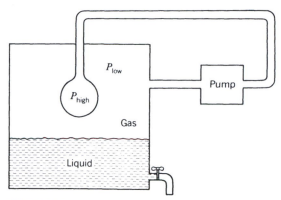

FIGURE 6.3
Schematic apparatus for liquefaction of a gas by throttling process. The pump maintains the pressure difference ($P_{high} - P_{low}$). The spherical termination of the high pressure pipe is a porous ceramic shell through which the gas expands in the throttling process.

(Fig. 6.2) that pushes this quantity of gas through the plug does an amount of work $P_i v_i$, in which v_i is the molar volume of the gas on the high pressure side of the plug. As the gas emerges from the plug, it does work on the piston that maintains the low pressure P_f, and this amount of work is $P_f v_f$. Thus the conservation of energy determines the final molar energy of the gas; it is the initial molar energy, plus the work $P_i v_i$ done on the gas, minus the work $P_f v_f$ done by the gas.

$$u_f = u_i + P_i v_i - P_f v_f \qquad (6.35)$$

or

$$u_f + P_f v_f = u_i + P_i v_i \qquad (6.36)$$

which can be written in terms of the molar enthalpy h as

$$h_f = h_i \qquad (6.37)$$

Although, on the basis of equation 6.37, we say that the Joule–Thomson process occurs at constant enthalpy, we stress that this simply implies that the final enthalpy is equal to the initial enthalpy. We do not imply anything about the enthalpy during the process; *the intermediate states of the gas are nonequilibrium states for which the enthalpy is not defined.*

The isenthalpic curves ("isenthalps") of nitrogen are shown in Fig. 6.4 The initial temperature and pressure in a throttling process determine a particular isenthalp. The final pressure then determines a point on this same isenthalp, thereby determining the final temperature.

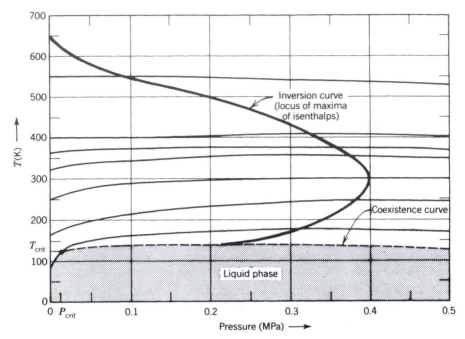

FIGURE 6.4

Isenthalps (solid), inversion temperature (dark), and coexistence curve for nitrogen; semiquantitative.

The isenthalps in Fig. 6.4 are concave, with maxima. If the initial temperature and pressure lie to the left of the maximum the throttling process necessarily cools the gas. If the initial temperature lies to the right of the maximum a small pressure drop heats the gas (though a large pressure drop may cross the maximum and can either heat or cool the gas). The maximum of the isenthalp therefore determines the inversion temperature, at which a small pressure change neither heats nor cools the gas.

The dark curve in Fig. 6.4 is a plot of inversion temperature as a function of pressure, obtained by connecting the maxima of the isenthalpic curves. Also shown on the figure is the curve of liquid–gas equilibrium. Points below the curve are in the liquid phase and those above are in the gaseous phase. This coexistence curve terminates in the "critical point." In the region of this point the "gas" and the "liquid" phases lose their distinguishability, as we shall study in some detail in Chapter 9.

If the change in pressure in a throttling process is sufficiently small we can employ the usual differential analysis.

$$dT = \left(\frac{\partial T}{\partial P} \right)_{H, N_1, N_2, \dots} dP \tag{6.38}$$

The derivative can be expressed in terms of standard measurable quantities (c_p, α, κ_T) by a procedure that may appear somewhat complicated on first reading, but that will be shown in Chapter 7 to follow a routine and straightforward recipe. By a now familiar mathematical identity (A.22),

$$dT = -\left[\left(\frac{\partial H}{\partial P}\right)_T \bigg/ \left(\frac{\partial H}{\partial T}\right)_P\right] dP \qquad (6.39)$$

where we suppress the subscripts $N_1, N_2, \ldots$ for simplicity, noting that the mole numbers remain constant throughout. However, $dH = T\,dS + V\,dP$ at constant mole numbers, so that

$$dT = -\frac{T(\partial S/\partial P)_T + V}{T(\partial S/\partial T)_P} dP \qquad (6.40)$$

The denominator is Nc_P. The derivative $(\partial S/\partial P)_T$ is equal to $-(\partial V/\partial T)_P$ by one of the class of "Maxwell relations," analogous to equations 3.62 or 3.65 (in the present case the two derivatives can be corroborated to be the two mixed second derivatives of the Gibbs potential). Identifying $(\partial S/\partial P)_T = -(\partial V/\partial T)_P = -V\alpha$ (equation 3.67) we finally find

$$dT = \frac{v}{c_P}(T\alpha - 1)\,dP \qquad (6.41)$$

This is a fundamental equation of the Joule–Thomson effect. As the change in pressure dP is negative, the sign of dT is opposite that of the quantity in parentheses. Thus if $T\alpha > 1$, a small decrease in pressure (in transiting the "throttling valve") cools the gas. The inversion temperature is determined by

$$\alpha T_{\text{inversion}} = 1 \qquad (6.42)$$

For an ideal gas the coefficient of thermal expansion α is equal to $1/T$, so that there is no change in temperature in a Joule–Thomson expansion. All gases approach ideal behavior at high temperature and low or moderate pressure, and the isenthalps correspondingly become "flat," as seen in Fig. 6.4. It is left to Example 2 to show that for real gases the temperature change is negative below the inversion temperature and positive above, and to evaluate the inversion temperature.

Example 2

Compute the inversion temperature of common gases, assuming them to be described by the van der Waals equation of state (3.41).

Solution

We must first evaluate the coefficient of expansion α. Differentiating the van der Waals equation of state (3.41) with respect to T, at constant P

$$\alpha = \frac{1}{v}\left(\frac{\partial v}{\partial T}\right)_P = \left[\frac{Tv}{v-b} - \frac{2a(v-b)}{Rv^2}\right]^{-1}$$

To express the right-hand side as a function of T and P is analytically difficult. An approximate solution follows from the recognition that molar volumes are on the order of 0.02 m^3, whence b/v is on the order of 10^{-3} and a/RTv is on the order of $10^{-3} - 10^{-4}$ (see Table 3.1). Hence a series expansion in b/v and a/RTv can reasonably be terminated at the lowest order term. Let

$$\varepsilon_1 \equiv \frac{b}{v} \qquad \varepsilon_2 \equiv \frac{a}{RTv}$$

Then

$$\alpha = \left[\frac{T}{1 - \varepsilon_1} - \frac{2T}{v}(v - b)\varepsilon_2 \right]^{-1}$$

$$= \frac{1}{T}\left[\frac{1}{1 - \varepsilon_1} - 2(1 - \varepsilon_1)\varepsilon_2 \right]^{-1}$$

Returning to equation 6.41

$$dT = \frac{v}{c_p}(T\alpha - 1)\, dP$$

from which we recall that

$$T_{\text{inv}}, \ \alpha = 1$$

It then follows that at the inversion temperature

$$[1 - \varepsilon_1 + 2\varepsilon_2 + \cdots] = 1$$

or

$$\varepsilon_1 = 2\varepsilon_2$$

The inversion temperature is now determined by

$$T_{\text{inv}} \simeq \frac{2a}{bR}$$

with cooling of the gas for temperature below T_{inv}, and heating above. From Table 3.1, we compute the inversion temperature of several gases: $T_{\text{inv}}(H_2) = 224$ K, $T_{\text{inv}}(Ne) = 302$ K, $T_{\text{inv}}(N_2) = 850$ K, $T_{\text{inv}}(O_2) = 1020$ K, $T_{\text{inv}}(CO_2) = 2260$ K. In fact the inversion temperature empirically depends strongly on the pressure —a dependence lost in our calculation by the neglect of higher-order terms. The observed inversion temperature at zero pressure for H_2 is 204 K, and for neon it is 228 K—in fair agreement with our crude calculation. For polyatomic gases the agreement is less satisfactory; the observed value for CO_2 is 1275 K whereas we have computed 2260 K.

PROBLEMS

6.3-1. A hole is opened in the wall separating two chemically identical single-component subsystems. Each of the subsystems is also in interaction with a pressure reservoir of pressure P'. Use the enthalpy minimum principle to show that the conditions of equilibrium are $T^{(1)} = T^{(2)}$ and $\mu^{(1)} = \mu^{(2)}$.

6.3-2. A gas has the following equations of state

$$P = \frac{U}{V} \qquad\qquad T = 3B\left(\frac{U^2}{NV}\right)^{1/3}$$

where B is a positive constant. The system obeys the Nernst postulate ($S \to 0$ as $T \to 0$). The gas, at an initial teperature T_i and initial pressure P_i, is passed through a "porous plug" in a Joule–Thomson process. The final pressure is P_f. Calculate the final temperature T_f.

6.3-3. Show that for an ideal van der Waals fluid

$$h = -\frac{2a}{v} + RT\left(c + \frac{v}{v - b}\right)$$

where h is the molar enthalpy. Assuming such a fluid to be passed through a porous plug and thereby expanded from v_i to v_f (with $v_f > v_i$), find the final temperature T_f in terms of the initial temperature T_i and the given data.

Evaluate the temperature change if the gas is CO_2, the mean temperature is $0°C$, the mean pressure is 10^7 Pa, and the change in pressure is 10^6 Pa. The molar heat capacity c_P of CO_2 at the relevant temperature and pressure is 29.5 J/mole-K. Carry calculation only to first order in b/v and a/RTv.

6.3-4. One mole of a monatomic ideal gas is in a cylinder with a movable piston on the other side of which is a pressure reservoir with $P_r = 1$ atm. How much heat must be added to the gas to increase its volume from 20 to 50 liters?

6.3-5. Assume that the gas of Problem 6.3-4 is an ideal van der Waals fluid with the van der Waals constants of argon (Table 3-1), and again calculate the heat required. Recall Problem 6.3-3.

6-4 THE GIBBS POTENTIAL; CHEMICAL REACTIONS

For a composite system in interaction with both thermal and pressure reservoirs the equilibrium state minimizes the Gibbs potential over the manifold of states of constant temperature and pressure (equal to those of the reservoirs).

The Gibbs potential is a natural function of the variables $T, P, N_1, N_2, \ldots$, and it is particularly convenient to use in the analysis of problems involving constant T and P. Innumerable processes of common experience occur in systems exposed to the atmosphere, and thereby maintained at constant temperature and pressure. And frequently a process of interest occurs in a small subsystem of a larger system that acts as both a thermal and a pressure reservoir (as in the fermentation of a grape in a large wine vat).

The Gibbs potential of a multicomponent system is related to the chemical potentials of the individual components, for $G = U - TS + PV$, and inserting the Euler relation $U = TS - PV + \mu_1 N_1 + \mu_2 N_2 + \cdots$ we find

$$G = \mu_1 N_1 + \mu_2 N_2 + \cdots \tag{6.43}$$

Thus, for a single component system the molar Gibbs potential is identical with μ

$$\frac{G}{N} = \mu \tag{6.44}$$

but for a multicomponent system

$$\frac{G}{N} = \mu_1 x_1 + \mu_2 x_2 + \cdots + \mu_r x_r \tag{6.45}$$

where x_j is the mole fraction (N_j/N) of the jth component. Accordingly, the chemical potential is often referred to as the *molar Gibbs potential* in single component systems or as the *partial molar Gibbs potential* in multicomponent systems.

The thermodynamics of chemical reactions is a particularly important application of the Gibbs potential.

Consider the chemical reaction

$$0 \rightleftarrows \sum_1^r \nu_j A_j \tag{6.46}$$

where the ν_j are the stoichiometric coefficients defined in Section 2.9. The change in Gibbs potential associated with virtual changes dN_j in the mole numbers is

$$dG = -S\,dT + V\,dP + \sum_j \mu_j\,dN_j \tag{6.47}$$

However the changes in the mole numbers must be in proportion to the stoichiometric coefficients, so that

$$\frac{dN_1}{\nu_1} = \frac{dN_2}{\nu_2} = \cdots \equiv d\tilde{N} \tag{6.48}$$

or, equivalently,

$$dN_j = \nu_j\,d\tilde{N} \tag{6.49}$$

where $d\tilde{N}$ is simply a proportionality factor defined by equation 6.48. If the chemical reaction is carried out at constant temperature and pressure (as in an open vessel) the condition of equilibrium then implies

$$dG = d\tilde{N}\sum_j \nu_j\mu_j = 0 \tag{6.50}$$

or

$$\sum_j \nu_j\mu_j = 0 \tag{6.51}$$

If the initial quantities of each of the chemical components is N_j^0 the chemical reaction proceeds to some extent and the mole numbers assume the new values

$$N_j = N_j^0 + \int dN_j = N_j^0 + \nu_j\Delta\tilde{N} \tag{6.52}$$

where $\Delta\tilde{N}$ is the factor of proportionality. The chemical potentials in equation 6.51 are functions of T, P, and the mole numbers, and hence of the single unknown parameter $\Delta\tilde{N}$. Solution of equation 6.51 for $\Delta\tilde{N}$ determines the equilibrium composition of the system.

The solution described is appropriate only providing that there is a sufficient quantity of each component present so that none is depleted before equilibrium is reached. That is, none of the quantities N_j in equation 6.52 can become negative. This consideration is most conveniently expressed in terms of the *degree of reaction*.

The maximum value of $\Delta\tilde{N}$ for which all N_j remain positive (in equation 6.52) defines the maximum permissible extent of the reaction. Similarly the minimum value of $\Delta\tilde{N}$ for which all N_j remain positive defines the maximum permissible extent of the reverse reaction. The actual value of $\Delta\tilde{N}$ in equilibrium may be anywhere between these two extremes. The *degree of reaction* ε is defined as

$$\varepsilon \equiv \frac{\Delta\tilde{N} - \Delta\tilde{N}_{\min}}{\Delta\tilde{N}_{\max} - \Delta\tilde{N}_{\min}} \tag{6.53}$$

It is possible that a straightforward solution of the equation of chemical equilibrium (6.51) may yield a value of $\Delta\tilde{N}$ that is larger than $\Delta\tilde{N}_{\max}$ or smaller than $\Delta\tilde{N}_{\min}$. In such a case the process is terminated by the depletion of one of its components. The physically relevant value of $\Delta\tilde{N}$ is then $\Delta\tilde{N}_{\max}$ (or $\Delta\tilde{N}_{\min}$). Although $\sum_j \nu_j\mu_j$ does not attain the value zero, it does attain the smallest absolute value accessible to the system.

Whereas the partial molar Gibbs potentials characterize the equilibrium condition, the enthalpy finds its expression in the *heat of reaction*. This fact follows from the general significance of the enthalpy as a "potential for heat flux" at constant pressure (equation 6.29). That is, the flux of heat from the surroundings to the system, during the chemical reaction, is equal to the change in the enthalpy. This change in enthalpy, in turn, can be related to the chemical potentials, for

$$H = G + TS = G - T\left(\frac{\partial G}{\partial T}\right)_{P, N_1, N_2, \ldots} \tag{6.54}$$

If an infinitesimal chemical reaction $d\tilde{N}$ occurs, both H and G change and

$$dH = \frac{dH}{d\tilde{N}} \, d\tilde{N} = \frac{dG}{d\tilde{N}} \, d\tilde{N} - T\frac{\partial}{\partial T}\left(\frac{dG}{d\tilde{N}}\right)_{P, N_1, N_2, \ldots} d\tilde{N} \tag{6.55}$$

But the change in Gibbs function is

$$dG = \sum_1^r \mu_j \, dN_j = \left(\sum_1^r \nu_j \mu_j\right) d\tilde{N} \tag{6.56}$$

whence

$$\frac{dG}{d\tilde{N}} = \sum_1^r \nu_j \mu_j \tag{6.57}$$

At equilibrium $dG/d\tilde{N}$ vanishes (but the temperature derivative of $dG/d\tilde{N}$ does not) so that in the vicinity of the equilibrium state equation 6.55 becomes

$$\frac{dH}{d\tilde{N}} = -T\frac{\partial}{\partial T}\left(\sum_1^r \nu_j \mu_j\right)_{P, N_1, N_2, \ldots} \tag{6.58}$$

The quantity $dH/d\tilde{N}$ is known as the *heat of reaction*; it is the heat absorbed per unit reaction in the vicinity of the equilibrium state. It is positive for *endothermic* reactions and negative for *exothermic* reactions.

We have assumed that the reaction considered is not one that goes to completion. If the reaction does go to completion, the summation in equation 6.57 does not vanish in the equilibrium state, and this summation appears as an additional term in equation 6.58.

As the summation in equation 6.58 vanishes at the equilibrium composition, it is intuitively evident that the temperature derivative of this quantity is related to the temperature dependence of the equilibrium concentrations. We shall find it convenient to develop this connection explicitly only in the special case of ideal gases, in Section 13.4. However, it is of interest here to note the plausibility of the relationship and to recognize that such a relationship permits the heat of reaction to be measured by determinations of equilibrium compositions at various temperatures rather than by relatively difficult calorimetric experiments.

The general methodology for the analysis of chemical reactions becomes specific and definite when applied to particular systems. To anchor the foregoing treatment in a fully explicit (and practically important) special case, the reader may well wish here to interpolate Chapter 13—and particularly Section 13.2 on chemical reactions in ideal gases.

Example 3

Five moles of H_2, 1 mole of CO_2, 1 mole of CH_4, and 3 moles of H_2O are allowed to react in a vessel maintained at a temperature T_0 and pressure P_0. The relevant reaction is

$$4H_2 + CO_2 \rightleftharpoons CH_4 + 2H_2O$$

Solution of the equilibrium condition gives the nominal solution $\Delta \tilde{N} = -\frac{1}{2}$. What are the mole numbers of each of the components? If the pressure is then increased to P_1 ($P_1 > P_0$) and the temperature is maintained constant ($= T_0$) the equilibrium condition gives a new nominal solution of $\Delta \tilde{N} = 1, 2$. What are the mole numbers of each of the components?

Solution

We first write the analogue of equation 6.52 for each component: $N_{H_2} = 5 - 4\Delta\tilde{N}$, $N_{CO_2} = 1 - \Delta\tilde{N}$, $N_{CH_4} = 1 + \Delta\tilde{N}$, $N_{H_2O} = 3 + 2\Delta\tilde{N}$. Setting each of these mole numbers equal to zero successively we find four roots for $\Delta\tilde{N}$: $\frac{5}{4}$, 1, -1, and $-\frac{3}{2}$. The positive and negative roots of smallest absolute values are, respectively,

$$\Delta\tilde{N}_{max} = 1 \qquad \Delta\tilde{N}_{min} = -1$$

These two bounds on $\Delta\tilde{N}$ correspond to depletion of CO_2 if the reaction proceeds too far in the "forward" direction, and to depletion of CH_4 if the reaction proceeds too far in the "reverse" direction.

The degree of reaction is now, by equation 6.53

$$\varepsilon = \frac{\Delta\tilde{N} + 1}{1 + 1} = \frac{1}{2}(\Delta\tilde{N} + 1)$$

If the nominal solution of the equilibrium condition gives $\Delta\tilde{N} = -\frac{1}{2}$ then $\varepsilon = \frac{1}{4}$ and $N_{H_2} = 3$, $N_{CO_2} = \frac{3}{2}$, $N_{CH_4} = \frac{1}{2}$ and $N_{H_2O} = 2$.

If the increase in pressure shifts the nominal solution for $\Delta\tilde{N}$ to $+1.2$ we reject this value as outside the acceptable range of $\Delta\tilde{N}$ (i.e., greater than $\Delta\tilde{N}_{max}$); it would lead to the nonphysical value of $\varepsilon = 1.1$ whereas ε must be between zero and unity. Hence the reaction is terminated at $\Delta\tilde{N} = \Delta\tilde{N}_{max} = \frac{1}{2}$ (or at $\varepsilon = 1$) by the depletion of CO_2. The final mole numbers are $N_{H_2} = 1$, $N_{CO_2} = \frac{1}{2}$, $N_{CH_4} = 2$, and $N_{H_2O} = 5$.

PROBLEMS

6.4-1. One half mole of H_2S, $\frac{3}{4}$ mole of H_2O, 2 moles of H_2, and 1 mole of SO_2 are allowed to react in a vessel maintained at a temperature of 300 K and a pressure of 10^4 Pa. The components can react by the chemical reaction

$$3H_2 + SO_2 \rightleftharpoons H_2S + 2H_2O$$

a) Write the condition of equilibrium in terms of the partial molar Gibbs potentials.

b) Show that

$$N_{H_2} = 2 - 3\Delta\tilde{N}$$

and similarly for the other components. For what value of $\Delta\tilde{N}$ does each N_j vanish?

c) Show that $\Delta\tilde{N}_{max} = \frac{2}{3}$ and $\Delta\tilde{N}_{min} = -\frac{3}{8}$. Which components are depleted in each of these cases?

d) Assume that the nominal solution of the equilibrium condition gives $\Delta\tilde{N} = \frac{1}{4}$. What is the degree of reaction ε? What are the mole fractions of each of the components in the equilibrium mixture?

e) Assume that the pressure is raised and that the nominal solution of the equilibrium condition now yields the value $\Delta\tilde{N} = 0.8$. What is the degree of reaction? What are the mole fractions of each of the components in the final state?

Answers:

c) H_2 and H_2O depleted

d) $\quad \varepsilon = \frac{3}{5}, \qquad x_{H_2O} = \frac{5}{16}$

e) $\quad \Delta\tilde{N} = \frac{2}{3}, \qquad x_{H_2O} = .59$

6-5 OTHER POTENTIALS

Various other potentials may occasionally become useful in particular applications. One such application will suffice to illustrate the general method.

Example 4

A bottle, of volume V, contains N_s moles of sugar, and it is filled with water and capped by a rigid lid. The lid though rigid is permeable to water but not to sugar. The bottle is immersed in a large vat of water. The pressure in the vat, at the position of the bottle, is P_v and the temperature is T. We seek the pressure P and the mole number N_w of water in the bottle.

Solution

We suppose that we are given the fundamental equation of a two-component mixture of sugar and water. Most conveniently, this fundamental equation will be cast in the representation $U[T, V, \mu_w, N_s]$; that is, in the representation in which S and N_w are replaced by their corresponding intensive parameters, but the volume V and the mole number of sugar N_s remain untransformed. The diathermal wall ensures that T has the value established by the vat (a thermal reservoir), and the semipermeable lid ensures that μ_w has the value established by the vat (a "water reservoir"). No problem remains! We know all the independent variables of the generalized potential $U[T, V, \mu_w, N_s]$. To find the pressure in the bottle we merely differentiate the potential:

$$P = -\frac{\partial U[T, V, \mu_w, N_s]}{\partial V} \tag{6.59}$$

It is left to the reader to compare this approach to the solution of the same problem in energy or entropy representations. Various unsought for variables enter into the analysis—such as the entropy of the contents of the bottle, or the entropy, energy, and mole number of the contents of the vat. And for each such extraneous variable, an additional equation is needed for its elimination. The choice of the appropriate representation clearly is the key to simplicity, and indeed to practicality, in thermodynamic calculations.

6-6 COMPILATIONS OF EMPIRICAL DATA; THE ENTHALPY OF FORMATION

In principle, thermodynamic data on specific systems would be most succinctly and conveniently given by a tabulation of the Gibbs potential as a function of temperature, pressure, and composition (mole fractions of the individual components). Such a tabulation would provide a fundamental equation in the representation most convenient to the experimentalist.

In practice it is customary to compile data on $h(T, P)$, $s(T, P)$, and $v(T, P)$, from which the molar Gibbs potential can be obtained ($g = h - Ts$). The tabulation of h, s, and v is redundant but convenient. For multicomponent systems analogous compilations must be made for each composition of interest.

Differences in the molar enthalpies of two states of a system can be evaluated experimentally by numerical integration of $dh = dQ/N + v\,dP$, for dQ as well as P and v can be measured along the path of integration.

The absolute scale of the enthalpy h, like that of the energy or of any other thermodynamic potential, is arbitrary, undetermined within an additive constant. For purposes of compilation of data, the scale of enthalpy is made definite by assigning the value zero to the molar enthalpy of each chemical element in its most stable form at a standard temperature and pressure, generally taken as

$$T_0 = 298.15 \text{ K} = 25°\text{C} \qquad P_0 = 0.1 \text{ MPa} \approx 1 \text{ atm}$$

The enthalpy defined by this choice of scale is called the *enthalpy of formation*.

The reference to the "most stable state" in the definition of the enthalpy of formation implies, for instance, that the value zero is assigned to the molecular form of oxygen (O_2) rather than to the atomic form (O); the molecular form is the most stable form at standard temperature and pressure.

If 1 mole of carbon and 1 mole of O_2 are chemically reacted to form 1 mole of CO_2, the reaction being carried out at standard temperature and pressure, it is observed that 393.52×10^3 J of heat are emitted. Hence the enthalpy of formation of CO_2 is taken as -393.52×10^3 J/mole in the standard state. This is the *standard enthalpy of formation* of CO_2. The enthalpy of formation of CO_2 at any other temperature and pressure is obtained by integration of $dh = dQ/N + v\, dP$.

The standard molar enthalpy of formation, the corresponding standard molar Gibbs potential, and the molar entropy in the standard state are tabulated for a wide range of compounds in the JANAF Thermochemical Tables (Dow Chemical Company, Midland, Michigan) and in various other similar compilations.

Tables of thermodynamic properties of a particular material can become very voluminous indeed if several properties (such as h, s, and v), or even a single property, are to be tabulated over wide ranges of the independent variables T and P. Nevertheless, for common materials such as water very extensive tabulations are readily available. In the case of water the tabulations are referred to as "Steam Tables." One form of steam table, referred to as a "superheated steam table," gives values of the molar volume v, energy u, enthalpy h, and entropy s as a function of temperature, for various values of pressure. An excerpt from such a table (by Sonntag and van Wilen), for a few values of the pressure, is given in Table 6.1. Another form, referred to as a "saturated steam table," gives values of the properties of the liquid and of the gaseous phases of water for values of P and T which lie on the gas-liquid coexistence curve. Such a "saturated steam table" will be given in Table 9.1.

Another very common technique for representation of thermodynamic data consists of "thermodynamic charts," or graphs. Such charts necessarily sacrifice precision, but they allow a large amount of data to be summarized succinctly and compactly. Conceptually, the simplest such chart would label the two coordinate axes by T and P. Then, for a single-component system one would draw families of curves of constant molar Gibbs potential μ. In principle that would permit evaluation of all desired data. Determination of the molar volume, for instance, would require reading the values of μ for two nearby pressures at the temperature of interest; this would permit numerical evaluation of the derivative $(\Delta\mu/\Delta P)_T$, and thence of the molar volume. Instead, a family of isochores is overlaid on the graph, with each isochore labeled by v. Similarly,

TABLE 6.1

Superheated Steam Table

The quantities u, h, and s are per unit mass (rather than molar); the units of u and h are Joules/kilogram, of v are m³/kilogram, and of s are Joules/kilogram-Kelvin. Temperatures are in degrees Celsius. The notation "Sat." under T refers to the temperature on the liquid-gas coexistence curve; this temperature is given in parentheses following each pressure value.

From R. E. Sonntag and G. Van Wylen, Introduction to Thermodynamics, Classical and Statistical, John Wiley & Sons, New York, 1982.

T	P = .010 MPa (45.81)				P = .050 MPa (81.33)				P = .10 MPa (99.63)			
	v	u	h	s	v	u	h	s	v	u	h	s
Sat.	14.674	2437.9	2584.7	8.1502	3.240	2483.9	2645.9	7.5939	1.6940	2506.1	2675.5	7.3594
50	14.869	2443.9	2592.6	8.1749								
100	17.196	2515.5	2687.5	8.4479	3.418	2511.6	2682.5	7.6947	1.6958	2506.7	2676.2	7.3614
150	19.512	2587.9	2783.0	8.6882	3.889	2585.6	2780.1	7.9401	1.9364	2582.8	2776.4	7.6134
200	21.825	2661.3	2879.5	8.9038	4.356	2659.9	2877.7	8.1580	2.172	2658.1	2875.3	7.8343
250	24.136	2736.0	2977.3	9.1002	4.820	2735.0	2976.0	8.3556	2.406	2733.7	2974.3	8.0333
300	26.445	2812.1	3076.5	9.2813	5.284	2811.3	3075.5	8.5373	2.639	2810.4	3074.3	8.2158
400	31.063	2968.9	3279.6	9.6077	6.209	2968.5	3278.9	8.8642	3.103	2967.9	3278.2	8.5435
500	35.679	3132.3	3489.1	9.8978	7.134	3132.0	3488.7	9.1546	3.565	3131.6	3488.1	8.8342
600	40.295	3302.5	3705.4	10.1608	8.057	3302.2	3705.1	9.4178	4.028	3301.9	3704.7	9.0976
700	44.911	3479.6	3928.7	10.4028	8.981	3479.4	3928.5	9.6599	4.490	3479.2	3928.2	9.3398
800	49.526	3663.8	4159.0	10.6281	9.904	3663.6	4158.9	9.8852	4.952	3663.5	4158.6	9.5652
900	54.141	3855.0	4396.4	10.8396	10.828	3854.9	4396.3	10.0967	5.414	3854.8	4396.1	9.7767
1000	58.757	4053.0	4640.6	11.0393	11.751	4052.9	4640.5	10.2964	5.875	4052.8	4640.3	9.9764
1100	63.372	4257.5	4891.2	11.2287	12.674	4257.4	4891.1	10.4859	6.337	4257.3	4891.0	10.1659
1200	67.987	4467.9	5147.8	11.4091	13.597	4467.8	5147.7	10.6662	6.799	4467.7	5147.6	10.3463
1300	72.602	4683.7	5409.7	11.5811	14.521	4683.6	5409.6	10.8382	7.260	4683.5	5409.5	10.5183

TABLE 6.1
(Continued)

	P = .20 MPa (120.23)				P = .30 MPa (133.55)				P = .40 MPa (143.63)			
Sat.	.8857	2529.5	2706.7	7.1272	.6058	2543.6	2725.3	6.9919	.4625	2553.6	2738.6	6.8959
150	.9596	2576.9	2768.8	7.2795	.6339	2570.8	2761.0	7.0778	.4708	2564.5	2752.8	6.9299
200	1.0803	2654.4	2870.5	7.5066	.7163	2650.7	2865.6	7.3115	.5342	2646.8	2860.5	7.1706
250	1.1988	2731.2	2971.0	7.7086	.7964	2728.7	2967.6	7.5166	.5951	2726.1	2964.2	7.3789
300	1.3162	2808.6	3071.8	7.8926	.8753	2806.7	3069.3	7.7022	.6548	2804.8	3066.8	7.5662
400	1.5493	2966.7	3276.6	8.2218	1.0315	2965.6	3275.0	8.0330	.7726	2964.4	3273.4	7.8985
500	1.7814	3130.8	3487.1	8.5133	1.1867	3130.0	3486.0	8.3251	.8893	3129.2	3484.9	8.1913
600	2.013	3301.4	3704.0	8.7770	1.3414	3300.8	3703.2	8.5892	1.0055	3300.2	3702.4	8.4558
700	2.244	3478.8	3927.6	9.0194	1.4957	3478.4	3927.1	8.8319	1.1215	3477.9	3926.5	8.6987
800	2.475	3663.1	4158.2	9.2449	1.6499	3662.9	4157.8	9.0576	1.2372	3662.4	4157.3	8.9244
900	2.706	3854.5	4395.8	9.4566	1.8041	3854.2	4395.4	9.2692	1.3529	3853.9	4395.1	9.1362
1000	2.937	4052.5	4640.0	9.6563	1.9581	4052.3	4639.7	9.4690	1.4685	4052.0	4639.4	9.3360
1100	3.168	4257.0	4890.7	9.8458	2.1121	4256.8	4890.4	9.6585	1.5840	4256.5	4890.2	9.5256
1200	3.399	4467.5	5147.3	10.0262	2.2661	4467.2	5147.1	9.8389	1.6996	4467.0	5146.8	9.7060
1300	3.630	4683.2	5409.3	10.1982	2.4201	4683.0	5409.0	10.0110	1.8151	4682.8	5408.8	9.8780

	P = .50 MPa (151.86)				P = .60 MPa (158.85)				P = .80 MPa (170.43)			
Sat.	.3749	2561.2	2748.7	6.8213	.3157	2567.4	2756.8	6.7600	.2404	2576.8	2769.1	6.6628
200	.4249	2642.9	2855.4	7.0592	.3520	2638.9	2850.1	6.9665	.2608	2630.6	2839.3	6.8158
250	.4744	2723.5	2960.7	7.2709	.3938	2720.9	2957.2	7.1816	.2931	2715.5	2950.0	7.0384
300	.5226	2802.9	3064.2	7.4599	.4344	2801.0	3061.6	7.3724	.3241	2797.2	3056.5	7.2328
350	.5701	2882.6	3167.7	7.6329	.4742	2881.2	3165.7	7.5464	.3544	2878.2	3161.7	7.4089
400	.6173	2963.2	3271.9	7.7938	.5137	2962.1	3270.3	7.7079	.3843	2959.7	3267.1	7.5716
500	.7109	3128.4	3483.9	8.0873	.5920	3127.6	3482.8	8.0021	.4433	3126.0	3480.6	7.8673
600	.8041	3299.6	3701.7	8.3522	.6697	3299.1	3700.9	8.2674	.5018	3297.9	3699.4	8.1333
700	.8969	3477.5	3925.9	8.5952	.7472	3477.0	3925.3	8.5107	.5601	3476.2	3924.2	8.3770
800	.9896	3662.1	4156.9	8.8211	.8245	3661.8	4156.5	8.7367	.6181	3661.1	4155.6	8.6033
900	1.0822	3853.6	4394.7	9.0329	.9017	3853.4	4394.4	8.9486	.6761	3852.8	4393.7	8.8153
1000	1.1747	4051.8	4639.1	9.2328	.9788	4051.5	4638.8	9.1485	.7340	4051.0	4638.2	9.0153
1100	1.2672	4256.3	4889.9	9.4224	1.0559	4256.1	4889.6	9.3381	.7919	4255.6	4889.1	9.2050
1200	1.3596	4466.8	5146.6	9.6029	1.1330	4466.5	5146.3	9.5185	.8497	4466.1	5145.9	9.3855
1300	1.4521	4682.5	5408.6	9.7749	1.2101	4682.3	5408.3	9.6906	.9076	4681.8	5407.9	9.5575

families of constant molar entropy s, of constant molar enthalpy h, of constant coefficient of thermal expansion α, of constant κ_T, and the like are also overlaid. The limit is set by readability of the chart.

It will be recognized that there is nothing unique about the variables assigned to the cartesian axes. Each family of curves serves as a (curvilinear) coordinate system. Thus a point of given v and s can be located as the intersection of the corresponding isochore and adiabat, and the value of any other plotted variable can then be read.

In practice there are many variants of thermodynamic charts in use. A popular type of chart is known as a Mollier chart —it assigns the molar enthalpy h and the molar entropy s to the cartesian axes; whereas the isochores and isobars appear as families of curves overlaid on the diagram. Another frequently used form of chart (a "temperature–entropy chart") assigns the temperature and the entropy to the coordinate axes, and overlays the molar enthalpy h and various other thermodynamic functions, the number again being limited mainly by readability (Figure 6.5).

Such full thermodynamic data is available for only a few systems, of relatively simple composition. For most systems only partial thermodynamic data are available. A very large scale international program on data compilation exists. The *International Journal of Thermophysics* (Plenum Press, New York and London) provides current reports of thermophysical measurements. The Center for Information and Numerical Data Analysis and Synthesis ("CINDAS"), located at Purdue University, publishes several series of data collections; of particular note is the *Thermophysical Properties Research Literature Retrieval Guide: 1900–1980*, (seven volumes) edited by J. F. Chancy and V. Ramdas (Plenum Publishing Corp., New York, 1982).

Finally, we briefly recall the procedure by which a fundamental equation for a single-component system can be constructed from minimal tabulated or measured data. The minimal information required is $\alpha(T, P)$, $c_p(T, P)$, and $\kappa_T(T, P)$, plus the values of v_0, s_0 in one reference state (and perhaps the enthalpy of formation). Given these data the molar Gibbs potential can be obtained by numerical integration of the Gibbs–Duhem relation $d(G/N) = -s\,dT + v\,dP$—but only after preliminary evaluations of $s(T, P)$ and $v(T, P)$ by numerical integration of the equations

$$ds = \left(\frac{\partial s}{\partial T}\right)_P dT + \left(\frac{\partial s}{\partial P}\right)_T dP = \frac{c_p}{T} dT - v\alpha\,dP$$

and

$$dv = v\alpha\,dT - v\kappa_T\,dP$$

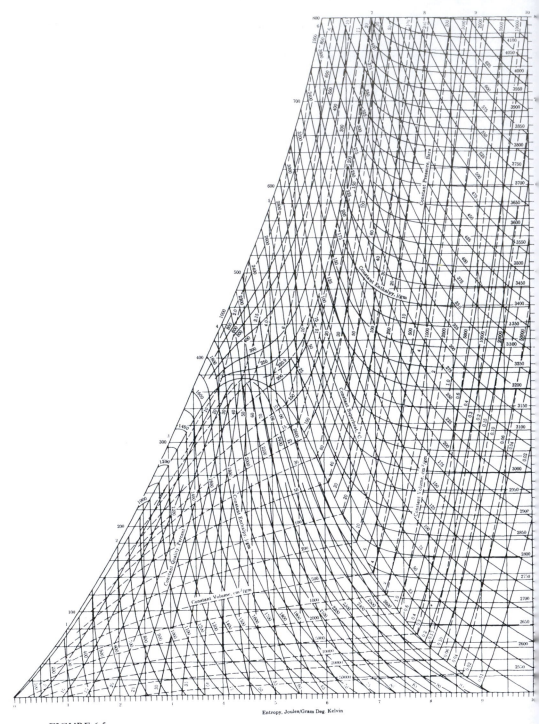

FIGURE 6.5
Temperature–entropy chart for water vapor ("steam"). From Keenan, Keyes, Hill and Moore, *Steam Tables*, copyright © 1969, John Wiley and Sons, Inc.

Note that "quality" is defined as the mole fraction in the gaseous state (in the two-phase region of the diagram).

Each of these integrations must be carried out over a network of paths covering the entire T–P plane—often a gigantic numerical undertaking.

6-7 THE MAXIMUM PRINCIPLES FOR THE MASSIEU FUNCTIONS

In the energy representation the energy is minimum for constant entropy, and from this it follows that each Legendre transform of the energy is minimum for constant values of the transformed (intensive) variables. Similarly, in the entropy representation the entropy is maximum for constant energy, and from this it follows that each Legendre transform of the entropy is maximum for constant values of the transformed (intensive) variables.

For two of the three common Massieu functions the maximum principles can be very easily obtained, for these functions are directly related to potentials (i.e., to transforms of the energy). By equation 5.61, we have

$$S\left[\frac{1}{T}\right] = -\frac{F}{T} \tag{6.60}$$

and, as F is minimum at constant temperature, $S[1/T]$ is clearly maximum. Again, by equation 5.63,

$$S\left[\frac{1}{T}, \frac{P}{T}\right] = -\frac{G}{T} \tag{6.61}$$

and, as G is minimum at constant pressure and temperature, $S[1/T, P/T]$ is clearly maximum.

For the remaining common Massieu function $S[P/T]$ we can repeat the logic of Section 6.1. We are concerned with a system in contact with a reservoir that maintains P/T constant, but permits $1/T$ to vary. It is readily recognized that such a reservoir is more of a mathematical fiction than a physically practical device, and the extremum principle for the function $S[P/T]$ is correspondingly artificial. Nevertheless, the derivation of this principle along the lines of Section 6.1 is an interesting exercise that I leave to the curious reader.

7

MAXWELL RELATIONS

7-1 THE MAXWELL RELATIONS

In Section 3.6 we observed that quantities such as the isothermal compressibility, the coefficient of thermal expansion, and the molar heat capacities describe properties of physical interest. Each of these is essentially a derivative $(\partial X / \partial Y)_{Z, W, \ldots}$ in which the variables are either extensive or intensive thermodynamic parameters. With a wide range of extensive and intensive parameters from which to choose, in general systems, the number of such possible derivatives is immense. But there are relations among such derivatives, so that a relatively small number of them can be considered as independent; all others can be expressed in terms of these few. Needless to say such relationships enormously simplify thermodynamic analyses. Nevertheless the relationships need not be memorized. There is a simple, straightforward procedure for producing the appropriate relationships as needed in the course of a thermodynamic calculation. That procedure is the subject of this chapter.

As an illustration of the existence of such relationships we recall equations 3.70 to 3.71

$$\frac{\partial^2 U}{\partial S \, \partial V} = \frac{\partial^2 U}{\partial V \, \partial S} \tag{7.1}$$

or

$$-\left(\frac{\partial P}{\partial S} \right)_{V, N_1, N_2, \ldots} = \left(\frac{\partial T}{\partial V} \right)_{S, N_1, N_2, \ldots} \tag{7.2}$$

This relation is the prototype of a whole class of similar equalities known as the *Maxwell relations*. These relations arise from the equality of the mixed partial derivatives of the fundamental relation expressed in any of the various possible alternative representations.

Given a particular thermodynamic potential, expressed in terms of its $(t + 1)$ natural variables, there are $t(t + 1)/2$ separate pairs of mixed second derivatives. Thus each potential yields $t(t + 1)/2$ Maxwell relations.

For a single-component simple system the internal energy is a function of three variables $(t = 2)$, and the three $[= (2 \cdot 3)/2]$ pairs of mixed second derivatives are $\partial^2 U/\partial S\, \partial V = \partial^2 U/\partial V\, \partial S$, $\partial^2 U/\partial S\, \partial N = \partial^2 U/\partial N\, \partial S$, and $\partial^2 U/\partial V\, \partial N = \partial^2 U/\partial N\, \partial V$. The complete set of Maxwell relations for a single-component simple system is given in the following listing, in which the first column states the potential from which the relation derives, the second column states the pair of independent variables with respect to which the mixed partial derivatives are taken, and the last column states the Maxwell relations themselves. A mnemonic diagram to be described in Section 7.2 provides a mental device for recalling relations of this form. In Section 7.3 we present a procedure for utilizing these relations in the solution of thermodynamic problems.

U	S, V	$\left(\dfrac{\partial T}{\partial V}\right)_{S,N} = -\left(\dfrac{\partial P}{\partial S}\right)_{V,N}$	(7.3)
$dU = T\,dS - P\,dV + \mu\,dN$	S, N	$\left(\dfrac{\partial T}{\partial N}\right)_{S,V} = \left(\dfrac{\partial \mu}{\partial S}\right)_{V,N}$	(7.4)
	V, N	$-\left(\dfrac{\partial P}{\partial N}\right)_{S,V} = \left(\dfrac{\partial \mu}{\partial V}\right)_{S,N}$	(7.5)
$U[T] \equiv F$	T, V	$\left(\dfrac{\partial S}{\partial V}\right)_{T,N} = \left(\dfrac{\partial P}{\partial T}\right)_{V,N}$	(7.6)
$dF = -S\,dT - P\,dV + \mu\,dN$	T, N	$-\left(\dfrac{\partial S}{\partial N}\right)_{T,V} = \left(\dfrac{\partial \mu}{\partial T}\right)_{V,N}$	(7.7)
	V, N	$-\left(\dfrac{\partial P}{\partial N}\right)_{T,V} = \left(\dfrac{\partial \mu}{\partial V}\right)_{T,N}$	(7.8)
$U[P] \equiv H$	S, P	$\left(\dfrac{\partial T}{\partial P}\right)_{S,N} = \left(\dfrac{\partial V}{\partial S}\right)_{P,N}$	(7.9)
$dH = T\,dS + V\,dP + \mu\,dN$	S, N	$\left(\dfrac{\partial T}{\partial N}\right)_{S,P} = \left(\dfrac{\partial \mu}{\partial S}\right)_{P,N}$	(7.10)
	P, N	$\left(\dfrac{\partial V}{\partial N}\right)_{S,P} = \left(\dfrac{\partial \mu}{\partial P}\right)_{S,N}$	(7.11)
$U[\mu]$	S, V	$\left(\dfrac{\partial T}{\partial V}\right)_{S,\mu} = -\left(\dfrac{\partial P}{\partial S}\right)_{V,\mu}$	(7.12)
$dU[\mu] = T\,dS - P\,dV - N\,d\mu$	S, μ	$\left(\dfrac{\partial T}{\partial \mu}\right)_{S,V} = -\left(\dfrac{\partial N}{\partial S}\right)_{V,\mu}$	(7.13)
	V, μ	$\left(\dfrac{\partial P}{\partial \mu}\right)_{S,V} = \left(\dfrac{\partial N}{\partial V}\right)_{S,\mu}$	(7.14)

$U[T, P] \equiv G$	T, P	$-\left(\dfrac{\partial S}{\partial P}\right)_{T,N} = \left(\dfrac{\partial V}{\partial T}\right)_{P,N}$ (7.15)
$dG = -S\,dT + V\,dP + \mu\,dN$	T, N	$-\left(\dfrac{\partial S}{\partial N}\right)_{T,P} = \left(\dfrac{\partial \mu}{\partial T}\right)_{P,N}$ (7.16)
	P, N	$\left(\dfrac{\partial V}{\partial N}\right)_{T,P} = \left(\dfrac{\partial \mu}{\partial P}\right)_{T,N}$ (7.17)
$U[T, \mu]$	T, V	$\left(\dfrac{\partial S}{\partial V}\right)_{T,\mu} = \left(\dfrac{\partial P}{\partial T}\right)_{V,\mu}$ (7.18)
$dU[T,\mu] = -S\,dT - P\,dV$	T, μ	$\left(\dfrac{\partial S}{\partial \mu}\right)_{T,V} = \left(\dfrac{\partial N}{\partial T}\right)_{V,\mu}$ (7.19)
$\qquad -N\,d\mu$		
	V, μ	$\left(\dfrac{\partial P}{\partial \mu}\right)_{T,V} = \left(\dfrac{\partial N}{\partial V}\right)_{T,\mu}$ (7.20)
$U[P, \mu]$	S, P	$\left(\dfrac{\partial T}{\partial P}\right)_{S,\mu} = \left(\dfrac{\partial V}{\partial S}\right)_{P,\mu}$ (7.21)
$dU[P,\mu] = T\,dS + V\,dP + N\,d\mu$	S, μ	$\left(\dfrac{\partial T}{\partial \mu}\right)_{S,P} = -\left(\dfrac{\partial N}{\partial S}\right)_{P,\mu}$ (7.22)
	P, μ	$\left(\dfrac{\partial V}{\partial \mu}\right)_{S,P} = -\left(\dfrac{\partial N}{\partial P}\right)_{S,\mu}$ (7.23)

7-2   A THERMODYNAMIC MNEMONIC DIAGRAM

A number of the most useful Maxwell relations can be remembered conveniently in terms of a simple mnemonic diagram.[1] This diagram, given in Fig. 7.1, consists of a square with arrows pointing upward along the two diagonals. The sides are labeled with the four common thermodynamic potentials, F, G, H, and U, in alphabetical order clockwise around the diagram, the Helmholtz potential F at the top. The two corners at the left are labeled with the extensive parameters V and S, and the two corners at the right are labeled with the intensive parameters T and P. ("**Valid Facts and Theoretical Understanding Generate Solutions to Hard Problems**" suggests the sequence of the labels.)

Each of the four thermodynamic potentials appearing on the square is flanked by its natural independent variables. Thus U is a natural function of V and S; F is a natural function of V and T; and G is a natural function of T and P. Each of the potentials also depends on the mole numbers, which are not indicated explicitly on the diagram.

---

[1] This diagram was presented by Professor Max Born in 1929 in a lecture heard by Professor Tisza. It appeared in the literature in a paper by F. O. Koenig, *J. Chem. Phys.* **3**, 29 (1935), and **56**, 4556 (1972). See also L. T. Klauder, *Am. Journ. Phys.* **36**, 556 (1968), and a number of other variants presented by a succession of authors in this journal.

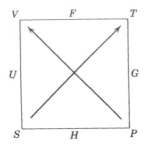

FIGURE 7.1
The thermodynamic square.

In the differential expression for each of the potentials, in terms of the differentials of its natural (flanking) variables, the associated algebraic sign is indicated by the diagonal arrow. An arrow pointing away from a natural variable implies a positive coefficient, whereas an arrow pointing toward a natural variable implies a negative coefficient. This scheme becomes evident by inspection of the diagram and of each of the following equations:

$$dU = T\,dS - P\,dV + \sum_k \mu_k\,dN_k \tag{7.24}$$

$$dF = -S\,dT - P\,dV + \sum_k \mu_k\,dN_k \tag{7.25}$$

$$dG = -S\,dT + V\,dP + \sum_k \mu_k\,dN_k \tag{7.26}$$

$$dH = T\,dS + V\,dP + \sum_k \mu_k\,dN_k \tag{7.27}$$

Finally the Maxwell relations can be read from the diagram. We then deal only with the corners of the diagram. The labeling of the four corners of the square can easily be seen to be suggestive of the relationship

$$\left(\frac{\partial V}{\partial S}\right)_P = \left(\frac{\partial T}{\partial P}\right)_S \qquad (\text{constant } N_1, N_2, \dots) \tag{7.28}$$

By mentally rotating the square on its side, we find, by exactly the same construction

$$\left(\frac{\partial S}{\partial P}\right)_T = -\left(\frac{\partial V}{\partial T}\right)_P \qquad (\text{constant } N_1, N_2, \dots) \tag{7.29}$$

The minus sign in this equation is to be inferred from the unsymmetrical placement of the arrows in this case. The two remaining rotations of the square give the two additional Maxwell relations

$$\left(\frac{\partial P}{\partial T}\right)_V = \left(\frac{\partial S}{\partial V}\right)_T \qquad \text{(constant } N_1, N_2, \dots) \qquad (7.30)$$

and

$$\left(\frac{\partial T}{\partial V}\right)_S = -\left(\frac{\partial P}{\partial S}\right)_V \qquad \text{(constant } N_1, N_2, \dots) \qquad (7.31)$$

These are the four most useful Maxwell relations in the conventional applications of thermodynamics.

The mnemonic diagram can be adapted to pairs of variables other than S and V. If we are interested in Legendre transformations dealing with S and N_j, the diagram takes the form shown in Fig. 7.2a. The arrow connecting N_j and μ_j has been reversed in relation to that which previously connected V and P to ake into account the fact that μ_j is analogous to $-P$. Equations 7.4, 7.7, 7.13, and 7.19 can be read directly from this diagram. Other diagrams can be constructed in a similar fashion, as indicated in the general case in Fig. 7.2b.

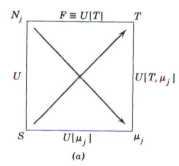

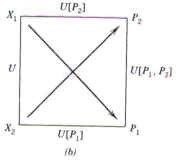

(a)   (b)

FIGURE 7.2

PROBLEMS

7.2-1. In the immediate vicinity of the state T_0, v_0 the volume of a particular system of 1 mole is observed to vary according to the relationship

$$v = v_0 + a(T - T_0) + b(P - P_0)$$

Calculate the transfer of heat dQ to the system if the molar volume is changed by a small increment $dv = v - v_0$ at constant temperature T_0.

Answer:
$$dQ = T\left(\frac{\partial S}{\partial V}\right)_T dV = T\left(\frac{\partial P}{\partial T}\right)_V dV = -\frac{aT}{b}dV$$

7.2-2. For a particular system of 1 mole, in the vicinity of a particular state, a change of pressure dP at constant T is observed to be accompanied by a heat flux $dQ = A\, dP$. What is the value of the coefficient of thermal expansion of this system, in the same state?

7.2-3. Show that the relation

$$\alpha = \frac{1}{T}$$

implies that c_P is independent of the pressure

$$\left(\frac{\partial c_P}{\partial P} \right)_T = 0$$

7-3 A PROCEDURE FOR THE REDUCTION OF DERIVATIVES IN SINGLE-COMPONENT SYSTEMS

In the practical applications of thermodynamics the experimental situation to be analyzed frequently dictates a partial derivative to be evaluated. For instance, we may be concerned with the analysis of the temperature change that is required to maintain the volume of a single-component system constant if the pressure is increased slightly. This temperature change is evidently

$$dT = \left(\frac{\partial T}{\partial P} \right)_{V,N} dP \qquad (7.32)$$

and consequently we are interested in an evaluation of the derivative $(\partial T / \partial P)_{V,N}$. A number of similar problems will be considered in Section 7.4. A general feature of the derivatives that arise in this way is that they are likely to involve constant mole numbers and that they generally involve both intensive and extensive parameters. *Of all such derivatives, only three can be independent, and any given derivative can be expressed in terms of an arbitrarily chosen set of three basic derivatives.* This set is conventionally chosen as c_P, α, and κ_T.

The choice of c_p, α, and κ_T is an implicit transformation to the Gibbs representation, for the three second derivatives in this representation are $\partial^2 g / \partial T^2$, $\partial^2 g / \partial T\, \partial P$, and $\partial^2 g / \partial P^2$; these derivatives are equal, respectively, to $-c_p / T$, $v\alpha$, and $-v\kappa_T$. For constant mole numbers these are the only independent second derivatives.

All first derivatives (involving both extensive and intensive parameters) can be written in terms of second derivatives of the Gibbs potential, of which we have now seen that c_p, α, and κ_T constitute a complete independent set (at constant mole numbers).

The procedure to be followed in this "reduction of derivatives" is straightforward in principle; the entropy S need only be replaced by

$-\partial G/\partial T$ and V must be replaced by $\partial G/\partial P$, thereby expressing the original derivative in terms of second derivatives of G with respect to T and P. In practice this procedure can become somewhat involved.

It is essential that the student of thermodynamics become thoroughly proficient in the "reduction of derivatives." To that purpose we present a procedure, based upon the "mnemonic square" and organized in a step by step recipe that accomplishes the reduction of any given derivative. Students are urged to do enough exercises of this type so that the procedure becomes automatic.

Consider a partial derivative involving constant mole numbers. It is desired to express this derivative in terms of c_P, α, and κ_T. We first recall the following identities which are to be employed in the mathematical manipulations (see Appendix A).

$$\left(\frac{\partial X}{\partial Y}\right)_Z = 1 \bigg/ \left(\frac{\partial Y}{\partial X}\right)_Z \tag{7.33}$$

and

$$\left(\frac{\partial X}{\partial Y}\right)_Z = \left(\frac{\partial X}{\partial W}\right)_Z \bigg/ \left(\frac{\partial Y}{\partial W}\right)_Z \tag{7.34}$$

$$\left(\frac{\partial X}{\partial Y}\right)_Z = -\left(\frac{\partial Z}{\partial Y}\right)_X \bigg/ \left(\frac{\partial Z}{\partial X}\right)_Y \tag{7.35}$$

The following steps are then to be taken in order:

1. **If the derivative contains any potentials, bring them one by one to the numerator and eliminate by the thermodynamic square** (equations 7.24 to 7.27).

Example

Reduce the derivative $(\partial P/\partial U)_{G,N}$.

$$\left(\frac{\partial P}{\partial U}\right)_{G,N} = \left[\left(\frac{\partial U}{\partial P}\right)_{G,N}\right]^{-1} \tag{by 7.33}$$

$$= \left[T\left(\frac{\partial S}{\partial P}\right)_{G,N} - P\left(\frac{\partial V}{\partial P}\right)_{G,N}\right]^{-1} \tag{by 7.24}$$

$$= \left[-T\left(\frac{\partial G}{\partial P}\right)_{S,N} \bigg/ \left(\frac{\partial G}{\partial S}\right)_{P,N} + P\left(\frac{\partial G}{\partial P}\right)_{V,N} \bigg/ \left(\frac{\partial G}{\partial V}\right)_{P,N}\right]^{-1} \tag{by 7.35}$$

$$= \left[-T\frac{-S(\partial T/\partial P)_{S,N} + V}{-S(\partial T/\partial S)_{P,N}} + P\frac{-S(\partial T/\partial P)_{V,N} + V}{-S(\partial T/\partial V)_{P,N}}\right]^{-1} \tag{by 7.26}$$

The remaining expression does not contain any potentials but may involve a number of derivatives. Choose these one by one and treat each according to the following procedure.

2. **If the derivative contains the chemical potential, bring it to the numerator and eliminate by means of the Gibbs–Duhem relation,** $d\mu = -s\,dT + v\,dP$.

Example
Reduce $(\partial\mu/\partial V)_{S,N}$.

$$\left(\frac{\partial\mu}{\partial V}\right)_{S,N} = -s\left(\frac{\partial T}{\partial V}\right)_{S,N} + v\left(\frac{\partial P}{\partial V}\right)_{S,N}$$

3. **If the derivative contains the entropy, bring it to the numerator. If one of the four Maxwell relations of the thermodynamic square now eliminates the entropy, invoke it. If the Maxwell relations do not eliminate the entropy put a ∂T under ∂S (employ equation 7.34 with $w = T$). The numerator will then be expressible as one of the specific heats** (either c_v or c_P).

Example
Consider the derivative $(\partial T/\partial P)_{S,N}$ appearing in the example of step 1:

$$\left(\frac{\partial T}{\partial P}\right)_{S,N} = -\left(\frac{\partial S}{\partial P}\right)_{T,N}\bigg/\left(\frac{\partial S}{\partial T}\right)_{P,N} \qquad \text{(by 7.35)}$$

$$= \left(\frac{\partial V}{\partial T}\right)_{P,N}\bigg/\frac{N}{T}c_P \qquad \text{(by 7.29)}$$

Example
Consider the derivative $(\partial S/\partial V)_{P,N}$. The Maxwell relation would give $(\partial S/\partial V)_{P,N} = (\partial P/\partial T)_{S,N}$ (equation 7.28), which would not eliminate the entropy. We therefore do not invoke the Maxwell relation but write

$$\left(\frac{\partial S}{\partial V}\right)_{P,N} = \frac{(\partial S/\partial T)_{P,N}}{(\partial V/\partial T)_{P,N}} = \frac{(N/T)c_P}{(\partial V/\partial T)_{P,N}} \qquad \text{(by 7.34)}$$

The derivative now contains neither any potential nor the entropy. It consequently contains only V, P, T (and N).

4. **Bring the volume to the numerator. The remaining derivative will be expressible in terms of α and κ_T.**

Example
Given $(\partial T/\partial P)_{V,N}$

$$\left(\frac{\partial T}{\partial P}\right)_{V,N} = -\left(\frac{\partial V}{\partial P}\right)_{T,N}\bigg/\left(\frac{\partial V}{\partial T}\right)_{P,N} = \frac{\kappa_T}{\alpha} \qquad \text{(by 7.35)}$$

5. **The originally given derivative has now been expressed in terms of the four quantities** c_v, c_P, α, **and** κ_T. **The specific heat at constant volume is eliminated by the equation**

$$c_v = c_P - Tv\alpha^2/\kappa_T \qquad (7.36)$$

This useful relation, which should be committed to memory, was alluded to in equation 3.75. The reader should be able to derive it as an exercise (see Problem 7.3-2).

This method of reduction of derivatives can be applied to multicomponent systems as well as to single-component systems, provided that the chemical potentials μ_j do not appear in the derivative (for the Gibbs–Duhem relation, which eliminates the chemical potential for single-component systems, merely introduces the chemical potentials of other components in multicomponent systems).

PROBLEMS

7.3-1. Thermodynamicists sometimes refer to the "first $T\,dS$ equation" and the "second $T\,dS$ equation";

$$T\,dS = Nc_v\,dT + (T\alpha/\kappa_T)\,dV \qquad (N \text{ constant})$$

$$T\,dS = Nc_P\,dT - TV\alpha\,dP \qquad (N \text{ constant})$$

Derive these equations.

7.3-2. Show that the second equation in the preceding problem leads directly to the relation

$$T\left(\frac{\partial s}{\partial T}\right)_v = c_P - Tv\alpha\left(\frac{\partial P}{\partial T}\right)_v$$

and so validates equation 7.36.

7.3-3. Calculate $(\partial H/\partial V)_{T,N}$ in terms of the standard quantities c_P, α, κ_T, T, and P.

Answer:
$$\left(\frac{\partial H}{\partial V}\right)_{T,N} = (T\alpha - 1)/\kappa_T$$

7.3-4. Reduce the derivative $(\partial v/\partial s)_P$.

7.3-5. Reduce the derivative $(\partial s/\partial f)_v$.

7.3-6. Reduce the derivative $(\partial s/\partial f)_P$.

7.3-7. Reduce the derivative $(\partial s/\partial v)_h$.

7-4   SOME SIMPLE APPLICATIONS

In this section we indicate several representative applications of the manipulations described in Section 7.3. In each case to be considered we first pose a problem. Typically, we are asked to find the change in one parameter when some other parameter is changed. Thus, in the simplest case, we might be asked to find the increase in the pressure of a system if its temperature is increased by ΔT, its volume being kept constant.

In the examples to be given we consider two types of solutions. First, the straightforward solution that assumes complete knowledge of the fundamental equation, and, second, the solution that can be obtained if c_P, α, and κ_T are assumed known and if the changes in parameters are small.

Adiabatic Compression

Consider a single-component system of some definite quantity of matter (characterized by the mole number N) enclosed within an adiabatic wall. The initial temperature and pressure of the system are known. The system is compressed quasi-statically so that the pressure increases from its initial value P_i to some definite final value P_f. We attempt to predict the changes in the various thermodynamic parameters (e.g., in the volume, temperature, internal energy, and chemical potential) of the system.

The essential key to the analysis of the problem is the fact that for a quasi-static process the adiabatic constraint implies constancy of the entropy. This fact follows, of course, from the quasi-static correspondence $dQ = T\,dS$.

We consider in particular the change in temperature. First, we assume the fundamental equation to be known. By differentiation, we can find the two equations of state $T = T(S, V, N)$ and $P = P(S, V, N)$. By knowing the initial temperature and pressure, we can thereby find the initial volume and entropy. Elimination of V between the two equations of state gives the temperature as a function of S, P, and N. Then, obviously,

$$\Delta T = T(S, P_f, N) - T(S, P_i, N) \tag{7.37}$$

If the fundamental equation is not known, but c_P, α, and κ_T are given, and if the pressure change is small, we have

$$dT = \left(\frac{\partial T}{\partial P} \right)_{S,N} dP \tag{7.38}$$

By the method of Section 7.3, we then obtain

$$dT = \frac{Tv\alpha}{c_P} dP \tag{7.39}$$

The change in chemical potential can be found similarly. Thus, for a small pressure change

$$d\mu = \left(\frac{\partial \mu}{\partial P}\right)_{S,N} dP \tag{7.40}$$

$$= \left(v - \frac{sTv\alpha}{c_P}\right) dP \tag{7.41}$$

The fractional change in volume associated with an (infinitesimal) adiabatic compression is characterized by the *adiabatic compressibility* κ_S, previously defined in equation 3.73. It was there stated that κ_S can be related to κ_T, c_p, and α (equation 3.76, and (see also Problem 3.9-5), an exercise that is now left to the reader in Problem 7.4-8.

Isothermal Compression

We now consider a system maintained at constant temperature and mole number and quasi-statically compressed from an initial pressure P_i to a final pressure P_f. We may be interested in the prediction of the changes in the values of U, S, V, and μ. By appropriate elimination of variables among the fundamental equation and the equations of state, any such parameter can be expressed in terms of T, P, and N, and the change in that parameter can then be computed directly.

For small changes in pressure we find

$$dS = \left(\frac{\partial S}{\partial P}\right)_{T,N} dP \tag{7.42}$$

$$= -\alpha V \, dP \tag{7.43}$$

also

$$dU = \left(\frac{\partial U}{\partial P}\right)_{T,N} dP \tag{7.44}$$

$$= (-T\alpha V + PV\kappa_T) \, dP \tag{7.45}$$

and similar equations exist for the other parameters.

One may inquire about the total quantity of heat that must be extracted from the system by the heat reservoir in order to keep the system at constant temperature during the isothermal compression. First, assume that the fundamental equation is known. Then

$$\Delta Q = T\Delta S = TS(T, P_f, N) - TS(T, P_i, N) \tag{7.46}$$

where $S(U, V, N)$ is reexpressed as a function of T, P, and N in standard fashion.

If the fundamental equation is not known we consider an infinitesimal isothermal compression, for which we have, from equation 7.43

$$dQ = -T\alpha V \, dP \qquad (7.47)$$

Finally, suppose that the pressure change is large, but that the fundamental equation is not known (so that the solution 7.46 is not available). Then, if α and V are known as functions of T and P, we integrate equation 7.47 at constant temperature

$$\Delta Q = -T \int_{P_i}^{P_f} \alpha V \, dP \qquad (7.48)$$

This solution must be equivalent to that given in equation 7.46.

Free Expansion

The third process we shall consider is a free expansion (recall Problems 3.4-8 and 4.2-3). The constraints that require the system to have a volume V_i are suddenly relaxed, allowing the system to expand to a volume V_f. If the system is a gas (which, of course, does not have to be the case), the expansion may be accomplished conveniently by confining the gas in one section of a rigid container, the other section of which is evacuated. If the septum separating the sections is suddenly fractured the gas spontaneously expands to the volume of the whole container. We seek to predict the change in the temperature and in the various other parameters of the system.

The total internal energy of the system remains constant during the free expansion. Neither heat nor work are transferred to the system by any external agency.

If the temperature is expressed in terms of U, V, and N, we find

$$T_f - T_i = T(U, V_f, N) - T(U, V_i, N) \qquad (7.49)$$

If the volume change is small

$$dT = \left(\frac{\partial T}{\partial V}\right)_{U, N} dV \qquad (7.50)$$

$$= \left(\frac{P}{Nc_v} - \frac{T\alpha}{Nc_v\kappa_T}\right) dV \qquad (7.51)$$

This process, unlike the two previously treated, is essentially irreversible and is not quasi-static (Problem 4.2-3).

Example

In practice the processes of interest rarely are so neatly defined as those just considered. No single thermodynamic parameter is apt to be constant in the process. More typically, measurements might be made of the temperature during the expansion stroke in the cylinder of an engine. The expansion is neither isothermal nor isentropic, for heat tends to flow uncontrolled through the cylinder walls. Nevertheless, the temperature can be evaluated empirically as a function of the volume, and this defines the process. Various other characterizations of real processes will occur readily to the reader, but the general methodology is well represented by the following particular example.

N moles of a material are expanded from V_1 to V_2 and the temperature is observed to decrease from T_1 to T_2, the temperature falling linearly with volume. Calculate the work done on the system and the heat transfer, expressing each result in terms of definite integrals of the tabulated functions c_p, α, and κ_T.

Solution

We first note that the tabulated functions $c_p(T, P)$, $\alpha(T, P)$, $\kappa_T(T, P)$, and $v(T, P)$ are redundant. The first three functions imply the last, as has already been shown in the example of Section 3.9.

Turning to the stated problem, the equation of the path in the T–V plane is

$$T = A + BV; \qquad A = (T_1 V_2 - T_2 V_1)/(V_2 - V_1); \qquad B = (T_2 - T_1)/(V_2 - V_1)$$

Furthermore, the pressure is known at each point on the path, for the known function $v(T, P)$ can be inverted to express P as a function of T and v, and thence of v alone

$$P = P(T, V) = P(A + BV, V)$$

The work done in the process is then

$$W = \int_{V_1}^{V_2} P(A + BV, V)\, dV$$

This integral must be performed numerically, but generally it is well within the capabilities of even a modest programmable hand calculator.

The heat input is calculated by considering S as a function of T and V.

$$dS = \left(\frac{\partial S}{\partial T}\right)_V dT + \left(\frac{\partial S}{\partial V}\right)_T dV$$

$$= \frac{N}{T} c_v \, dT + \left(\frac{\partial P}{\partial T}\right)_V dV$$

$$= \left(\frac{Nc_p}{T} - \frac{V\alpha^2}{\kappa_T}\right) dT + \frac{\alpha}{\kappa_T} dV$$

But on the path, $dT = B\, dV$, so that

$$dS = \left(NB\frac{c_p}{T} - \frac{BV\alpha^2}{\kappa_T} + \frac{\alpha}{\kappa_T}\right) dV$$

Thus the heat input is

$$Q = \int_{V_1}^{V_2} \left[N B c_p - (A + BV)(BV\alpha - 1)\alpha/\kappa_T \right] dV$$

Again the factors in the integral must be evaluated at the appropriate values of P and T corresponding to the point V on the path, and the integral over V must then be carried out numerically.

It is often convenient to approximate the given data by polynomial expressions in the region of interest; numerous packaged computer programs for such "fits" are available. Then the integrals can be evaluated either numerically or analytically.

Example

In the P–v plane of a particular substance, two states, A and D, are defined by

$$P_A = 10^5 \text{ Pa} \qquad v_A = 2 \times 10^{-2} \text{ m}^3/\text{mole}$$

$$P_D = 10^4 \text{ Pa} \qquad v_D = 10^{-1} \text{ m}^3/\text{mole}$$

and it is also ascertained that $T_A = 350.9$ K. If 1 mole of this substance is initially in the state A, and if a thermal reservoir at temperature 150 K is available, how much work can be delivered to a reversible work source in a process that leaves the system in the state D?

The following data are available. The adiabats of the system are of the form

$$Pv^2 = \text{constant} \qquad (\text{for } s = \text{constant})$$

Measurements of c_p and α are known only at the pressure of 10^5 Pa.

$$c_p = Bv^{2/3} \qquad (\text{for } P = 10^5 \text{ Pa});$$

$$B = 10^{8/3} = 464.2 \text{ J/m}^2\text{K}$$

$$\alpha = 3/T \qquad (\text{for } P = 10^5 \text{ Pa})$$

and no measurements of κ_T are available.

The reader is strongly urged to analyze this problem independently before reading the following solution.

Solution

In order to assess the maximum work that can be delivered in a reversible process $A \rightarrow D$ it is necessary only to know $u_D - u_A$ and $s_D - s_A$.

The adiabat that passes through the state D is described by $Pv^2 = 10^2$ Pa $\cdot$ m^6; it intersects the isobar $P = 10^5$ Pa at a point C for which

$$P_C = 10^5 \text{ Pa} \qquad v_c = 10^{-3/2} \text{ m}^3 = 3.16 \times 10^{-2} \text{ m}^3$$

As a two-step quasi-static process joining A and D we choose the isobaric process $A \to C$ followed by the isentropic process $C \to D$. By considering these two processes in turn we seek to evaluate first $u_C - u_A$ and $s_C - s_A$ and then $u_D - u_C$ and $s_D - s_C$, yielding finally $u_D - u_A$ and $s_D - s_A$.

We first consider the isobaric process $A \to C$.

$$du = T\,ds - P\,dv = \left(\frac{c_p}{v\alpha} - P \right) dv = \left(\frac{1}{3} Bv^{-1/3}T - P_A \right) dv$$

We cannot integrate this directly for we do not yet know $T(v)$ along the isobar. To calculate $T(v)$ we write

$$\left(\frac{\partial T}{\partial v} \right)_P = \frac{1}{v\alpha} = \frac{T}{3v} \qquad (\text{for } P = P_A)$$

or integrating

$$\ln\left(\frac{T}{T_A} \right) = \frac{1}{3}\ln\left(\frac{v}{v_A} \right)$$

and

$$T = 350.9 \times (50v)^{1/3} \qquad (\text{on } P = 10^5 \text{ Pa isobar})$$

Returning now to the calculation of $u_C - u_A$

$$du = \left[\tfrac{1}{3}B \times 350.9 \times (50)^{1/3} - 10^5 \right] dv \approx 10^5\, dv$$

or

$$u_C - u_A = 10^5 \times (v_C - v_A) = 1.16 \times 10^3 \text{ J}$$

We now require the difference $u_D - u_C$. Along the adiabat we have

$$u_D - u_C = -\int_{v_C}^{v_D} P\,dv = -10^2 \int_{v_C}^{v_D} \frac{dv}{v^2} = 10^2 \left[v_D^{-1} - v_C^{-1} \right] = -2.16 \times 10^3 \text{ J}$$

Finally, then, we have the required energy difference

$$u_D - u_A = -10^3 \text{ J}$$

We now turn our attention to the entropy difference $s_D - s_A = s_C - s_A$. Along the isobar AC

$$ds = \left(\frac{\partial s}{\partial v} \right)_P dv = \frac{c_p}{Tv\alpha}\,dv = \frac{1}{3}Bv^{-1/3}\,dv$$

and

$$s_D - s_A = s_C - s_A = \tfrac{1}{2}B\left[v_C^{2/3} - v_A^{2/3} \right] = 6.1 \text{ J/K}$$

Knowing Δu and Δs for the process, we turn to the problem of delivering maximum work. The increase in entropy of the system permits us to *extract* energy from the thermal reservoir.

$$(-Q_{\text{res}}) = T_{\text{res}}\Delta s = 150 \times 6.1 = 916 \text{ J}$$

The total energy that can then be delivered to the reversible work source is $(-\Delta u) + (-Q_{\text{res}})$, or

$$\text{work delivered} = 1.92 \times 10^3 \text{ J}$$

PROBLEMS

7.4-1. In the analysis of a Joule–Thomson experiment we may be given the initial and final molar volumes of the gas, rather than the initial and final pressures. Express the derivative $(\partial T/\partial v)_h$ in terms of c_p, α, and κ_T.

7.4-2. The adiabatic bulk modulus is defined by

$$\beta_S = -v\left(\frac{\partial P}{\partial v}\right)_s = -V\left(\frac{\partial P}{\partial V}\right)_{S,N}$$

Express this quantity in terms of c_p, c_v, α, and κ_T (do not eliminate c_p). What is the relation of your result to the identity $\kappa_s/\kappa_T = c_v/c_p$ (recall Problem 3.9-5)?

7.4-3. Evaluate the change in temperature in an infinitesimal free expansion of a simple ideal gas (equation 7.51). Does this result also hold if the change in volume is comparable to the initial volume? Can you give a more general argument for a simple ideal gas, not based on equation 7.51?

7.4-4. Show that equation 7.46 can be written as

$$Q = U_f[P,\mu] - U_i[P,\mu]$$

so that $U[P,\mu]$ can be interpreted as a "potential for heat at constant T and N."

7.4-5. A 1% decrease in volume of a system is carried out adiabatically. Find the change in the chemical potential in terms of c_p, α, and κ_T (and the state functions P, T, u, v, s, etc).

7.4-6. Two moles of an imperfect gas occupy a volume of 1 liter and are at a temperature of 100 K and a pressure of 2 MPa. The gas is allowed to expand freely into an additional volume, initially evacuated, of 10 cm³. Find the change in enthalpy.

At the initial conditions $c_p = 0.8$ J/mole · K, $\kappa_T = 3 \times 10^{-6}$ Pa⁻¹, and $\alpha = 0.002$ K⁻¹.

Answer:

$$\Delta H = -\left[\frac{P - (c_p - Pv\alpha)}{(c_p\kappa_T - Tv\alpha^2)}\right]\Delta v = 15 \text{ J}$$

7.4-7. Show that $(\partial c_v/\partial v)_T = T(\partial^2 P/\partial T^2)_v$ and evaluate this quantity for a system obeying the van der Waals equation of state.

7.4-8. Show that

$$\left(\frac{\partial c_p}{\partial P}\right)_T = -Tv\left[\alpha^2 + \left(\frac{\partial \alpha}{\partial T}\right)_P\right]$$

Evaluate this quantity for a system obeying the equation of state

$$P\left(v + \frac{A}{T^2}\right) = RT$$

7.4-9. One mole of the system of Problem 7.4-8 is expanded isothermally from an initial pressure P_0 to a final pressure P_f. Calculate the heat flux to the system in this process.

Answer:

$$Q = -RT \ln\left(\frac{P_f}{P_i}\right) - 2A(P_f - P_i)/T^2$$

7.4-10. A system obeys the van der Waals equation of state. One mole of this system is expanded isothermally at temperature T from an initial volume v_0 to a final volume v_f. Find the heat transfer to the system in this expansion.

7.4-11. Two moles of O_2 are initially at a pressure of 10^5 Pa and a temperature of $0°C$. An adiabatic compression is carried out to a final temperature of $300°C$. Find the final pressure by integration of equation 7.39. Assume that O_2 is a simple ideal gas with a molar heat capacity c_p which can be represented by

$$c_p = 26.20 + 11.49 \times 10^{-3}T - 3.223 \times 10^{-6}T^2$$

where c_p is in J/mole and T is in kelvins.

Answer:
$$P_f \simeq 15 \times 10^5 \text{ Pa}$$

7.4-12. A ball bearing of mass 10 g just fits in a vertical glass tube of cross-sectional area 2 cm². The bottom of the tube is connected to a vessel of volume 5 liters, filled with oxygen at a temperature of 30°C. The top of the tube is open to the atmosphere, which is at a pressure of 10^5 Pa and a temperature of 30°C. What is the period of vertical oscillation of the ball? Assume that the compressions and expansions of the oxygen are slow enough to be essentially quasi-static but fast enough to be adiabatic. Assume that O_2 is a simple ideal gas with a molar heat capacity as given in Problem 7.4-11.

7.4-13. Calculate the change in the molar internal energy in a throttling process in which the pressure change is dP, expressing the result in terms of standard parameters.

7.4-14. Assuming that a gas undergoes a free expansion and that the temperature is found to change by dT, calculate the difference dP between the initial and final pressure.

7.4-15. One mole of an ideal van der Waals fluid is contained in a vessel of volume V_i at temperature T_i. A valve is opened, permitting the fluid to expand into an initially evacuated vessel, so that the final volume is V_f. The walls of the vessels are adiabatic. Find the final temperature T_f.

Evaluate your result for $V_i = 2 \times 10^{-3}$ m³, $V_f = 5 \times 10^{-3}$ m³, $N = 1$, $T_i = 300$ K, and the van der Waals constants are those of argon (Table 3.1). What was the initial pressure of the gas?

7.4-16. Assuming the expansion of the ideal van der Waals fluid of Problem 7.4-15 to be carried out quasi-statically and adiabatically, again find the final temperature T_f.

Evaluate your result with the numerical data specified in Problem 7.4-15.

7.4-17. It is observed that an adiabatic decrease in molar volume of 1% produces a particular change in the chemical potential μ. What percentage change in molar volume, carried out isothermally, produces the same change in μ?

7.4-18. A cylinder is fitted with a piston, and the cylinder contains helium gas. The sides of the cylinder are adiabatic, impermeable, and rigid, but the bottom of the cylinder is thermally conductive, permeable to helium, and rigid. Through this permeable wall the system is in contact with a reservoir of constant T and μ_{He} (the chemical potential of He). Calculate the compressibility of the system $[-(1/V)(dV/dP)]$ in terms of the properties of helium (c_p, v, α, κ_T, etc.) and thereby demonstrate that this compressibility diverges. Discuss the physical reason for this divergence.

7.4-19. The cylinder in Problem 7.4-18 is initially filled with $\frac{1}{10}$ mole of Ne. Assume both He and Ne to be monatomic ideal gases. The bottom of the cylinder is again permeable to He, but not to Ne. Calculate the pressure in the cylinder and the compressibility $(-1/V)(dV/dP)$ as functions of T, V, and μ_{He}.
Hint: Recall Problems 5.3-1, 5.3-10, and 6.2-3.

7.4-20. A system is composed of 1 mole of a particular substance. In the P–v plane two states (A and B) lie on the locus $Pv^2 =$ constant, so that $P_A v_A^2 = P_B v_B^2$. The following properties of the system have been measured *along this locus*: $c_p = Cv^2$, $\alpha = D/v$, and $\kappa_T = Ev$, where C, D, and E are constants. Calculate the temperature T_B in terms of T_A, P_A, v_A, v_B, and the constants C, D, and E.

Answer:

$$T_B = T_A + (v_B - v_A)/D + 2EP_A v_A^2 D^{-1} \ln(v_B/v_a)$$

7.4-21. A system is composed of 1 mole of a particular substance. Two thermodynamic states, designated as A and B, lie on the locus $Pv =$ constant. The following properties of the system have been measured *along this locus*; $c_p = Cv$, $\alpha = D/v^2$, and $\kappa_T = Ev$, where C, D, and E are constants. Calculate the difference in molar energies $(u_B - u_A)$ in terms of T_A, P_A, v_A, v_B, and the constants C, D, and E.

7.4-22. The constant-volume heat capacity of a particular simple system is

$$c_v = AT^3 \qquad (A = \text{constant})$$

In addition the equation of state is known to be of the form

$$(v - v_0)P = B(T)$$

where $B(T)$ is an unspecified function of T. Evaluate the permissible functional form of $B(T)$.

In terms of the undetermined constants appearing in your functional representation of $B(T)$, evaluate α, c_p, and κ_T as functions of T and v.
Hint: Examine the derivative $\partial^2 s / \partial T\, \partial v$.

Answer:

$$c_p = AT^3 + (T^3/DT + E), \text{ where } D \text{ and } E \text{ are constants.}$$

7.4-23. A system is expanded along a straight line in the P–v plane, from the initial state (P_0, v_0) to the final state (P_f, v_f). Calculate the heat transfer per mole to the system in this process. It is to be assumed that α, κ_T, and c_p are known only along the isochore $v = v_0$ and the isobar $P = P_f$; in fact it is sufficient to specify that the quantity $(c_v \kappa_T / \alpha)$ has the value AP on the isochore $v = v_0$, and the quantity $(c_p / v\alpha)$ has the value Bv on the isobar $P = P_f$, where A and B are known constants. That is

$$\frac{c_v \kappa_T}{\alpha} = AP \qquad \left(\text{for } v = v_0\right)$$

$$\frac{c_p}{v\alpha} = Bv \qquad \left(\text{for } P = P_f\right)$$

Answer:

$$Q = \tfrac{1}{2}A(P_f^2 - P_0^2) + \tfrac{1}{2}B(v_f^2 - v_0^2) + \tfrac{1}{2}(P_0 - P_f)(v_f - v_0)$$

7.4-24. A nonideal gas undergoes a throttling process (i.e., a Joule–Thomson expansion) from an initial pressure P_0 to a final pressure P_f. The initial temperature is T_0 and the initial molar volume is v_0. Calculate the final temperature T_f if it is given that

$$\kappa_T = \frac{A}{v^2} \text{ along the } T = T_0 \text{ isotherm } (A > 0)$$

$$\alpha = \alpha_0 \text{ along the } T = T_0 \text{ isotherm}$$

and

$$c_p = c_p^0 \text{ along the } P = P_f \text{ isobar}$$

What is the condition on T_0 in order that the temperature be lowered by the expansion?

7-5 GENERALIZATIONS: MAGNETIC SYSTEMS

For systems other than simple systems there exists a complete parallelism to the formalism of Legendre transformation, of Maxwell relations, and of reduction of derivatives by the mnemonic square.

The fundamental equation of a magnetic system is of the form (recall Section 3.8 and Appendix B)

$$U = U(S, V, I, N) \tag{7.52}$$

Legendre transformations with respect to S, V, and N simply retain the magnetic moment I as a parameter. Thus the enthalpy is a function of S,

P, I, and N.

$$H \equiv U[P] = U + PV = H(S, P, I, N) \tag{7.53}$$

An analogous transformation can be made with respect to the magnetic coordinate

$$U[B_e] = U - B_e I \tag{7.54}$$

and this potential is a function of S, V, B_e, and N. The condition of equilibrium for a system at constant external field is that this potential be minimum.

Various other potentials result from multiple Legendre transformations, as depicted in the mnemonic squares of Fig. 7.3. Maxwell relations and the relationships between potentials can be read from these squares in a completely straightforward fashion.

$$\left(\frac{\partial V}{\partial I}\right)_{S,P} = \left(\frac{\partial B_e}{\partial P}\right)_{S,I}$$

$$\left(\frac{\partial I}{\partial P}\right)_{S,B_e} = -\left(\frac{\partial V}{\partial B_e}\right)_{S,P}$$

$$\left(\frac{\partial V}{\partial I}\right)_{T,P} = \left(\frac{\partial B_e}{\partial P}\right)_{T,I}$$

$$\left(\frac{\partial I}{\partial P}\right)_{T,B_e} = -\left(\frac{\partial V}{\partial B_e}\right)_{T,P}$$

$$\left(\frac{\partial S}{\partial I}\right)_{V,T} = -\left(\frac{\partial B_e}{\partial T}\right)_{V,I}$$

$$\left(\frac{\partial T}{\partial I}\right)_{V,S} = \left(\frac{\partial B_e}{\partial S}\right)_{V,I}$$

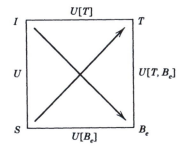

FIGURE 7.3

The "magnetic enthalpy" $U[P, B_e] \equiv U + PV - B_e I$ is an interesting and useful potential. It is minimum for systems maintained at constant pressure and constant external field. Furthermore, as in equation 6.29 for the enthalpy, $dU[P, B_e] = T dS = dQ$ at constant P, B_e, and N. Thus the magnetic enthalpy $U[P, B_e]$ acts as a "potential for heat" for systems maintained at constant pressure and magnetic field.

Example

A particular material obeys the fundamental equation of the "paramagnetic model" (equation 3.66), with $T_0 = 200$ K and $I_0^2/2R = 10$ Tesla^2 K/m^{2}J. Two moles of this material are maintained at constant pressure in an external field of $B_e = 0.2$ Tesla (or 2000 gauss), and the system is heated from an initial temperature of 5 K to a final temperature of 10 K. What is the heat input to the system?

Solution

The heat input is the change in the "magnetic enthalpy" $U[P, B_e]$. For a system in which the fundamental relation is independent of volume, $P \equiv \partial U/\partial V = 0$, so that $U[P, B_e]$ degenerates to $U - B_e I = U[B_e]$. Furthermore for the paramagnetic model (equation 3.66), $U = NRT$ and $I = (NI_0^2/2RT)B_e$, so that $U[P, B_e] = U[B_e] = NRT - (NI_0^2/2RT)B_e^2$. Thus

$$Q = N\left[R \Delta T - \frac{I_0^2}{2R} B_e^2 \Delta\left(\frac{1}{T}\right)\right]$$

$$= 2[8.314 \times 5 + 10 \times 0.04 \times 0.1] J = 83.22J$$

(Note that the magnetic contribution, arising from the second term, is small compared to the nonmagnetic first-term contribution; in reality the nonmagnetic contribution to the heat capacity of real solids falls rapidly at low temperatures and would be comparably small. Recall Problem 3.9-6.)

PROBLEMS

7.5-1. Calculate the "magnetic Gibbs potential" $U[T, B_e]$ for the paramagnetic model of equation 3.66. Corroborate that the derivative of this potential with respect to B_e at constant T has its proper value.

7.5-2. Repeat Problem 7.5-1 for the system with the fundamental equation given in Problem 3.8-2.

Answer:

$$U[T, B_e] = \frac{1}{2} N \frac{\chi}{\mu_0} B_e^2 - \frac{1}{2} NRT \ln(k_B T/2\varepsilon)$$

7.5-3. Calculate $(\partial I/\partial T)_s$ for the paramagnetic model of equation 3.66. Also calculate $(\partial S/\partial B_e)_I$. What is the relationship between these derivatives, as read from the mnemonic square?

7.5-4. Show that

$$C_{B_e} - C_I = \frac{\mu_0^2 T}{\chi_T^2} \left(\frac{\partial I}{\partial T} \right)_{B_e}$$

and

$$\frac{C_{B_e}}{C_I} = \frac{\chi_T}{\chi_S}$$

where C_{B_e} and C_I are heat capacities and χ_T and χ_s are susceptibilities: $\chi_T \equiv \mu_0 (\partial I / \partial B_e)_T$.

8

STABILITY OF THERMODYNAMIC SYSTEMS

8-1 INTRINSIC STABILITY OF THERMODYNAMIC SYSTEMS

The basic extremum principle of thermodynamics implies both that $dS = 0$ and that $d^2S < 0$, the first of these conditions stating that the entropy is an extremum and the second stating that the extremum is, in particular, a maximum. We have not yet fully exploited the second condition, which determines the *stability* of predicted equilibrium states. Similarly, in classical mechanics the stable equilibrium of a rigid pendulum is at the position of minimum potential energy. A so-called "unstable equilibrium" exists at the inverted point where the potential energy is maximum.

Considerations of stability lead to some of the most interesting and significant predictions of thermodynamics. In this chapter we investigate the conditions under which a system is stable. In Chapter 9 we consider phase transitions, which are the consequences of instability.

Consider two identical subsystems, each with a fundamental equation $S = S(U, V, N)$, separated by a totally restrictive wall. Suppose the dependence of S on U to be qualitatively as sketched in Fig. 8.1. If we were to remove an amount of energy ΔU from the first subsystem and transfer it to the second subsystem the total entropy would change from its initial value of $2S(U, V, N)$ to $S(U + \Delta U, V, N) + S(U - \Delta U, V, N)$. With the shape of the curve shown in the figure the resultant entropy would be larger than the initial entropy! If the adiabatic restraint were removed in such a system energy would flow spontaneously across the wall; one subsystem thereby would increase its energy (and its temperature) at the expense of the other. Even within one subsystem the system would find it advantageous to transfer energy from one region to another, developing internal inhomogeneities. Such a loss of homogeneity is the hallmark of a phase transition.

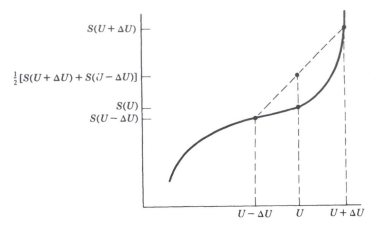

FIGURE 8.1
For a convex fundamental relation, as shown, the average entropy is increased by transfer
of energy between two subsystems; such a system is unstable.

It is evident from Fig. 8.1 that the condition of stability is the *concavity*
of the entropy.[1]

$$S(U + \Delta U, V, N) + S(U - \Delta U, V, N) \leq 2S(U, V, N) \qquad \text{(for all } \Delta\text{)}$$

$$(8.1)$$

For $\Delta U \to 0$ this condition reduces to its differential form

$$\left(\frac{\partial^2 S}{\partial U^2} \right)_{V, N} \leq 0 \qquad (8.2)$$

However this differential form is less restrictive than the concavity condition (8.1), which must hold for *all* ΔU rather than for $\Delta U \to 0$ only.
It is evident that the same considerations apply to a transfer of volume

$$S(U, V + \Delta V, N) + S(U, V - \Delta V, N) \leq 2S(U, V, N) \qquad (8.3)$$

or in differential form

$$\left(\frac{\partial^2 S}{\partial V^2} \right)_{U, N} \leq 0 \qquad (8.4)$$

A fundamental equation that does not satisfy the concavity conditions
might be obtained from a statistical mechanical calculation or from

---

[1] R. B. Griffiths, *J. Math. Phys.* **5**, 1215 (1964). L. Galgani and A. Scotti, *Physica* **40**, 150 (1968);
42, 242 (1969); *Pure and Appl. Chem.* **22**, 229 (1970).

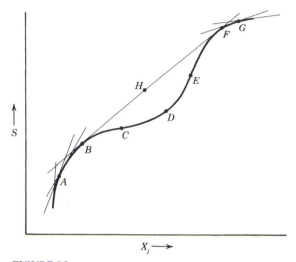

FIGURE 8.2
The underlying fundamental relation *ABCDEFG* is unstable. The stable fundamental
relation is *ABHFG*. Points on the straight line *BHF* correspond to inhomogeneous
combinations of the two phases at *B* and *F*.

extrapolation of experimental data. The stable thermodynamic fundamental equation is then obtained from this "underlying fundamental equation" by the construction shown in Fig. 8.2. The family of tangent lines
that lie everywhere *above* the curve (the superior tangents) are drawn; *the
thermodynamic fundamental equation is the envelope of these superior tangent lines.*

In Fig. 8.2 the portion *BCDEF* of the underlying fundamental relation
is unstable and is replaced by the straight line *BHF*. It should be noted
that only the portion *CDE* fails to satisfy the differential (or "local") form
of the stability condition (8.2), whereas the entire portion *BCDEF* violates
the global form (8.1). The portions of the curve *BC* and *EF* are said to be
"locally stable" but "globally unstable."

A point on a straight portion (*BHF* in Fig. 8.2) of the fundamental
relation corresponds to a phase separation in which part of the system is
in state *B* and part in state *F*, as we shall see in some detail in Chapter 9.

In the three-dimensional *S–U–V* subspace the global condition of
stability requires that the entropy surface $S(U, V, \ldots)$ lie everywhere
below its tangent *planes*. That is, for arbitrary ΔU and ΔV

$$S(U + \Delta U, V + \Delta V, N) + S(U - \Delta U, V - \Delta V, N) \leq 2S(U, V, N)$$

$$(8.5)$$

from which equations 8.2 and 8.4 again follow, as well as the additional requirement (see Problem 8.1-1) that

$$\frac{\partial^2 S}{\partial U^2} \frac{\partial^2 S}{\partial V^2} - \left(\frac{\partial^2 S}{\partial U \, \partial V} \right)^2 \geq 0 \qquad (8.6)$$

We shall soon obtain this equation by an alternative method, by applying the analogue of the simple curvature condition 8.2 to the Legendre transforms of the entropy.

To recapitulate, stability requires that the entropy surface lie everywhere below its family of tangent planes. The local conditions of stability are weaker conditions. They require not only that $(\partial^2 S/\partial U^2)_{V,N}$ and $(\partial^2 S/\partial V^2)_{U,N}$ be negative, but that $[(\partial^2 S/\partial U^2)(\partial^2 S/\partial V^2)] - (\partial^2 S/\partial U \, \partial V)^2$ must be positive. The condition $\partial^2 S / \partial U^2 \leq 0$ ensures that the curve of intersection of the entropy surface with the plane of constant V (passing through the equilibrium point) have negative curvature. The condition $\partial^2 S/\partial V^2 < 0$ similarly ensures that the curve of intersection of the entropy surface with the plane of constant U have negative curvature. These two "partial curvatures" are not sufficient to ensure concavity, for the surface could be "fluted," curving downward along the four directions $\pm U$ and $\pm V$, but curving upward along the four diagonal directions (between the U and V axes). It is this fluted structure that is forbidden by the third differential stability criterion (8.6).

In physical terms the local stability conditions ensure that inhomogeneities of either u or v separately do not increase the entropy, and also that a coupled inhomogeneity of u and v together does not increase the entropy.

For magnetic systems analogous relations hold, with the magnetic moment replacing the volume.[2]

Before turning to the full physical implications of these stability conditions it is useful first (Section 8.2) to consider their analogues for other thermodynamic potentials. We here take note only of the most easily interpreted inequality (equation 8.3), which suggests the type of information later to be inferred from all the stability conditions. Equation 8.2 requires that

$$\left(\frac{\partial^2 S}{\partial U^2} \right)_{V,N} = -\frac{1}{T^2} \left(\frac{\partial T}{\partial U} \right)_{V,N} = -\frac{1}{NT^2 c_v} \leq 0 \qquad (8.7)$$

whence *the molar heat capacity must be positive in a stable system.* The remaining stability conditions will place analogous restrictions on other physically significant observables.

Finally, and in summary, in an $r + 2$ dimensional thermodynamic space $(S, X_0, X_1, \ldots, X_r)$ *stability requires that the entropy hyper-surface lie everywhere below its family of tangent hyper-planes.*

[2] R. B. Griffiths, *J. Math. Phys.* **5**, 1215 (1964).

PROBLEMS

8.1-1. To establish the inequality 8.6 expand the left-hand side of 8.5 in a Taylor series to second order in ΔU and ΔV. Show that this leads to the condition

$$S_{UU}(\Delta U)^2 + 2S_{UV}\Delta U\Delta V + S_{VV}(\Delta V)^2 \le 0$$

Recalling that $S_{UU} \equiv \partial^2 S/\partial U^2 \le 0$, show that this can be written in the form

$$\left(S_{UU}\Delta U + S_{UV}\Delta V\right)^2 + \left(S_{UU}S_{VV} - S_{UV}^2\right)(\Delta V)^2 \ge 0$$

and that this condition in turn leads to equation 8.6.

8.1-2. Consider the fundamental equation of a monatomic ideal gas and show that S is a concave function of U and V, and also of N.

8-2 STABILITY CONDITIONS FOR THERMODYNAMIC POTENTIALS

The reformulation of the stability criteria in energy representation requires only a straightforward transcription of language. Whereas the entropy is maximum, the energy is minimum; thus the concavity of the entropy surface is replaced by *convexity* of the energy surface.

The stable energy surface lies *above* its tangent planes

$$U(S + \Delta S, V + \Delta V, N) + U(S - \Delta S, V - \Delta V, N) \ge 2U(S, V, N)$$

$$(8.8)$$

The local conditions of convexity become

$$\frac{\partial^2 U}{\partial S^2} = \frac{\partial T}{\partial S} \ge 0 \qquad \frac{\partial^2 U}{\partial V^2} = -\frac{\partial P}{\partial V} \ge 0 \qquad (8.9)$$

and for cooperative variations of S and V

$$\frac{\partial^2 U}{\partial S^2}\frac{\partial^2 U}{\partial V^2} - \left(\frac{\partial^2 U}{\partial S\,\partial V}\right)^2 \ge 0 \qquad (8.10)$$

This result can be extended easily to the Legendre transforms of the energy, or of the entropy. We first recall the properties of Legendre transformations (equation 5.31)

$$P = \frac{\partial U}{\partial X} \qquad \text{and} \qquad X = -\frac{\partial U[P]}{\partial P} \qquad (8.11)$$

whence

$$\frac{\partial X}{\partial P} = -\frac{\partial^2 U[P]}{\partial P^2} = \frac{1}{\dfrac{\partial^2 U}{\partial X^2}} \tag{8.12}$$

Hence the sign of $\partial^2 U[P]/\partial P^2$ is the negative of the sign of $\partial^2 U/\partial X^2$. If U is a convex function of X then $U[P]$ is a concave function of P. It follows that the Helmholtz potential is a concave function of the temperature and a convex function of the volume

$$\left(\frac{\partial^2 F}{\partial T^2}\right)_{V,N} \leq 0 \qquad \left(\frac{\partial^2 F}{\partial V^2}\right)_{T,N} \geq 0 \tag{8.13}$$

The enthalpy is a convex function of the entropy and a concave function of the pressure

$$\left(\frac{\partial^2 H}{\partial S^2}\right)_{P,N} \geq 0 \qquad \left(\frac{\partial^2 H}{\partial P^2}\right)_{S,N} \leq 0 \tag{8.14}$$

The Gibbs potential is a concave function of both temperature and pressure

$$\left(\frac{\partial^2 G}{\partial T^2}\right)_{P,N} \leq 0 \qquad \left(\frac{\partial^2 G}{\partial P^2}\right)_{T,N} \leq 0 \tag{8.15}$$

In summary, for constant N the *thermodynamic potentials* (the energy and its Legendre transforms) *are convex functions of their extensive variables and concave functions of their intensive variables*. Similarly for constant N the *Massieu functions* (the entropy and its Legendre transforms) *are concave functions of their extensive variables and convex functions of their intensive variables*.

PROBLEMS

8.2-1. *a*) Show that in the region $X > 0$ the function $Y = X^n$ is concave for $0 < n < 1$ and convex for $n < 0$ or $n > 1$.

The following four equations are asserted to be fundamental equations of physical systems. ___

(*b*) $\quad F = A\left(\dfrac{N^5 T}{V^3}\right)^{\frac{1}{2}}$ (*c*) $\quad G = BT^{\frac{1}{2}}P^2 N$

(*d*) $\quad H = \dfrac{CS^2 P^{\frac{1}{2}}}{N}$ (*e*) $\quad U = D\left(\dfrac{S^3 V^4}{N^5}\right)^{\frac{1}{2}}$

Which of these equations violate the criteria of stability? Assume A, B, C, and D to be positive constants. Recall the "fluting condition" (equation 8.10).

8.2-2. Prove that

$$\left(\frac{\partial^2 F}{\partial V^2}\right)_T = \frac{\dfrac{\partial^2 U}{\partial S^2}\dfrac{\partial^2 U}{\partial V^2} - \left(\dfrac{\partial^2 U}{\partial S\, \partial V}\right)^2}{\dfrac{\partial^2 U}{\partial S^2}}$$

Hint: Note that $(\partial^2 F/\partial V^2)_T = -(\partial P/\partial V)_T$, and consider P formally to be a function of S and V.

This identity casts an interesting perspective on the formalism. The quantity in the numerator, being positive, ensures that the energy surface lies above its local tangent planes (recall the discussion of "fluting" after equation 8.6). The primary curvature condition on F, along the V axis, is redundant with the "fluting" condition on U. *Only primary curvature conditions need be invoked if all potentials are considered.*

8.2-3. Show that stability requires equations 8.15 and

$$\left(\frac{\partial^2 G}{\partial T^2}\right)\left(\frac{\partial^2 G}{\partial P^2}\right) - \left(\frac{\partial^2 G}{\partial T\, \partial P}\right)^2 \geq 0$$

(Recall Problem 8.1-1.)

8-3 PHYSICAL CONSEQUENCES OF STABILITY

We turn finally to a direct interpretation of the local stability criteria in terms of limitations on the signs of quantities such as c_v, c_p, α, and κ_T. The first such inference was obtained in equations 8.2 or 8.7, where we found that $c_v \geq 0$. Similarly, the convexity of the Helmholtz potential with respect to the volume gives

$$\left(\frac{\partial^2 F}{\partial V^2}\right)_T = -\left(\frac{\partial P}{\partial V}\right)_T = \frac{1}{V\kappa_T} \geq 0 \tag{8.16}$$

or

$$\kappa_T > 0 \tag{8.17}$$

The fact that both c_v and κ_T are positive (equations 8.7 and 8.17) has further implications which become evident when we recall the identities of

Problem 3.9-5

$$c_p - c_v = \frac{Tv\alpha^2}{\kappa_T} \tag{8.18}$$

and

$$\frac{\kappa_s}{\kappa_T} = \frac{c_v}{c_p} \tag{8.19}$$

From these it follows that stability requires

$$c_p \geq c_v \geq 0 \tag{8.20}$$

and

$$\kappa_T \geq \kappa_s \geq 0 \tag{8.21}$$

Thus both heat capacities and both compressibilities must be positive in a stable system. *Addition of heat, either at constant pressure or at constant volume, necessarily increases the temperature of a stable system—the more so at constant volume than at constant pressure. And decreasing the volume, either isothermally or isentropically, necessarily increases the pressure of a stable system—the less so isothermally than isentropically.*

PROBLEMS

8.3-1. Explain on intuitive grounds why $c_p \geq c_v$ and why $\kappa_T \geq \kappa_S$.
Hint: Consider the energy input and the energy output during constant-pressure and constant-volume heating processes.

8.3-2. Show that the fundamental equation of a monatomic ideal gas satisfies the criteria of intrinsic stability.

8.3-3. Show that the van der Waals equation of state does not satisfy the criteria of intrinsic stability for all values of the parameters. Sketch the curves of P versus V for constant T (the isotherms of the gas) and show the region of local instability.

8-4   LE CHATELIER'S PRINCIPLE; THE QUALITATIVE EFFECT OF FLUCTUATIONS

The physical content of the stability criteria is known at *Le Chatelier's Principle*. According to this principle the criterion for stability is that *any inhomogeneity that somehow develops in a system should induce a process that tends to eradicate the inhomogeneity.*

As an example, suppose that a container of fluid is in equilibrium and an incident photon is suddenly absorbed at some point within it, locally heating the fluid slightly. Heat flows away from this heated region and, by the stability condition (that the specific heat is positive), this flow of heat tends to *lower* the local temperature toward the ambient value. The initial homogeneity of the system thereby is restored.

Similarly, a longitudinal vibrational wave in a fluid system induces local regions of alternately high and low density. The regions of increased density, and hence of increased pressure, tend to expand, and the regions of low density contract. The stability condition (that the compressibility is positive) ensures that these responses tend to restore the local pressure toward homogeneity.

In fact local inhomogeneities always occur in physical systems even in the absence of incident photons or of externally induced vibrations. In a gas, for instance, the individual molecules move at random, and by pure chance this motion produces regions of high density and other regions of low density.

From the perspective of statistical mechanics all systems undergo continual local fluctuations. The equilibrium state, static from the viewpoint of classical thermodynamics, is incessantly dynamic. Local inhomogeneities continually and spontaneously generate, only to be attenuated and dissipated in accordance with the Le Chatelier principle.

An informative analogy exists between a thermodynamic system and a model of a marble rolling within a "potential well." The stable state is at the minimum of the surface. The criterion of stability is that the surface be convex.

In a slightly more sophisticated viewpoint we can conceive of the marble as being subject to Brownian motion—perhaps being buffeted by some type of random collisions. These are the mechanical analogues of the spontaneous fluctuations that occur in all real systems. The potential minimum does not necessarily coincide with the instantaneous position of the system, but rather with its "expected value"; it is this "expected value" that enters thermodynamic descriptions. The curvature of the potential well then plays a crucial and continual role, restoring the system toward the "expected state" after each Brownian impact (fluctuation). This "induced restoring force" is the content of the Le Chatelier principle.

We note in passing that in the atypical but important case in which the potential well is both shallow and asymmetric, the time-averaged position may deviate measurably from the "expected state" at the potential minimum. In such a case classical thermodynamics makes spurious predictions which deviate from observational data, for thermodynamic measurements yield *average* values (recall Chapter 1). Such a pathological case arises at higher-order phase transitions—the correct theory of which was developed in the 1970s. We shall explore that area in Chapter 11.

8-5   THE LE CHATELIER–BRAUN PRINCIPLE

Returning to the physical interpretation of the stability criteria, a more subtle insight than that given by the Le Chatelier principle is formulated in the Le Chatelier–Braun principle.

Consider a system that is taken out of equilibrium by some action or fluctuation. According to the Le Chatelier principle the perturbation directly induces a process that attenuates the perturbation. But various other secondary processes are also induced, indirectly. The content of the Le Chatelier–Braun principle is that these indirectly induced processes also act to attenuate the initial perturbation.

A simple example may clarify the principle. Consider a subsystem contained within a cylinder with diathermal walls and a loosely fitting piston, all immersed within a "bath" (a thermal and pressure reservoir). The piston is moved outward slightly, either by an external agent or by a fluctuation. The primary effect is that the internal pressure is decreased—the pressure difference across the piston then acts to push it inward; this is the Le Chatelier principle. A second effect is that the initial expansion dV alters the temperature of the subsystem; $dT = (\partial T/\partial V)_S \, dV = -(T\alpha/Nc_v\kappa_T)\, dV$. This change of temperature may have either sign, depending on the sign of α. Consequently there is a flow of heat through the cylinder walls, inward if α is positive and outward if α is negative (sign dQ = sign α). This flow of heat, in turn, tends to change the pressure of the system: $dP = (1/T)(\partial P/\partial S)_V \, dQ = (\alpha/NT^2c_v\kappa_T)\, dQ$. The pressure is increased for either sign of α. Thus a secondary induced process (heat flow) also acts to diminish the initial perturbation. This is the Le Chatelier–Braun principle.

To demonstrate both the Le Chatelier and the Le Chatelier–Braun principles formally, let a spontaneous fluctuation dX_1^f occur in a composite system. This fluctuation is accompanied by a change in the intensive parameter P_1 of the subsystem

$$dP_1^f = \frac{\partial P_1}{\partial X_1} \, dX_1^f \tag{8.22}$$

The fluctuation dX_1^f also alters the intensive parameter P_2

$$dP_2^f = \frac{\partial P_2}{\partial X_1} \, dX_1^f \tag{8.23}$$

Now we can inquire as to the changes in X_1 and X_2 which are driven by these two deviations dP_1^f and dP_2^f. We designate the driven change in dX_j by dX_j^r, the superscript indicating "response." The signs of dX_1^r and dX_2^r are determined by the minimization of the total energy (at constant total entropy)

$$d(U + U^{\text{res}}) = (P_1 - P_1^{\text{res}}) \, dX_1^r + (P_2 - P_2^{\text{res}}) \, dX_2^r \leq 0 \quad (8.24)$$

$$= dP_1^f dX_1^r + dP_2^f dX_2^r \leq 0 \quad (8.25)$$

Hence, since dX_1^r and dX_2^r are independent

$$dP_1^f dX_1^r \leq 0 \quad (8.26)$$

and

$$dP_2^f dX_2^r \leq 0 \quad (8.27)$$

From the first of these and equation 8.22

$$\frac{dP_1}{dX_1} \, dX_1^f dX_1^r \leq 0 \quad (8.28)$$

and similarly

$$\frac{dP_2}{dX_1} \, dX_1^f dX_2^f \leq 0 \quad (8.29)$$

We examine these two results in turn. The first, equation 8.28, is the formal statement of the Le Chatelier principle. For multiplying by dP_1/dX_1, which is positive by virtue of the convexity criterion of stability,

$$\frac{dP_1}{dX_1} \, dX_1^f \cdot \frac{dP_1}{dX_1} \, dX_1^r \leq 0 \quad (8.30)$$

or

$$dP_1^f dP_1^{r(1)} \leq 0 \quad (8.31)$$

That is, the response dX_1^r produces a change $dP_1^{r(1)}$ in the intensive parameter P_1 that is opposite in sign to the change dP_1^f induced by the initial fluctuation.

The second inequality, (8.29), can be rewritten by the Maxwell relation

$$\frac{\partial P_2}{\partial X_1} = \frac{\partial P_1}{\partial X_2} \tag{8.32}$$

in the form

$$dX_1^f \cdot \left(\frac{\partial P_1}{\partial X_2} \, dX_2^r \right) \leq 0 \tag{8.33}$$

Then, multiplying by the positive quantity dP_1/dX_1

$$\left(\frac{\partial P_1}{\partial X_1} \, dX_1^f \right)\left(\frac{dP_1}{dX_2} \, dX_2^r \right) \leq 0 \tag{8.34}$$

or

$$\left(dP_1^f \right)\left(dP_1^{r(2)} \right) \leq 0 \tag{8.35}$$

That is, the response dX_2^r produces a change $dP_1^{r(2)}$ in the intensive parameter P_1 which is opposite in sign to the change in P_1 directly induced by the initial fluctuation. This is the Le Chatelier–Braun principle.

Finally, it is of some interest to note that equation 8.33 is subject to another closely correlated interpretation. Multiplying by the positive quantity dP_2/dX_2

$$\left(\frac{\partial P_2}{\partial X_1} \, dX_1^f \right)\left(\frac{\partial P_2}{\partial X_2} \, dX_2^r \right) \leq 0 \tag{8.36}$$

or

$$\left(dP_2^f \right)\left(dP_2^{r(2)} \right) \leq 0 \tag{8.37}$$

That is, the response in X_2 produces a change in P_2 opposite in sign to the change induced by the initial fluctuation in X_1.

PROBLEMS

8.5-1. A system is in equilibrium with its environment at a common temperature and a common pressure. The entropy of the system is increased slightly (by a fluctuation in which heat flows into the system, or by the purposeful injection of heat into the system). Explain the implications of both the Le Chatelier and the Le Chatelier–Braun principles to the ensuing processes, proving your assertions in detail.

9

FIRST-ORDER PHASE TRANSITIONS

9-1 FIRST–ORDER PHASE TRANSITIONS IN SINGLE COMPONENT SYSTEMS

Ordinary water is liquid at room temperature and atmospheric pressure, but if cooled below 273.15 K it solidifies; and if heated above 373.15 K it vaporizes. At each of these temperatures the material undergoes a precipitous change of properties—a "phase transition." At high pressures water undergoes several additional phase transitions from one solid form to another. These distinguishable solid phases, designated as "ice I," "ice II," "ice III," ..., differ in crystal structure and in essentially all thermodynamic properties (such as compressibility, molar heat capacity, and various molar potentials such as u or f). The "phase diagram" of water is shown in Fig. 9.1.

Each transition is associated with a linear region in the thermodynamic fundamental relation (such as BHF in Fig. 8.2), and each can be viewed as the result of failure of the stability criteria (convexity or concavity) in the underlying fundamental relation.

In this section we shall consider systems for which the underlying fundamental relation is unstable. By a qualitative consideration of fluctuations in such systems we shall see that *the fluctuations are profoundly influenced by the details of the underlying fundamental relation*. In contrast, *the average values of the extensive parameters reflect only the stable thermodynamic fundamental relation*.

Consideration of the manner in which the form of the underlying fundamental relation influences the thermodynamic fluctuations will provide a physical interpretation of the stability considerations of Chapter 8 and of the construction of Fig. 8.2 (in which the thermodynamic fundamental relation is constructed as the envelope of tangent planes).

A simple mechanical model illustrates the considerations to follow by an intuitively transparent analogy. Consider a semicircular section of pipe, closed at both ends. The pipe stands vertically on a table, in the form of

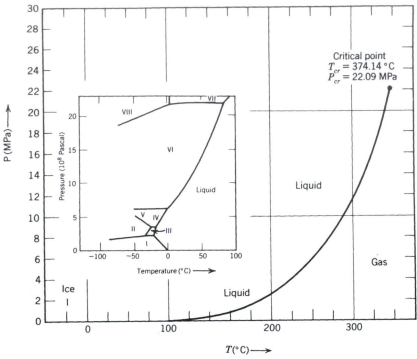

FIGURE 9.1
Phase diagram of water. The region of gas-phase stability is represented by an indiscernibly narrow horizontal strip above the positive temperature axis in the phase diagram (small figure). The background graph is a magnification of the vertical scale to show the gas phase and the gas–liquid coexistence curve.

an inverted U (Fig. 9.2). The pipe contains a freely-sliding internal piston separating the pipe into two sections, each of which contains one mole of a gas. The symmetry of the system will prove to have important consequences, and to break this symmetry we consider that each section of the pipe contains a small metallic "ball bearing" (i.e., a small metallic sphere). The two ball bearings are of dissimilar metals, with different coefficients of thermal expansion.

At some particular temperature, which we designate as T_c, the two spheres have equal radii; at temperatures above T_c the right-hand sphere is the larger.

The piston, momentarily brought to the apex of the pipe, can fall into either of the two legs, compressing the gas in that leg and expanding the gas in the other leg. In either of these competing equilibrium states the pressure difference exactly compensates the effect of the weight of the piston.

In the absence of the two ball bearings the two competing equilibrium states would be fully equivalent. But with the ball bearings present the

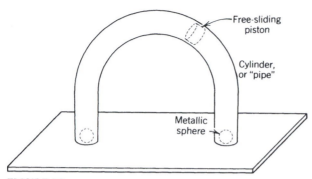

FIGURE 9.2
A simple mechanical model.

more stable equilibrium position is that to the left if $T > T_c$, and it is that to the right if $T < T_c$.

From a thermodynamic viewpoint the Helmholtz potential of the system is $F = U - TS$, and the energy U contains the gravitational potential energy of the piston as well as the familiar thermodynamic energies of the two gases (and, of course, the thermodynamic energies of the two ball bearings, which we assume to be small and/or equal). Thus the Helmholtz potential of the system has two local minima, the lower minimum corresponding to the piston being on the side of the smaller sphere.

As the temperature is lowered through T_c the two minima of the Helmholtz potential shift, the absolute minimum changing from the left-hand to the right-hand side.

A similar shift of the equilibrium position of the piston from one side to the other can be induced at a given temperature by tilting the table—or, in the thermodynamic analogue, by adjustment of some thermodynamic parameter other than the temperature.

The shift of the equilibrium state from one local minimum to the other constitutes a *first-order phase transition*, induced either by a change in temperature or by a change in some other thermodynamic parameter.

The two states between which a first-order phase transition occurs are distinct, occurring at separate regions of the thermodynamic configuration space.

To anticipate "critical phenomena" and "second-order phase transitions" (Chapter 10) it is useful briefly to consider the case in which the ball bearings are identical or absent. Then at low temperatures the two competing minima are equivalent. However as the temperature is increased the two equilibrium positions of the piston rise in the pipe, approaching the apex. Above a particular temperature T_{cr}, there is only one equilibrium position, with the piston at the apex of the pipe. Inversely, lowering the temperature from $T > T_{cr}$ to $T < T_{cr}$, the single equilibrium state bifurcates into two (symmetric) equilibrium states. The temperature T_{cr} is the "critical temperature," and the transition at T_{cr} is a "second-order phase transition."

The states between which a second-order phase transition occurs are contiguous states in the thermodynamic configuration space.

In this chapter we consider first-order phase transitions. Second-order transitions will be discussed in Chapter 10. We shall there also consider the "mechanical model" in quantitative detail, whereas we here discuss it only qualitatively.

Returning to the case of dissimilar spheres, consider the piston residing in the higher minimum—that is, in the same side of the pipe as the larger ball bearing. Finding itself in such a minimum of the Helmholtz potential, the piston will remain temporarily in that minimum though undergoing thermodynamic fluctuations ("Brownian motion"). After a sufficiently long time a giant fluctuation will carry the piston "over the top" and into the stable minimum. It then will remain in this deeper minimum until an even larger (and enormously less probable) fluctuation takes it back to the less stable minimum, after which the entire scenario is repeated. The probability of fluctuations falls so rapidly with increasing amplitude (as we shall see in Chapter 19) that *the system spends almost all of its time in the more stable minimum.* All of this dynamics is ignored by macroscopic thermodynamics, which concerns itself only with the stable equilibrium state.

To discuss the dynamics of the transition in a more thermodynamic context it is convenient to shift our attention to a familiar thermodynamic system that again has a thermodynamic potential with two local minima separated by an unstable intermediate region of concavity. Specifically we consider a vessel of water vapor at a pressure of 1 atm and at a temperature somewhat above 373.15 K (i.e., above the "normal boiling point" of water). We focus our attention on a small subsystem—a spherical region of such a (variable) radius that at any instant it contains one milligram of water. This subsystem is effectively in contact with a thermal reservoir and a pressure reservoir, and the condition of equilibrium is that the Gibbs potential $G(T, P, N)$ of the small subsystem be minimum. The two independent variables which are determined by the equilibrium conditions are the energy U and the volume V of the subsystem.

If the Gibbs potential has the form shown in Fig. 9.3, where X_j is the volume, the system is stable in the lower minimum. This minimum corresponds to a considerably larger volume (or a smaller density) than does the secondary local minimum.

Consider the behavior of a fluctuation in volume. Such fluctuations occur continually and spontaneously. The slope of the curve in Fig. 9.3 represents an intensive parameter (in the present case a difference in pressure) which acts as a restoring "force" driving the system back toward density homogeneity in accordance with Le Chatelier's principle. Occa-

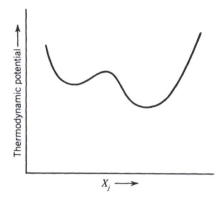

FIGURE 9.3
Thermodynamic potential with multiple minima.

sionally a fluctuation may be so large that it takes the system over the maximum, to the region of the secondary minimum. The system then settles in the region of this secondary minimum—but only for an instant. A relatively small (and therefore much more frequent) fluctuation is all that is required to overcome the more shallow barrier at the secondary minimum. The system quickly returns to its stable state. Thus very small droplets of high density (liquid phase!) occasionally form in the gas, live briefly, and evanesce.

If the secondary minimum were far removed from the absolute minimum, with a very high intermediate barrier, the fluctuations from one minimum to another would be very improbable. In Chapter 19 it will be shown that the probability of such fluctuations decreases exponentially with the height of the intermediate free-energy barrier. In solid systems (in which interaction energies are high) it is not uncommon for multiple minima to exist with intermediate barriers so high that transitions from one minimum to another take times on the order of the age of the universe! Systems trapped in such secondary "metastable" minima are *effectively* in stable equilibrium (as if the deeper minimum did not exist at all).

Returning to the case of water vapor at temperatures somewhat above the "boiling point," let us suppose that we lower the temperature of the entire system. The form of the Gibbs potential varies as shown schematically in Fig. 9.4. At the temperature T_4 the two minima become equal, and below this temperature the high density (liquid) phase becomes absolutely stable. Thus T_4 is the temperature of the phase transition (at the prescribed pressure).

If the vapor is cooled very gently through the transition temperature the system finds itself in a state that had been absolutely stable but that is now metastable. Sooner or later a fluctuation within the system will "discover" the truly stable state, forming a nucleus of condensed liquid. This nucleus then grows rapidly, and the entire system suddenly undergoes the transition. In fact the time required for the system to discover the

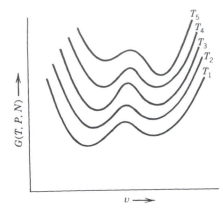

FIGURE 9.4
Schematic variation of Gibbs potential with volume (or reciprocal density) for various temperatures ($T_1 < T_2 < T_3 < T_4 < T_5$). The temperature T_4 is the transition temperature. The high density phase is stable below the transition temperature.

preferable state by an "exploratory" fluctuation is unobservably short in the case of the vapor to liquid condensation. But in the transition from liquid to ice the delay time is easily observed in a pure sample. The liquid so cooled below its solidification (freezing) temperature is said to be "supercooled." A slight tap on the container, however, sets up longitudinal waves with alternating regions of "condensation" and "rarefaction," and these externally induced fluctuations substitute for spontaneous fluctuations to initiate a precipitous transition.

A useful perspective emerges when the values of the Gibbs potential at each of its minima are plotted against temperature. The result is as shown schematically in Fig. 9.5. If these minimum values were taken from Fig. 9.4 there would be only two such curves, but any number is possible. At equilibrium the smallest minimum is stable, so the true Gibbs potential is the lower envelope of the curves shown in Fig. 9.5. The discontinuities in the entropy (and hence the latent heat) correspond to the discontinuities in slope of this envelope function.

Figure 9.5 should be extended into an additional dimension, the additional coordinate P playing a role analogous to T. The Gibbs potential is then represented by the lower envelope *surface*, as each of the three

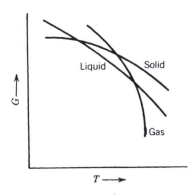

FIGURE 9.5
Minima of the Gibbs potential as a function of T.

single-phase surfaces intersect. The projection of these curves of intersection onto the $P-T$ plane is the now familiar phase diagram (e.g., Fig. 9.1).

A phase transition occurs as the state of the system passes from one envelope surface, across an intersection curve, to another envelope surface.

The variable X_j, or V in Fig. 9.4, can be any extensive parameter. In a transition from paramagnetic to ferromagnetic phases X_j is the magnetic moment. In transitions from one crystal form to another (e.g., from cubic to hexagonal) the relevant parameter X_j is a crystal symmetry variable. In a solubility transition it may be the mole number of one component. We shall see examples of such transitions subsequently. All conform to the general pattern described.

At a first-order phase transition the molar Gibbs potential of the two phases are equal, but other molar potentials (u, f, h, etc.) are discontinuous across the transition, as are the molar volume and the molar entropy. The two phases inhabit different regions in "thermodynamic space," and equality of any property other than the Gibbs potential would be a pure coincidence. The discontinuity in the molar potentials is the defining property of a first-order transition.

As shown in Fig. 9.6, as one moves along the liquid–gas coexistence curve away from the solid phase (i.e., toward higher temperature), the discontinuities in molar volume and molar energy become progressively smaller. The two phases become more nearly alike. Finally, at the terminus of the liquid–gas coexistence curve, the two phases become indistinguishable. The first-order transition degenerates into a more subtle transition, a *second-order transition*, to which we shall return in Chapter 10. The terminus of the coexistence curve is called a *critical point*.

The existence of the critical point precludes the possibility of a sharp distinction between the generic term *liquid* and the generic term *gas*. In crossing the liquid–gas coexistence curve in a first-order transition we distinguish two phases, one of which is "clearly" a gas and one of which is

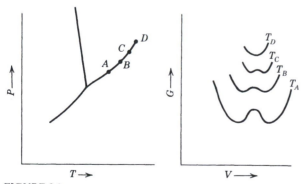

FIGURE 9.6

The two minima of G corresponding to four points on the coexistence curve. The minima coalesce at the critical point D.

"clearly" a liquid. But starting at one of these (say the liquid, immediately above the coexistence curve) we can trace an alternate path that skirts around the critical point and arrives at the other state (the "gas") without ever encountering a phase transition! Thus the terms *gas* and *liquid* have more intuitive connotation than strictly defined denotation. Together liquids and gases constitute the *fluid phase*. Despite this we shall follow the standard usage and refer to "the liquid phase" and "the gaseous phase" in a liquid–gas first-order transition.

There is another point of great interest in Fig. 9.1: the opposite terminus of the liquid–gas coexistence curve. This point is the coterminus of three coexistence curves, and it is a unique point at which gaseous, liquid, and solid phases coexist. Such a state of three-phase compatibility is a "triple point"—in this case the triple point of water. The uniquely defined temperature of the triple point of water is assigned the (arbitrary) value of 273.16 K to define the Kelvin scale of temperature (recall Section 2.6).

PROBLEM

9.1-1. The slopes of all three curves in Fig. 9.5 are shown as negative. Is this necessary? Is there a restriction on the curvature of these curves?

9-2 THE DISCONTINUITY IN THE ENTROPY—LATENT HEAT

Phase diagrams, such as Fig. 9.1, are divided by coexistence curves into regions in which one or another phase is stable. At any point on such a curve the two phases have precisely equal molar Gibbs potentials, and both phases can coexist.

Consider a sample of water at such a pressure and temperature that it is in the "ice" region of Fig. 9.1a. To increase the temperature of the ice one must supply roughly 2.1 kJ/kg for every kelvin of temperature increase (the specific heat capacity of ice). If heat is supplied at a constant rate the temperature increases at an approximately constant rate. But when the temperature reaches the "melting temperature," on the solid–liquid coexistence line, the temperature ceases to rise. As additional heat is supplied ice melts, forming liquid water at the same temperature. It requires roughly 335 kJ to melt each kg of ice. At any moment the amount of liquid water in the container depends on the quantity of heat that has entered the container since the arrival of the system at the coexistence curve (i.e., at the melting temperature). When finally the requisite amount of heat has been supplied, and the ice has been entirely melted, continued heat input again results in an increase in temperature—now at a rate determined by the specific heat capacity of liquid water ($\simeq 4.2$ kJ/kg-K).

The quantity of heat required to melt one mole of solid is the *heat of fusion* (or the *latent heat of fusion*). It is related to the difference in molar entropies of the liquid and the solid phase by

$$\ell_{LS} = T\left[s^{(L)} - s^{(S)}\right] \tag{9.1}$$

where T is the melting temperature at the given pressure.

More generally, the latent heat in any first-order transition is

$$\ell = T\Delta s \tag{9.2}$$

where T is the temperature of the transition and Δs is the difference in molar entropies of the two phases. Alternatively, the latent heat can be written as the difference in the molar enthalpies of the two phases

$$\ell = \Delta h \tag{9.3}$$

which follows immediately from the identity $h = Ts + \mu$ (and the fact that μ, the molar Gibbs function, is equal in each phase). The molar enthalpies of each phase are tabulated for very many substances.

If the phase transition is between liquid and gaseous phases the latent heat is called the *heat of vaporization*, and if it is between solid and gaseous phases it is called the *heat of sublimation*.

At a pressure of one atmosphere the liquid–gas transition (boiling) of water occurs at 373.15 K, and the latent heat of vaporization is then 40.7 kJ/mole (540 cal/g).

In each case the latent heat must be put *into* the system as it makes a transition from the low-temperature phase to the high-temperature phase. Both the molar entropy and the molar enthalpy are greater in the high-temperature phase than in the low-temperature phase.

It should be noted that the method by which the transition is induced is irrelevant—the latent heat is independent thereof. Instead of heating the ice at constant pressure (crossing the coexistence curve of Fig. 9.1*a* "horizontally"), the pressure could be increased at constant temperature (crossing the coexistence curve "vertically"). In either case the same latent heat would be drawn from the thermal reservoir.

The functional form of the liquid–gas coexistence curve for water is given in "saturated steam tables"—the designation "saturated" denoting that the steam is in equilibrium with the liquid phase. ("Superheated steam tables" denote compilations of the properties of the vapor phase alone, at temperatures above that on the coexistence curve at the given pressure). An example of such a saturated steam table is given in Table 9.1, from Sonntag and Van Wylen. The properties s, u, v and h of each

TABLE 9.1
"Steam Table"; Properties of the Gaseous and Liquid Phases on the Coexistence Curve of Water[a]

Temp. °C T	Press. kPa P	Specific Volume		Internal Energy			Enthalpy			Entropy		
		Sat. Liquid v_f	Sat. Vapor v_g	Sat. Liquid u_f	Evap. u_{fg}	Sat. Vapor u_g	Sat. Liquid h_f	Evap. h_{fg}	Sat. Vapor h_g	Sat. Liquid s_f	Evap. s_{fg}	Sat. Vapor s_g
0.01	0.6113	0.001 000	206.14	.00	2375.3	2375.3	.01	2501.3	2501.4	.0000	9.1562	9.1562
5	0.8721	0.001 000	147.12	20.97	2361.3	2382.3	20.98	2489.6	2510.6	.0761	8.9496	9.0257
10	1.2276	0.001 000	106.38	42.00	2347.2	2389.2	42.01	2477.7	2519.8	.1510	8.7498	8.9008
15	1.7051	0.001 001	77.93	62.99	2333.1	2396.1	62.99	2465.9	2528.9	.2245	8.5569	8.7814
20	2.339	0.001 002	57.79	83.95	2319.0	2402.9	83.96	2454.1	2538.1	.2966	8.3706	8.6672
25	3.169	0.001 003	43.36	104.88	2304.9	2409.8	104.89	2442.3	2547.2	.3674	8.1905	8.5580
30	4.246	0.001 004	32.89	125.78	2290.8	2416.6	125.79	2430.5	2556.3	.4369	8.0164	8.4533
35	5.628	0.001 006	25.22	146.67	2276.7	2423.4	146.68	2418.6	2565.3	.5053	7.8478	8.3531
40	7.384	0.001 008	19.52	167.56	2262.6	2430.1	167.57	2406.7	2574.3	.5725	7.6845	8.2570
45	9.593	0.001 010	15.26	188.44	2248.4	2436.8	188.45	2394.8	2583.2	.6387	7.5261	8.1648
50	12.349	0.001 012	12.03	209.32	2234.2	2443.5	209.33	2382.7	2592.1	.7038	7.3725	8.0763
55	15.758	0.001 015	9.568	230.21	2219.9	2450.1	230.23	2370.7	2600.9	.7679	7.2234	7.9913
60	19.940	0.001 017	7.671	251.11	2205.5	2456.6	251.13	2358.5	2609.6	.8312	7.0784	7.9096
65	25.03	0.001 020	6.197	272.02	2191.1	2463.1	272.06	2346.2	2618.3	.8935	6.9375	7.8310
70	31.19	0.001 023	5.042	292.95	2176.6	2469.6	292.98	2333.8	2626.8	.9549	6.8004	7.7553
75	38.58	0.001 026	4.131	313.90	2162.0	2475.9	313.93	2321.4	2635.3	1.0155	6.6669	7.6824
80	47.39	0.001 029	3.407	334.86	2147.4	2482.2	334.91	2308.8	2643.7	1.0753	6.5369	7.6122
85	57.83	0.001 033	2.828	355.84	2132.6	2488.4	355.90	2296.0	2651.9	1.1343	6.4102	7.5445
90	70.14	0.001 036	2.361	376.85	2117.7	2494.5	376.92	2283.2	2660.1	1.1925	6.2866	7.4791
95	84.55	0.001 040	1.982	397.88	2102.7	2500.6	397.96	2270.2	2668.1	1.2500	6.1659	7.4159

	Press. in MPa											
100	0.101 35	0.001 044	1.6729	418.94	2087.6	2506.5	419.04	2257.0	2676.1	1.3069	6.0480	7.3549
105	0.120 82	0.001 048	1.4194	440.02	2072.3	2512.4	440.15	2243.7	2683.8	1.3630	5.9328	7.2958
110	0.143 27	0.001 052	1.2102	461.14	2057.0	2518.1	461.30	2230.2	2691.5	1.4185	5.8202	7.2387
115	0.169 06	0.001 056	1.0366	482.30	2041.4	2523.7	482.48	2216.5	2699.0	1.4734	5.7100	7.1833
120	0.198 53	0.001 060	0.8919	503.50	2025.8	2529.3	503.71	2202.6	2706.3	1.5276	5.6020	7.1296
125	0.2321	0.001 065	0.7706	524.74	2009.9	2534.6	524.99	2188.5	2713.5	1.5813	5.4962	7.0775
130	0.2701	0.001 070	0.6685	546.02	1993.9	2539.9	546.31	2174.2	2720.5	1.6344	5.3925	7.0269
135	0.3130	0.001 075	0.5822	567.35	1977.7	2545.0	567.69	2159.6	2727.3	1.6870	5.2907	6.9777
140	0.3613	0.001 080	0.5089	588.74	1961.3	2550.0	589.13	2144.7	2733.9	1.7391	5.1908	6.9299
145	0.4154	0.001 085	0.4463	610.18	1944.7	2554.9	610.63	2129.6	2740.3	1.7907	5.0926	6.8833
150	0.4758	0.001 091	0.3928	631.68	1927.9	2559.5	632.20	2114.3	2746.5	1.8418	4.9960	6.8379
155	0.5431	0.001 096	0.3468	653.24	1910.8	2564.1	653.84	2098.6	2752.4	1.8925	4.9010	6.7935
160	0.6178	0.001 102	0.3071	674.87	1893.5	2568.4	675.55	2082.6	2758.1	1.9427	4.8075	6.7502
165	0.7005	0.001 108	0.2727	696.56	1876.0	2572.5	697.34	2066.2	2763.5	1.9925	4.7153	6.7078
170	0.7917	0.001 114	0.2428	718.33	1858.1	2576.5	719.21	2049.5	2768.7	2.0419	4.6244	6.6663
175	0.8920	0.001 121	0.2168	740.17	1840.0	2580.2	741.17	2032.4	2773.6	2.0909	4.5347	6.6256
180	1.0021	0.001 127	0.194 05	762.09	1821.6	2583.7	763.22	2015.0	2778.2	2.1396	4.4461	6.5857
185	1.1227	0.001 134	0.179 09	784.10	1802.9	2587.0	785.37	1997.1	2782.4	2.1879	4.3586	6.5465
190	1.2544	0.001 141	0.156 54	806.19	1783.8	2590.0	807.62	1978.8	2786.4	2.2359	4.2720	6.5079
195	1.3978	0.001 149	0.141 05	828.37	1764.4	2592.8	829.98	1960.0	2790.0	2.2835	4.1863	6.4698
200	1.5538	0.001 157	0.127 36	850.65	1744.7	2595.3	852.45	1940.7	2793.2	2.3309	4.1014	6.4323
205	1.7230	0.001 164	0.115 21	873.04	1724.5	2597.5	875.04	1921.0	2796.0	2.3780	4.0172	6.3952
210	1.9062	0.001 173	0.104 41	895.53	1703.9	2599.5	897.76	1900.7	2798.5	2.4248	3.9337	6.3585
215	2.104	0.001 181	0.094 79	918.14	1682.9	2601.1	920.62	1879.9	2800.5	2.4714	3.8507	6.3221

TABLE 9.1 (continued)

Temp. °C T	Press. MPa P	Specific Volume		Internal Energy			Enthalpy			Entropy		
		Sat. Liquid v_f	Sat. Vapor v_g	Sat. Liquid u_f	Evap. u_{fg}	Sat. Vapor u_g	Sat. Liquid h_f	Evap. h_{fg}	Sat. Vapor h_g	Sat. Liquid s_f	Evap. s_{fg}	Sat. Vapor s_g
220	2.318	0.001 190	0.086 19	940.87	1661.5	2602.4	943.62	1858.5	2802.1	2.5178	3.7683	6.2861
225	2.548	0.001 199	0.078 49	963.73	1639.6	2603.3	966.78	1836.5	2803.3	2.5639	3.6863	6.2503
230	2.795	0.001 209	0.071 58	986.74	1617.2	2603.9	990.12	1813.8	2804.0	2.6099	3.6047	6.2146
235	3.060	0.001 219	0.065 37	1009.89	1594.2	2604.1	1013.62	1790.5	2804.2	2.6558	3.5233	6.1791
240	3.344	0.001 229	0.059 76	1033.21	1570.8	2604.0	1037.32	1766.5	2803.8	2.7015	3.4422	6.1437
245	3.648	0.001 240	0.054 71	1056.71	1546.7	2603.4	1061.23	1741.7	2803.0	2.7472	3.3612	6.1083
250	3.973	0.001 251	0.050 13	1080.39	1522.0	2602.4	1085.36	1716.2	2801.5	2.7927	3.2802	6.0730
255	4.319	0.001 263	0.045 98	1104.28	1496.7	2600.9	1109.73	1689.8	2799.5	2.8383	3.1992	6.0375
260	4.688	0.001 276	0.042 21	1128.39	1470.6	2599.0	1134.37	1662.5	2796.9	2.8838	3.1181	6.0019
265	5.081	0.001 289	0.038 77	1152.74	1443.9	2596.6	1159.28	1634.4	2793.6	2.9294	3.0368	5.9662
270	5.499	0.001 302	0.035 64	1177.36	1416.3	2593.7	1184.51	1605.2	2789.7	2.9751	2.9551	5.9301
275	5.942	0.001 317	0.032 79	1202.25	1387.9	2590.2	1210.07	1574.9	2785.0	3.0208	2.8730	5.8938
280	6.412	0.001 332	0.030 17	1227.46	1358.7	2586.1	1235.99	1543.6	2779.6	3.0668	2.7903	5.8571
285	6.909	0.001 348	0.027 77	1253.00	1328.4	2581.4	1262.31	1511.0	2773.3	3.1130	2.7070	5.8199
290	7.436	0.001 366	0.025 57	1278.92	1297.1	2576.0	1289.07	1477.1	2766.2	3.1594	2.6227	5.7821
295	7.993	0.001 384	0.023 54	1305.2	1264.7	2569.9	1316.3	1441.8	2758.1	3.2062	2.5375	5.7437
300	8.581	0.001 404	0.021 67	1332.0	1231.0	2563.0	1344.0	1404.9	2749.0	3.2534	2.4511	5.7045
305	9.202	0.001 425	0.019 948	1359.3	1195.9	2555.2	1372.4	1366.4	2738.7	3.3010	2.3633	5.6643
310	9.856	0.001 447	0.018 350	1387.1	1159.4	2546.4	1401.3	1326.0	2727.3	3.3493	2.2737	5.6230

315	10.547	0.001 472	1415.5	1121.1	2536.6	1431.0	1283.5	2714.5	3.3982	2.1821	5.5804
320	11.274	0.001 499	1444.6	1080.9	2525.5	1461.5	1238.6	2700.1	3.4480	2.0882	5.5362
330	12.845	0.001 561	1505.3	993.7	2498.9	1525.3	1140.6	2665.9	3.5507	1.8909	5.4417
340	14.586	0.001 638	1570.3	894.3	2464.6	1594.2	1027.9	2622.0	3.6594	1.6763	5.3357
350	16.513	0.001 740	1641.9	776.6	2418.4	1670.6	893.4	2563.9	3.7777	1.4335	5.2112
360	18.651	0.001 893	1725.2	626.3	2351.5	1760.5	720.5	2481.0	3.9147	1.1379	5.0526
370	21.03	0.002 213	1844.0	384.5	2228.5	1890.5	441.6	2332.1	4.1106	.6865	4.7971
374.14	22.09	0.003 155	2029.6	0	2029.6	2099.3	0	2099.3	4.4298	0	4.4298

[a] From R. E. Sonntag and G. J. Van Wylen, *Introduction to Thermodynamics, Classical and Statistical*, John Wiley & Sons, New York, 1982 (adapted from J. H. Keenan, F. G. Keyes, P. G. Hill, and J. G. Moore, *Steam Tables*, John Wiley & Sons, New York, 1978).

phase are conventionally listed in such tables; the latent heat of the transition is the difference in the molar enthalpies of the two phases, or it can also be obtained as $T\Delta s$.

Similar data are compiled in the thermophysical data literature for a wide variety of other materials.

The molar volume, like the molar entropy and the molar energy, is discontinuous across the coexistence curve. For water this is particularly interesting in the case of the solid–liquid coexistence curve. It is common experience that ice floats in liquid water. The molar volume of the solid (ice) phase accordingly is *greater* than the molar volume of the liquid phase—an uncommon attribute of H_2O. The much more common situation is that in which the solid phase is more compact, with a smaller molar volume. One mundane consequence of this peculiar property of H_2O is the proclivity of frozen plumbing to burst. A compensating consequence, to which we shall return in Section 9.3, is the possibility of ice skating. And, underlying all, this peculiar property of water is essential to the very possibility of life on earth. If ice were more dense than liquid water the frozen winter surfaces of lakes and oceans would sink to the bottom; new surface liquid, unprotected by an ice layer, would again freeze (and sink) until the entire body of water would be frozen solid ("frozen under" instead of "frozen over").

PROBLEMS

9.2-1. In a particular solid–liquid phase transition the point P_0, T_0 lies on the coexistence curve. The latent heat of vaporization at this point is ℓ_0. A nearby point on the coexistence curve has pressure $P_0 + p$ and temperature $T_0 + t$; the local slope of the coexistence curve in the $P-T$ plane is p/t. Assuming v, c_p, α, and κ_T to be known in each phase in the vicinity of the states of interest, find the latent heat at the point $P_0 + p$, $T_0 + t$.

9.2-2. Discuss the equilibrium that eventually results if a solid is placed in an initially evacuated closed container and is maintained at a given temperature. Explain why the solid–gas coexistence curve is said to define the "vapor pressure of the solid" at the given temperature.

9-3   THE SLOPE OF COEXISTENCE CURVES; THE CLAPEYRON EQUATION

The coexistence curves illustrated in Fig. 9.1 are less arbitrary than is immediately evident; the slope dP/dT of a coexistence curve is fully determined by the properties of the two coexisting phases.

The slope of a coexistence curve is of direct physical interest. Consider cubes of ice at equilibrium in a glass of water. Given the ambient pressure, the temperature of the mixed system is determined by the liquid–solid coexistence curve of water; if the temperature were not on the coexistence curve some ice would melt, or some liquid would freeze, until the temperature would again lie on the coexistence curve (or one phase would become depleted). At 1 atm of pressure the temperature would be 273.15 K. If the ambient pressure were to decrease—perhaps by virtue of a change in altitude (the glass of water is to be served by the flight attendant in an airplane), or by a variation in atmospheric conditions (approach of a storm)—then the temperature of the glass of water would appropriately adjust to a new point on the coexistence curve. If ΔP were the change in pressure then the change in temperature would be $\Delta T = \Delta P / (dP/dT)_{cc}$, where the derivative in the denominator is the slope of the coexistence curve.

Ice skating, to which we have made an earlier allusion, presents another interesting example. The pressure applied to the ice directly beneath the blade of the skate shifts the ice across the solid–liquid coexistence curve (vertically upward in Fig. 9.1*a*), providing a lubricating film of liquid on which the skate slides.

The possibility of ice skating depends on the negative slope of the liquid–solid coexistence curve of water. The existence of the ice on the upper surface of the lake, rather than on the bottom, reflects the larger molar volume of the solid phase of water as compared to that of the liquid phase. The connection of these two facts, which are not independent, lies in the Clapeyron equation, to which we now turn.

Consider the four states shown in Fig. 9.7. States A and A' are on the coexistence curve, but they correspond to different phases (to the left-hand and right-hand regions respectively.) Similarly for the states B and B'. The pressure difference $P_B - P_A$ (or, equivalently, $P_{B'} - P_{A'}$) is assumed to be infinitesimal $(= dP)$, and similarly for the temperature difference $T_B - T_A$ $(= dT)$. The slope of the curve is dP/dT.

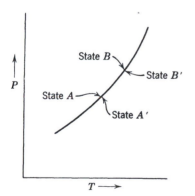

FIGURE 9.7
Four coexistence states.

Phase equilibrium requires that

$$\mu_A = \mu_{A'} \tag{9.4}$$

and

$$\mu_B = \mu_{B'} \tag{9.5}$$

whence

$$\mu_B - \mu_A = \mu_{B'} - \mu_{A'} \tag{9.6}$$

But

$$\mu_B - \mu_A = -s\,dT + v\,dP \tag{9.7}$$

and

$$\mu_{B'} - \mu_{A'} = -s'\,dT + v'\,dP \tag{9.8}$$

in which s and s' are the molar entropies and v and v' are the molar volumes in each of the phases. By inserting equations 9.7 and 9.8 in equation 9.6 and rearranging the terms, we easily find

$$\frac{dP}{dT} = \frac{s' - s}{v' - v} \tag{9.9}$$

$$\frac{dP}{dT} = \frac{\Delta s}{\Delta v} \tag{9.10}$$

in which Δs and Δv are the discontinuities in molar entropy and molar volume associated with the phase transition. According to equation 9.2 the latent heat is

$$\ell = T\Delta s \tag{9.11}$$

whence

$$\frac{dP}{dT} = \frac{\ell}{T\Delta v} \tag{9.12}$$

This is the Clapeyron equation.

The Clapeyron equation embodies the Le Chatelier principle. Consider a solid–liquid transition with a positive latent heat ($s_\ell > s_s$) and a positive difference of molar volumes ($v_\ell > v_s$). The slope of the phase curve is correspondingly positive. Then an increase in pressure at constant temperature tends to drive the system to the more dense (solid) phase (alleviating the pressure increase), and an increase in temperature tends to drive the system to the more entropic (liquid) phase. Conversely, if $s_\ell > s_s$ but $v_\ell < v_s$, then the slope of the coexistence curve is negative, and an increase of the pressure (at constant T) tends to drive the system to the liquid phase—again the more dense phase.

In practical problems in which the Clapeyron equation is applied it is often sufficient to neglect the molar volume of the liquid phase relative to the molar volume of the gaseous phase ($v_g - v_\ell \simeq v_g$), and to approximate the molar volume of the gas by the ideal gas equation ($v_g \simeq RT/P$). This "Clapeyron–Clausius approximation" may be used where appropriate in the problems at the end of this section.

Example

A light rigid metallic bar of rectangular cross section lies on a block of ice, extending slightly over each end. The width of the bar is 2 mm and the length of the bar in contact with the ice is 25 cm. Two equal masses, each of mass M, are hung from the extending ends of the bar. The entire system is at atmospheric pressure and is maintained at a temperature of $T = -2°C$. What is the minimum value of M for which the bar will pass through the block of ice by "regelation"? The given data are that the latent heat of fusion of water is 80 cal/gram, that the density of liquid water is 1 gram/cm^3, and that ice cubes float with $\simeq 4/5$ of their volume submerged.

Solution

The Clapeyron equation permits us to find the pressure at which the solid-liquid transition occurs at $T = -2°C$. However we must first use the "ice cube data" to obtain the difference Δv in molar volumes of liquid and solid phases. The data given imply that the density of ice is 0.8 g/cm^3. Furthermore $v_{liq} \simeq 18$ cm^3/mole, and therefore $v_{solid} \simeq 22.5 \times 10^{-6}$ m^3/mole. Thus

$$\left. \frac{dP}{dT} \right)_{cc} = \frac{\ell}{T\Delta v} = \frac{(80 \times 4.2 \times 18) \text{ J/mole}}{271 \times (-4.5 \times 10^{-6}) \text{ K} - \text{m}^3/\text{mole}} = -5 \times 10^6 \text{ Pa/K}$$

so that the pressure difference required is

$$P \simeq -5 \times 10^6 \times (-2) \simeq 10^7 \text{ Pa}$$

This pressure is to be obtained by a weight $2Mg$ acting on the area $A = 5 \times 10^{-5}$ m^2,

$$M = \frac{1}{2} \Delta P \frac{A}{g}$$

$$= \frac{1}{2} (10^7 \text{ Pa}) (5 \times 10^{-5} \text{ m}^2) \left/ \left(9.8 \frac{\text{m}}{\text{s}^2} \right) \right. = 2.6 \text{ Kg}$$

PROBLEMS

9.3-1. A particular liquid boils at $127°C$ at a pressure of 800 mm Hg. It has a heat of vaporization of 1000 cal/mole. At what temperature will it boil if the pressure is raised to 810 mm Hg?

9.3-2. A long vertical column is closed at the bottom and open at the top; it is partially filled with a particular liquid and cooled to $-5°C$. At this temperature the fluid solidifies below a particular level, remaining liquid above this level. If the temperature is further lowered to $-5.2°C$ the solid–liquid interface moves upward by 40 cm. The latent heat (per unit mass) is 2 cal/g, and the density of the liquid phase is 1 g/cm^3. Find the density of the solid phase. Neglect thermal expansion of all materials.

Hint: Note that the pressure at the original position of the interface remains constant.

Answer:

2.6 g/cm^3

9.3-3. It is found that a certain liquid boils at a temperature of $95°C$ at the top of a hill, whereas it boils at a temperature of $105°C$ at the bottom. The latent heat is 1000 cal/mole. What is the approximate height of the hill?

9.3-4. Two weights are hung on the ends of a wire, which passes over a block of ice. The wire gradually passes through the block of ice, but the block remains intact even after the wire has passed completely through it. Explain why less mass is required if a semi-flexible wire is used, rather than a rigid bar as in the Example.

9.3-5. In the vicinity of the triple point the vapor pressure of liquid ammonia (in Pascals) is represented by

$$\ln P = 24.38 - \frac{3063}{T}$$

This is the equation of the liquid–vapor boundary curve in a P–T diagram. Similarly, the vapor pressure of solid ammonia is

$$\ln P = 27.92 - \frac{3754}{T}$$

What are the temperature and pressure at the triple point? What are the latent heats of sublimation and vaporization? What is the latent heat of fusion at the triple point?

9.3-6. Let x be the mole fraction of solid phase in a solid–liquid two-phase system. If the temperature is changed at constant total volume, find the rate of change of x; that is, find dx/dT. Assume that the standard parameters v, α, κ_T, c_P are known for each phase.

9.3-7. A particular material has a latent heat of vaporization of 5×10^3 J/mole, constant along the coexistence curve. One mole of this material exists in two-phase (liquid–vapor) equilibrium in a container of volume $V = 10^{-2}$ m^3, at a temperature of 300 K and a pressure of 10^5 Pa. The system is heated at constant volume, increasing the pressure to 2.0×10^5 Pa. (Note that this is *not* a small ΔP.) The vapor phase can be treated as a monatomic ideal gas, and the molar volume of the liquid can be neglected relative to that of the gas. Find the initial and final mole fractions of the vapor phase $[x \equiv N_g/(N_g + N_\ell)]$.

9.3-8. Draw the phase diagram, in the B_e–T plane, for a simple ferromagnet; assume no magnetocrystalline anisotropy and assume the external field B_e to be always parallel to a fixed axis in space. What is the slope of the coexistence curve? Explain this slope in terms of the Clapeyron equation.

9.3-9. A system has coexistence curves similar to those shown in Fig. 9.6a, but with the liquid–solid coexistence curve having a positive slope. Sketch the isotherms in the P–v plane for temperature T such that
(a) $T < T_t$, (b) $T = T_t$, (c) $T_t \leq T < T_{crit}$, (d) $T_t < T \leq T_{crit}$, (e) $T = T_{crit}$, (f) $T \geq T_{crit}$.
Here T_t and T_{crit} denote the triple point and critical temperatures, respectively.

9-4 UNSTABLE ISOTHERMS AND FIRST-ORDER PHASE TRANSITIONS

Our discussion of the origin of first-order phase transitions has focused, quite properly, on the multiple minima of the Gibbs potential. But although the Gibbs potential may be the fundamental entity at play, a more common description of a thermodynamic system is in terms of the form of its isotherms. For many gases the shape of the isotherms is well represented (at least semiquantitatively) by the van der Waals equation of state (recall Section 3.5)

$$P = \frac{RT}{(v - b)} - \frac{a}{v^2} \tag{9.13}$$

The shape of such van der Waals isotherms is shown schematically in the P–v diagram of Fig. 9.8.

As pointed out in Section 3.5 the van der Waals equation of state can be viewed as an "underlying equation of state," obtained by curve fitting, by inference based on plausible heuristic reasoning, or by statistical mechanical calculations based on a simple molecular model. Other empirical or semiempirical equations of state exist, and they all have isotherms that are similar to those shown in Fig. 9.8.

We now explore the manner in which isotherms of the general form shown reveal and define a phase transition.

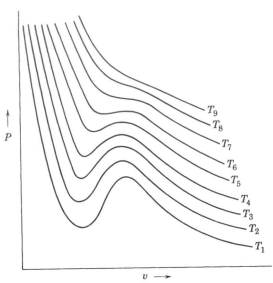

FIGURE 9.8
van der Waals isotherms (schematic). $T_1 < T_2 < T_3 \ldots$

It should be noted immediately that the isotherms of Fig. 9.8 do not satisfy the criteria of intrinsic stability everywhere, for one of these criteria (equation 8.21) is $\kappa_T > 0$, or

$$\left(\frac{\partial P}{\partial V} \right)_T < 0 \tag{9.14}$$

This condition clearly is violated over the portion *FKM* of a typical isotherm (which, for clarity, is shown separately in Fig. 9.9). Because of this violation of the stability condition a portion of the isotherm must be unphysical, superseded by a phase transition in a manner which will be explored shortly.

The molar Gibbs potential is essentially determined by the form of the isotherm. From the Gibbs–Duhem relation we recall that

$$d\mu = -s\,dT + v\,dP \tag{9.15}$$

whence, integrating at constant temperature

$$\mu = \int v\,dP + \phi(T) \tag{9.16}$$

where $\phi(T)$ is an undetermined function of the temperature, arising as the "constant of integration." The integrand $v(P)$, for constant temperature, is given by Fig. 9.9, which is most conveniently represented with P as

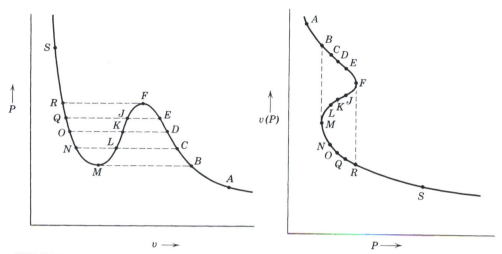

FIGURE 9.9
A particular isotherm of the van der Waals shape.

abscissa and v as ordinate. By arbitrarily assigning a value to the chemical potential at the point A, we can now compute the value of μ at any other point on the same isotherm, such as B, for from equation 9.16

$$\mu_B - \mu_A = \int_A^B v(P)\, dP \qquad (9.17)$$

In this way we obtain Fig. 9.10. This figure, representing μ versus P, can be considered as a plane section of a three-dimensional representation of μ versus P and T, as shown in Fig. 9.11. Four different constant-temperature sections of the μ-surface, corresponding to four isotherms, are shown. It is also noted that the closed loop of the μ versus P curves, which results from the fact that $v(P)$ is triple valued in P (see Fig. 9.9), disappears for high temperatures in accordance with Fig. 9.8.

Finally, we note that the relation $\mu = \mu(T, P)$ constitutes a fundamental relation for one mole of the material, as the chemical potential μ is the Gibbs function per mole. It would then appear from Fig. 9.11 that we have almost succeeded in the construction of a fundamental equation from a single given equation of state, but it should be recalled that although each of the traces of the μ-surface (in the various constant temperature planes of Fig. 9.11) has the proper form, each contains an additive "constant" $\phi(T)$, which varies from one temperature plane to another. Consequently, we do not know the complete form of the $\mu(T, P)$-surface, although we certainly are able to form a rather good mental picture of its essential topological properties.

With this qualitative picture of the fundamental relation implied by the van der Waals equation, we return to the question of stability.

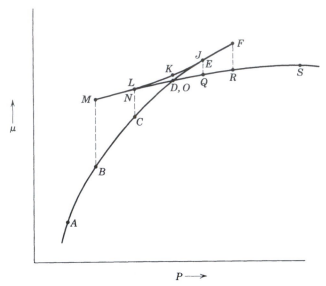

FIGURE 9.10
Isothermal dependence of the molar Gibbs potential on pressure.

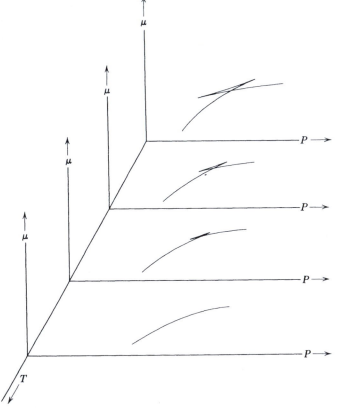

FIGURE 9.11
Functional dependence of the molar Gibbs potential.

Consider a system in the state A of Fig. 9.9 and in contact with thermal and pressure reservoirs. Suppose the pressure of the reservoir to be increased quasi-statically, maintaining the temperature constant. The system proceeds along the isotherm in Fig. 9.9 from the point A in the direction of point B. For pressures less than P_B we see that the volume of the system (for given pressure and temperature) is single valued and unique. As the pressure increases above P_B, however, three states of equal P and T become available to the system, as, for example, the states designated by C, L, and N. Of these three states L is unstable, but at both C and N the Gibbs potential is a (local) minimum. These two local minimum values of the Gibbs potential (or of μ) are indicated by the points C and N in Fig. 9.10. Whether the system actually selects the state C or the state N depends upon which of these two local minima of the Gibbs potential is the lower, or absolute, minimum. It is clear from Fig. 9.10 that the state C is the true physical state for this value of the pressure and temperature.

As the pressure is further slowly increased, the unique point D is reached. At this point the μ-surface intersects itself, as shown in Fig. 9.10, and the absolute minimum of μ or G thereafter comes from the other branch of the curve. Thus at the pressure $P_E = P_Q$, which is greater than P_D, the physical state is Q. Below P_D the right-hand branch of the isotherm in Fig. 9.9a is the physically significant branch, whereas above P_D the left-hand branch is physically significant. *The physical isotherm thus deduced from the hypothetical isotherm of Fig. 9.9 is therefore shown in Fig. 9.12.*

The isotherm of Fig. 9.9 belongs to an "underlying fundamental relation"; that of Fig. 9.12 belongs to the stable "thermodynamic fundamental relation."

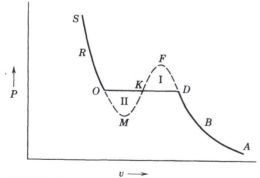

FIGURE 9.12
The physical van der Waals isotherm. The "underlying" isotherm is $SOMKFDA$, but the equal-area construction converts it to the physical isotherm $SOKDA$.

The points D and O are determined by the condition that $\mu_D = \mu_O$ or, from equation 9.17

$$\int_D^O v(P)\, dP = 0 \qquad (9.18)$$

where the integral is taken along the hypothetical isotherm. Referring to Fig. 9.9, we see that this condition can be given a direct graphical interpretation by breaking the integral into several portions

$$\int_D^F v\, dP + \int_F^K v\, dP + \int_K^M v\, dP + \int_M^O v\, dP = 0 \qquad (9.19)$$

and rearranging as follows

$$\int_D^F v\, dP - \int_K^F v\, dP = \int_M^K v\, dP - \int_M^O v\, dP \qquad (9.20)$$

Now the integral $\int_D^F v\, dP$ is the area under the arc DF in Fig. 9.12 and the integral $\int_K^F v\, dP$ is the area under the arc KF. The difference in these integrals is the area in the closed region $DFKD$, or the area marked I in Fig. 9.12. Similarly, the right-hand side of equation 9.20 represents the area II in Fig. 9.12, and the unique points O and D are therefore determined by the graphical condition

$$\text{area I} = \text{area II} \qquad (9.21)$$

It is only after the nominal (non-monotonic) isotherm has been truncated by this equal area construction that it represents a true physical isotherm.

Not only is there a nonzero change in the molar volume at the phase transition, but there are associated nonzero changes in the molar energy and the molar entropy as well. The change in the entropy can be computed by integrating the quantity

$$ds = \left(\frac{\partial s}{\partial v}\right)_T dv \qquad (9.22)$$

along the hypothetical isotherm $OMKFD$. Alternatively, by the thermodynamic mnemonic diagram, we can write

$$\Delta s = s_D - s_O = \int_{OMKFD} \left(\frac{\partial P}{\partial T}\right)_v dv \qquad (9.23)$$

A geometrical interpretation of this entropy difference, in terms of the area between neighboring isotherms, is shown in Fig. 9.13.

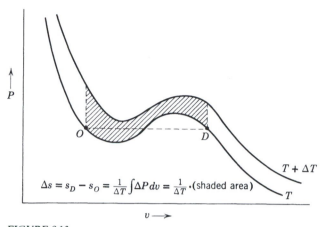

FIGURE 9.13

The discontinuity in molar entropy. The area between adjacent isotherms is related to the entropy discontinuity and thence to the latent heat.

As the system is transformed at fixed temperature and pressure from the pure phase O to the pure phase D, it absorbs an amount of heat per mole equal to $l_{DO} = T\Delta s$. The volume change per mole is $\Delta v = v_D - v_O$, and this is associated with a transfer of work equal to $P\Delta v$. Consequently, the total change in the molar energy is

$$\Delta u = u_D - u_O = T\Delta s - P\Delta v \qquad (9.24)$$

Each isotherm, such as that of Fig. 9.12, has now been classified into three regions. The region SO is in the liquid phase. The region DA is in the gaseous phase. The flat region OKD corresponds to a mixture of the two phases. Thereby the entire $P-v$ plane is classified as to phase, as shown in Fig. 9.14. The mixed liquid-plus-gas region is bounded by the inverted parabola-like curve joining the extremities of the flat regions of each isotherm.

Within the two-phase region any given point denotes a mixture of the two phases at the extremities of the flat portion of the isotherm passing through that point. The fraction of the system that exists in each of the two phases is governed by the "lever rule." Let us suppose that the molar volumes at the two extremities of the flat region of the isotherm are v_ℓ and v_g (suggesting but not requiring that the two phases are liquid and gas, for definiteness). Let the molar volume of the mixed system be $v = V/N$. Then if x_ℓ and x_g are the mole fractions of the two phases

$$V = Nv = Nx_\ell v_\ell + Nx_g v_g \qquad (9.25)$$

from which one easily finds

$$x_\ell = \frac{v_g - v}{v_g - v_\ell} \qquad (9.26)$$

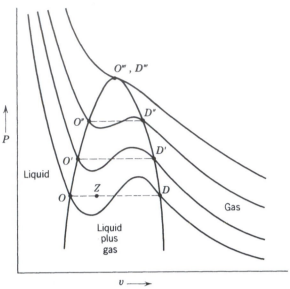

FIGURE 9.14
Phase classification of the $P - v$ plane.

and

$$x_g = \frac{v - v_\ell}{v_g - v_\ell} \tag{9.27}$$

That is, an intermediate point on the flat portion of the isotherm implies a mole fraction of each phase that is equal to the fractional distance of the point from the *opposite* end of the flat region. Thus the point Z in Fig. 9.14 denotes a mixed liquid–gas system with a mole fraction of liquid phase equal to the "length" ZD divided by the "length" OD. This is the very convenient and pictorial lever rule.

The vertex of the two-phase region, or the point at which O'' and D'' coincide in Fig. 9.14, corresponds to the *critical point*—the termination of the gas–liquid coexistence curve in Fig. 9.1a. For temperatures above the critical temperature the isotherms are monotonic (Fig. 9.14) and the molar Gibbs potential no longer is reentrant (Fig. 9.10).

Just as a P–v diagram exhibits a two-phase region, associated with the discontinuity in the molar volume, so a T–s diagram exhibits a two-phase region associated with the discontinuity in the molar entropy.

Example 1

Find the critical temperature T_{cr} and critical pressure P_{cr} for a system described by the van der Waals equation of state. Write the van der Waals equation of state in terms of the reduced variables $\tilde{T} \equiv T/T_{cr}$, $\tilde{P} \equiv P/P_{cr}$ and $\tilde{v} \equiv v/v_{cr}$.

Solution

The critical state coincides with a point of horizontal inflection of the isotherm, or

$$\left(\frac{\partial P}{\partial v}\right)_{T_{cr}} = \left(\frac{\partial^2 P}{\partial v^2}\right)_{T_{cr}} = 0.$$

(Why?) Solving these two simultaneous equations gives

$$v_{cr} = 3b \qquad P_{cr} = \frac{a}{27b^2}, \qquad RT_{cr} = \frac{8a}{27b}$$

from which we can write the van der Waals equation in reduced variables:

$$\tilde{P} = \frac{8\tilde{T}}{3\tilde{v} - 1} - \frac{3}{\tilde{v}^2}$$

Example 2

Calculate the functional form of the boundary of the two-phase region in the P–T plane for a system described by the van der Waals equation of state.

Solution

We work in reduced variables, as defined in the preceding example. We consider a fixed temperature and we carry out a Gibbs equal area construction on the corresponding isotherm. Let the extremities of the two-phase region, corresponding to the reduced temperature $\tilde{T}$, be $\tilde{v}_g$ and $\tilde{v}_\ell$. The equal area construction corresponding to equations 9.20 and 9.21 is

$$\int_{v_\ell}^{v_g} \tilde{P} \, d\tilde{v} = \tilde{P}_\ell(\tilde{v}_g - \tilde{v}_\ell)$$

where $\tilde{P}_\ell = \tilde{P}_g$ is the reduced pressure at which the phase transition occurs (at the given reduced temperature $\tilde{T}$). The reader should draw the isotherm, identify the significance of each side of the preceding equation, and reconcile this form of the statement with that in equations 9.20 and 9.21; he or she should also justify the use of reduced variables in the equation. Direct evaluation of the integral gives

$$\ln(3\tilde{v}_g - 1) + \frac{9}{4\tilde{T}}\frac{1}{\tilde{v}_g} - \frac{1}{3\tilde{v}_g - 1} = \ln(3\tilde{v}_\ell - 1) + \frac{9}{4\tilde{T}}\frac{1}{\tilde{v}_\ell} - \frac{1}{3\tilde{v}_\ell - 1}$$

Simultaneous solution of this equation and of the van der Waals equations for $\tilde{v}_g(\tilde{P}, \tilde{T})$ and $\tilde{v}_\ell(\tilde{P}, \tilde{T})$ gives $\tilde{v}_g$, $\tilde{v}_\ell$ and $\tilde{P}$ for each value of $\tilde{T}$.

PROBLEMS

9.4-1. Show that the difference in molar volumes across a coexistence curve is given by $\Delta v = -P^{-1}\Delta f$.

9.4-2. Derive the expressions for v_c, P_c and T_c given in Example 1.

9.4-3. Using the van der Waals constants for H_2O, as given in Table 3.1, calculate the critical temperature and pressure of water. How does this compare with the observed value $T_c = 647.05$ K (Table 10.1)?

9.4-4. Show that for sufficiently low temperature the van der Waals isotherm intersects the $P = 0$ axis, predicting a region of negative pressure. Find the temperature below which the isotherm exhibits this unphysical behavior.

Hint: Let $\check{P} = 0$ in the reduced van der Waals equation and consider the condition that the resultant quadratic equation for the variable $\tilde{v}^{-1}$ have two real roots.

Answer:
$$\tilde{T} = \tfrac{27}{32} \simeq 0.84$$

9.4-5. Is the fundamental equation of an ideal van der Waals fluid, as given in Section 3.5, an "underlying fundamental relation" or a "thermodynamic fundamental relation?" Why?

9.4-6. Explicitly derive the relationship among $\tilde{v}_g$, $\tilde{v}_\ell$ and $\tilde{T}$, as given in Example 2.

9.4-7. A particular substance satisfies the van der Waals equation of state. The coexistence curve is plotted in the $\check{P}, \tilde{T}$ plane, so that the critical point is at $(1, 1)$. Calculate the reduced pressure of the transition for $\tilde{T} = 0.95$. Calculate the reduced molar volumes for the corresponding gas and liquid phases.

Answer:

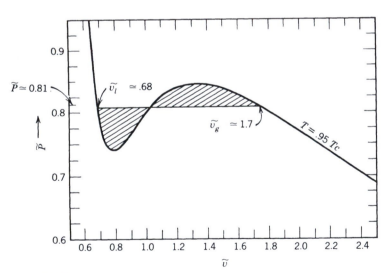

FIGURE 9.15
The $\tilde{T} = 0.95$ isotherm.

The $\tilde{T} = 0.95$ isotherm is shown in Fig. 9.15.
Counting squares permits the equal area construction

shown, giving the approximate roots indicated on the figure. Refinement of these roots by the analytic method of Example 2 yields $\tilde{P} = 0.814$, $\tilde{v}_g = 1.71$ and $\tilde{v}_\ell = 0.683$

9.4-8. Using the two points at $\tilde{T} = 0.95$ and $\tilde{T} = 1$ on the coexistence curve of a fluid obeying the van der Waals equation of state (Problem 9.4-7), calculate the average latent heat of vaporization over this range. Specifically apply this result to H_2O.

9.4-9. Plot the van der Waals isotherm, in reduced variables, for $T = 0.9T_c$. Make an equal area construction by counting squares on the graph paper. Corroborate and refine this estimate by the method of Example 2.

9.4-10. Repeat problem 9.4-8 in the range $0.90 \leq \tilde{T} \leq 0.95$, using the results of problems 9.4-7 and 9.4-9. Does the latent heat vary as the temperature approaches T_c? What is the expected value of the latent heat precisely at T_c? The latent heat of vaporization of water at atmospheric pressure is $\simeq 540$ calories per gram. Is this value qualitatively consistent with the trend suggested by your results?

9.4-11. Two moles of a van der Waals fluid are maintained at a temperature $T = 0.95T_c$ in a volume of 200 cm³. Find the mole number and volume of each phase. Use the van der Waals constants of oxygen.

9-5 GENERAL ATTRIBUTES OF FIRST-ORDER PHASE TRANSITIONS

Our discussion of first-order transitions has been based on the general shape of realistic isotherms, of which the van der Waals isotherm is a characteristic representative. The problem can be viewed in a more general perspective based on the convexity or concavity of thermodynamic potentials.

Consider a general thermodynamic potential, $U[P_s, \ldots, P_t]$, that is a function of $S, X_1, X_2, \ldots, X_{s-1}, P_s, \ldots, P_t$. The criterion of stability is that $U[P_s, \ldots, P_t]$ must be a convex function of its extensive parameters and a concave function of its intensive parameters. Geometrically, the function must lie above its tangent hyperplanes in the $X_1, \ldots, X_{s-1}$ subspace and below its tangent hyperplanes in the $P_s, \ldots, P_t$ subspace.

Consider the function $U[P_s, \ldots, P_t]$ as a function of X_j, and suppose it to have the form shown in Fig. 9.16a. A tangent line DO is also shown. It will be noted that the function lies above this tangent line. It also lies above all tangent lines drawn at points to the left of D or to the right of O. The function does not lie above tangent lines drawn to points intermediate between D and O. The local curvature of the potential is positive for all points except those between points F and M. Nevertheless a phase

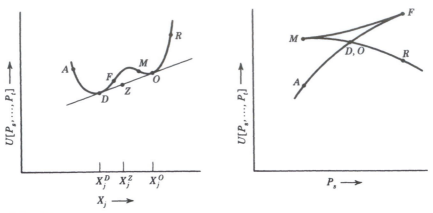

FIGURE 9.16
Stability reconstruction for a general potential.

transition occurs from the phase at D to the phase at O. Global curvature fails (becomes negative) at D before local curvature fails at F.

The "amended" thermodynamic potential $U[P_s, \ldots, P_t]$ consists of the segment AD in Fig. 9.15a, the straight line two-phase segment DO, and the original segment OR.

An intermediate point on the straight line segment, such as Z, corresponds to a mixture of phases D and O. The mole fraction of phase D varies linearly from unity to zero as Z moves from point D to point O, from which it immediately follows that

$$X = \frac{\left(X_j^0 - X_j^Z \right)}{\left(X_j^0 - X_j^D \right)}$$

This is again the "lever rule."

The value of the thermodynamic potential $U[P_s, \ldots, P_t]$ in the mixed state (i.e., at Z) clearly is less than that in the pure state (on the initial curve corresponding to X_j^z). Thus the mixed state given by the straight line construction does minimize $U[P_s, \ldots, P_t]$ and does correspond to the physical equilibrium state of the system.

The dependence of $U[P_s, \ldots, P_t]$ on an *intensive* parameter P_s is subject to similar considerations, which should now appear familiar. The Gibbs potential $U[T, P] = N\mu(T, P)$ is the particular example studied in the preceding section. The local curvature is negative except for the segment MF (Fig. 9.16b). But the segment MD lies above, rather than below, the tangent drawn to the segment ADF at D. Only the curve $ADOR$ lies everywhere below the tangent lines, thereby satisfying the conditions of global stability.

Thus the particular results of the preceding section are of very general applicability to all thermodynamic potentials.

9-6   FIRST-ORDER PHASE TRANSITIONS IN MULTICOMPONENT SYSTEMS—GIBBS PHASE RULE

If a system has more than two phases, as does water (recall Fig. 9.1), the phase diagram can become quite elaborate. In multicomponent systems the two-dimensional phase diagram is replaced by a multidimensional space, and the possible complexity would appear to escalate rapidly. Fortunately, however, the permissible complexity is severely limited by the "Gibbs phase rule." This restriction on the form of the boundaries of phase stability applies to single-component systems as well as to multi-component systems, but it is convenient to explore it directly in the general case.

The criteria of stability, as developed in Chapter 8, apply to multicom-ponent systems as well as to single-component systems. It is necessary only to consider the various mole numbers of the components as extensive parameters that are completely analogous to the volume V and the entropy S. Specifically, for a single-component system the fundamental relation is of the form

$$U = U(S, V, N) \tag{9.28}$$

or, in molar form

$$u = u(s, v) \tag{9.29}$$

For a multicomponent system the fundamental relation is

$$U = U(S, V, N_1, N_2, \ldots, N_r) \tag{9.30}$$

and the molar form is

$$u = u(s, v, x_1, x_2, \ldots, x_{r-1}) \tag{9.31}$$

The mole fractions $x_j = N_j/N$ sum to unity, so that only $r - 1$ of the x_j are independent, and only $r - 1$ of the mole fractions appear as indepen-dent variables in equation 9.31. All of this is (or should be) familiar, but it is repeated here to stress that the formalism is completely symmetric in the variables $s, v, x_1, \ldots, x_{r-1}$, and that the stability criteria can be interpreted accordingly. At the equilibrium state the energy, the enthalpy, and the Helmholtz and Gibbs potentials are convex functions of the mole fractions $x_1, x_2, \ldots, x_{r-1}$ (see Problems 9.6-1 and 9.6-2).

If the stability criteria are not satisfied in multicomponent systems a phase transition again occurs. The mole fractions, like the molar entropies and the molar volumes, differ in each phase. Thus the phases generally are different in gross composition. A mixture of salt (NaCl) and water brought to the boiling temperature undergoes a phase transition in which the gaseous phase is almost pure water, whereas the coexistent liquid phase contains both constituents—the difference in composition between the two phases in this case is the basis of purification by distillation.

Given the fact that a phase transition does occur, in either a single or multicomponent system, we are faced with the problem of how such a multiphase system can be treated within the framework of thermodynamic theory. The solution is simple indeed, for we need only consider each separate phase as a simple system and the given system as a composite system. The "wall" between the simple systems or phases is then completely nonrestrictive and may be analyzed by the methods appropriate to nonrestrictive walls.

As an example consider a container maintained at a temperature T and a pressure P and enclosing a mixture of two components. The system is observed to contain two phases: a liquid phase and a solid phase. We wish to find the composition of each phase.

The chemical potential of the first component in the liquid phase is $\mu_1^{(L)}(T, P, x_1^{(L)})$, and in the solid phase it is $\mu_1^{(S)}(T, P, x_1^{(S)})$; it should be noted that different functional forms for μ_1 are appropriate to each phase. The condition of equilibrium with respect to the transfer of the first component from phase to phase is

$$\mu_1^{(L)}\left(T, P, x_1^{(L)}\right) = \mu_1^{(S)}\left(T, P, x_1^{(S)}\right) \tag{9.32}$$

Similarly, the chemical potentials of the second component are $\mu_2^{(L)}(T, P, x_1^{(L)})$ and $\mu_2^{(S)}(T, P, x_1^{(S)})$; we can write these in terms of x_1 rather than x_2 because $x_1 + x_2$ is unity in each phase. Thus equating $\mu_2^{(L)}$ and $\mu_2^{(S)}$ gives a second equation, which, with equation 9.32, determines $x_1^{(L)}$ and $x_1^{(S)}$.

Let us suppose that three coexistent phases are observed in the foregoing system. Denoting these by I, II, and III, we have for the first component

$$\mu_1^{I}\left(T, P, x_1^{I}\right) = \mu_1^{II}\left(T, P, x_1^{II}\right) = \mu_1^{III}\left(T, P, x_1^{III}\right) \tag{9.33}$$

and a similar pair of equations for the second component. Thus we have four equations and only three composition variables: x_1^{I}, x_1^{II}, and x_1^{III}. This means that we are *not* free to specify both T and P a priori, but if T is specified then the four equations determine P, x_1^{I}, x_1^{II}, and x_1^{III}. Although it is possible to select both a temperature and a pressure arbitrarily, and then to find a two-phase state, a three-phase state can exist only for one particular pressure if the temperature is specified.

In the same system we might inquire about the existence of a state in which four phases coexist. Analogous to equation 9.33, we have three equations for the first component and three for the second. Thus we have six equations involving T, P, x_1^I, x_1^{II}, x_1^{III}, and x_1^{IV}. This means that we can have four coexistent phases only for a uniquely defined temperature and pressure, neither of which can be arbitrarily preselected by the experimenter but which are unique properties of the system.

Five phases cannot coexist in a two-component system, for the eight resultant equations would then overdetermine the seven variables $(T, P, x_1^I, \ldots, x_1^V)$, and no solution would be possible in general.

We can easily repeat the foregoing counting of variables for a multi-component, multiphase system. In a system with r components the chemical potentials in the first phase are functions of the variables, $T, P, x_1^I, x_2^I, \ldots, x_{r-1}^I$. The chemical potentials in the second phase are functions of $T, P, x_1^{II}, x_2^{II}, \ldots, x_{r-1}^{II}$. If there are M phases, the complete set of independent variables thus consists of T, P, and $M(r-1)$ mole fractions; $2 + M(r-1)$ variables in all. There are $M-1$ equations of chemical potential equality for each component, or a total of $r(M-1)$ equations. Therefore the number f of variables, which can be arbitrarily assigned, is $[2 + M(r-1)] - r(M-1)$, or

$$f = r - M + 2 \tag{9.34}$$

The fact that $r - M + 2$ variables from the set $(T, P, x_1^I, x_2^I, \ldots, x_{r-1}^M)$ can be assigned arbitrarily in a system with r components and M phases is the *Gibbs phase rule*.

The quantity f can be interpreted alternatively as the number of *thermodynamic degrees of freedom*, previously introduced in Section 3.2 and defined as the number of *intensive parameters* capable of independent variation. To justify this interpretation we now count the number of thermodynamic degrees of freedom in a straightforward way, and we show that this number agrees with equation 9.34.

For a single-component system in a single phase there are two degrees of freedom, the Gibbs–Duhem relation eliminating one of the three variables T, P, μ. For a single-component system with two phases there are three intensive parameters (T, P, and μ, each constant from phase to phase) and there are two Gibbs–Duhem relations. There is thus one degree of freedom. In Fig. 9.1 pairs of phases accordingly coexist over one-dimensional regions (curves).

If we have three coexistent phases of a single-component system, the three Gibbs–Duhem relations completely determine the three intensive parameters T, P, and μ. The three phases can coexist only in a unique zero-dimensional region, or point; the several "triple points" in Fig. 9.1.

For a multicomponent, multiphase system the number of degrees of freedom can be counted easily in similar fashion. If the system has r components, there are $r + 2$ intensive parameters: $T, P, \mu_1, \mu_2, \ldots, \mu_r$. Each of these parameters is a constant from phase to phase. But in each of the M phases there is a Gibbs–Duhem relation. These M relations reduce the number of independent parameters to $(r + 2) - M$. The number of degrees of freedom f is therefore $r - M + 2$, as given in equation 9.34.

The Gibbs phase rule therefore can be stated as follows. *In a system with r components and M coexistent phases it is possible arbitrarily to preassign $r - M + 2$ variables from the set $(T, P, x_1^1, x_2^1, \ldots, x_{r-1}^M)$ or from the set $(T, P, \mu_1, \mu_2, \ldots, \mu_r)$.*

It is now a simple matter to corroborate that the Gibbs phase rule gives the same results for single-component and two-component systems as we found in the preceding several paragraphs. For single-component systems $r = 1$ and $f = 0$ if $M = 3$. This agrees with our previous conclusion that the triple point is a unique state for a single-component system. Similarly, for the two-component system we saw that four phases coexist in a unique point ($f = 0$, $r = 2$, $M = 4$), that the temperature could be arbitrarily assigned for the three-phase system ($f = 1$, $r = 2$, $M = 3$), and that both T and P could be arbitrarily assigned for the two-phase system ($f = 2$, $r = 2$, $M = 2$).

PROBLEMS

9.6-1. In a particular system, solute A and solute B are each dissolved in solvent C.

a) What is the dimensionality of the space in which the phase regions exist?
b) What is the dimensionality of the region over which two phases coexist?
c) What is the dimensionality of the region over which three phases coexist?
d) What is the maximum number of phases that can coexist in this system?

9.6-2. If g, the molar Gibbs function, is a convex function of $x_1, x_2, \ldots, x_{r-1}$, show that a change of variables to $x_2, x_3, \ldots, x_r$ results in g being a convex function of $x_2, x_3, \ldots, x_r$. That is, show that the convexity condition of the molar Gibbs potential is independent of the choice of the "redundant" mole fraction.

9.6-3. Show that the conditions of stability in a multicomponent system require that the partial molar Gibbs potential μ_j of any component be an increasing function of the mole fraction x_j of that component, both at constant v and at constant P, and both at constant s and at constant T.

9-7   PHASE DIAGRAMS FOR BINARY SYSTEMS

The Gibbs phase rule (equation 9.34) provides the basis for the study of the possible forms assumed by phase diagrams. These phase diagrams, particularly for binary (two-component) or ternary (three-component) systems, are of great practical importance in metallurgy and physical chemistry, and much work has been done on their classification. To illustrate the application of the phase rule, we shall discuss two typical diagrams for binary systems.

For a single-component system the Gibbs function per mole is a function of temperature and pressure, as in the three-dimensional representation in Fig. 9.11. The "phase diagram" in the two-dimensional $T-P$ plane (such as Fig. 9.1) is a projection of the curve of intersection (of the μ-surface with itself) onto the $T-P$ plane.

For a binary system the molar Gibbs function $G/(N_1 + N_2)$ is a function of the *three* variables T, P, and x_1. The analogue of Fig. 9.11 is then four-dimensional, and the analogue of the $T-P$ phase diagram is three-dimensional. It is obtained by projection of the "hypercurve" of intersection onto the P, T, x_1 "hyperplane."

The three-dimensional phase diagram for a simple but common type of binary gas–liquid system is shown in Fig. 9.17. For obvious reasons of graphic convenience the three-dimensional space is represented by a series of two-dimensional constant-pressure sections. At a fixed value of the mole fraction x_1 and fixed pressure the gaseous phase is stable at high temperature and the liquid phase is stable at low temperature. At a temperature such as that labeled C in Figure 9.17 the system separates into two phases—a liquid phase at A and a gaseous phase at B. The

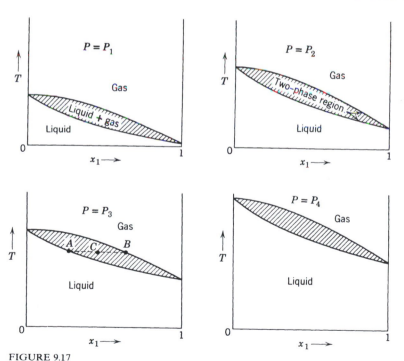

FIGURE 9.17

The three-dimensional phase diagram of a typical gas-liquid binary system. The two-dimensional sections are constant pressure planes, with $P_1 < P_2 < P_3 < P_4$.

composition at point C in Figure 9.17 is analogous to the volume at point Z in Figure 9.14 and a form of the lever rule is clearly applicable.

The region marked "gas" in Figure 9.17 is a three-dimensional region, and T, P, and x_1 can be independently varied within this region. This is true also for the region marked "liquid." In each case $r = 2$, $M = 1$, and $f = 3$.

The state represented by point C in Figure 9.17 is really a two-phase state, composed of A and B. Thus only A and B are physical points, and the shaded region occupied by point C is a sort of nonphysical "hole" in the diagram. The two-phase region is the surface enclosing the shaded volume in Figure 9.17. This surface is two-dimensional ($r = 2$, $M = 2$, $f = 2$). Specifying T and P determines x_1^A and x_1^B uniquely.

If a binary liquid with the mole fraction x_1^A is heated at atmospheric pressure, it will follow a vertical line in the appropriate diagram in Fig. 9.17. When it reaches point A, it will begin to boil. The vapor that escapes will have the composition appropriate to point B.

A common type of phase diagram for a liquid–solid, two-component system is indicated schematically in Fig. 9.18 in which only a single constant-pressure section is shown. Two distinct solid phases, of different crystal structure, exist: One is labeled α and the other is labeled β. The curve $BDHA$ is called the *liquidus* curve, and the curves BEL and ACJ are called *solidus* curves. Point G corresponds to a two-phase system—some liquid at H and some solid at F. Point K corresponds to α-solid at J plus β-solid at L.

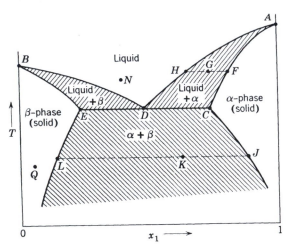

FIGURE 9.18
Typical phase diagram for a binary system at constant pressure.

If a liquid with composition x_H is cooled, the first solid to precipitate out has composition x_F. If it is desired to have the solid precipitate with the same composition as the liquid, it is necessary to start with a liquid of composition x_D. A liquid of this composition is called a *eutectic* solution. A eutectic solution freezes sharply and homogeneously, producing good alloy castings in metallurgical practice.

The liquidus and solidus curves are the traces of two-dimensional surfaces in the complete $T–x_1–P$ space. The eutectic point D is the trace of a curve in the full $T–x_1–P$ space. The eutectic is a three-phase region, in which liquid at D, β-solid at E, and α-solid at C can coexist. The fact that a three-phase system can exist over a one-dimensional curve follows from the phase rule ($r = 2$, $M = 3$, $f = 1$).

Suppose we start at a state such as N in the liquid phase. Keeping T and x_1 constant, we decrease the pressure so that we follow a straight line perpendicular to the plane of Fig. 9.18 in the $T–x_1–P$ space. We eventually come to a two-phase surface, which represents the liquid–gas phase transition. This phase transition occurs at a particular pressure for the given temperature and the given composition. Similarly, there is another particular pressure which corresponds to the temperature and composition of point Q and for which the solid β is in equilibrium with its own vapor. To each point T, x_1 we can associate a particular pressure P in this way. Then a phase diagram can be drawn, as shown in Fig. 9.19. This phase diagram differs from that of Fig. 9.18 in that the pressure at each point is different, and each point represents at least a two-phase system (of which one phase is the vapor). The curve $B'D'$ is now a one-dimensional curve ($M = 3$, $f = 1$), and the eutectic point D' is a unique point ($M = 4$, $f = 0$). Point B' is the triple point of the pure first component and point A' is the triple point of the pure second component.

Although Figs. 9.18 and 9.19 are very similar in general appearance, they are clearly very different in meaning, and confusion can easily arise

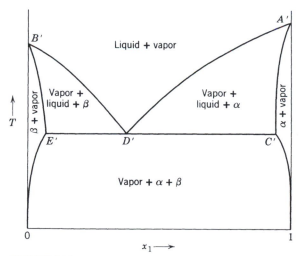

FIGURE 9.19

Phase diagram for a binary system in equilibrium with its vapor phase.

from failure to distinguish carefully between these two types of phase diagrams. The detailed forms of phase diagrams can take on a myriad of differences in detail, but the dimensionality of the intersections of the various multiphase regions is determined entirely by the phase rule.

PROBLEMS

9.7-1. The phase diagram of a solution of A in B, at a pressure of 1 atm, is as shown. The upper bounding curve of the two-phase region can be represented by

$$T = T_0 - (T_0 - T_1)x_A^2$$

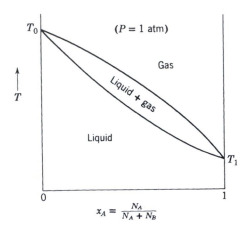

The lower bounding curve can be represented by

$$T = T_0 - (T_0 - T_1)x_A(2 - x_A)$$

A beaker containing equal mole numbers of A and B is brought to its boiling temperature. What is the composition of the vapor as it first begins to boil off? Does boiling tend to increase or decrease the mole fraction of A in the remaining liquid?

Answer:
$x_A(\text{vapor}) = 0.866$

9.7-2. Show that if a small fraction $(-dN/N)$ of the material is boiled off the system referred to in Problem 9.7-1, the change in the mole fraction in the remaining liquid is

$$dx_A = -\left[(2x_A - x_A^2)^{\frac{1}{2}} - x_A\right]\left(\frac{-dN}{N}\right)$$

9.7-3. The phase diagram of a solution of A in B, at a pressure of 1 atm and in the region of small mole fraction ($x_A \ll 1$), is as shown. The upper bounding

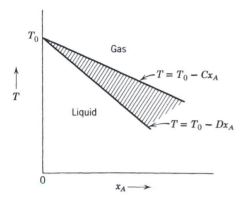

curve of the two-phase region can be represented by

$$T = T_0 - Cx_A$$

and the lower bounding curve by

$$T = T_0 - Dx_A$$

in which C and D are positive constants ($D > C$).

Assume that a liquid of mole fraction x_A^0 is brought to a boil and kept boiling until only a fraction (N_f/N_i) of the material remains; derive an expression for the final mole fraction of A.

Show that if $D = 3C$ and if $N_f/N_i = \frac{1}{2}$, the final mole fraction of component A is one fourth its initial value.

10

CRITICAL PHENOMENA

10-1 THERMODYNAMICS IN THE NEIGHBORHOOD OF THE CRITICAL POINT

The entire structure of thermodynamics, as described in the preceding chapters, appeared at mid-century to be logically complete, but the structure foundered on one ostensibly minor detail. That "detail" had to do with the properties of systems in the neighborhood of the critical point. Classical thermodynamics correctly predicted that various "generalized susceptibilities" (heat capacities, compressibilities, magnetic susceptibilities, etc.) should diverge at the critical point, and the general structure of classical thermodynamics strongly suggested the analytic form (or "shape") of those divergences. The generalized susceptibilities do diverge, but the analytic form of the divergences is not as expected. In addition the divergences exhibit regularities indicative of an underlying integrative principle inexplicable by classical thermodynamics.

Observations of the enormous fluctuations at critical points date back to 1869, when T. Andrews[1] reported the "critical opalescence" of fluids. The scattering of light by the huge density fluctuations renders water "milky" and opaque at or very near the critical temperature and pressure (647.29 K, 22.09 MPa). Warming or cooling the water a fraction of a Kelvin restores it to its normal transparent state.

Similarly, the magnetic susceptibility diverges for a magnetic system near its critical transition, and again the fluctuations in the magnetic moment are divergent.

A variety of other types of systems exhibit critical or "second-order" transitions; several are listed in Table 10.1 along with the corresponding "order parameter" (the thermodynamic quantity that exhibits divergent fluctuations, analogous to the magnetic moment).

[1] T. Andrews, *Phil. Trans. Royal Soc.* **159**, 575 (1869).

TABLE 10.1
Examples of Critical Points and Their Order Parameters*

Critical Point	Order Parameter	Example	T_{cr} (K)
Liquid–gas	Molar volume	H_2O	647.05
Ferromagnetic	Magnetic moment	Fe	1044.0
Antiferromagnetic	Sublattice magnetic moment	FeF_2	78.26
λ-line in ^{4}He	^{4}He quantum mechanical amplitude	^{4}He	1.8–2.1
Superconductivity	Electron pair amplitude	Pb	7.19
Binary fluid mixture	Fractional segregation of components	CCl_4–C_7F_{14}	301.78
Binary alloy	Fraction of one atomic species on one sublattice	Cu–Zn	739
Ferroelectric	Electric dipole moment	Triglycine sulfate	322.5

*Adapted from Shang-Keng Ma, *Modern Theory of Critical Phenomena* (Addison-Wesley Advanced Book Program, CA, 1976. Used by permission).

In order to fix these preliminary ideas in a specific way we focus on the gas–liquid transition in a fluid. Consider first a point P, T on the coexistence curve; two local minima of the *underlying Gibbs potential* then compete, as in Fig. 10.1 (page 205). If the point of interest were to move off the coexistence curve in either direction then one or the other of the two minima would become the lower. The two physical states, corresponding to the two minima, have very different values of molar volume,

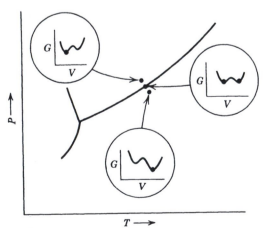

FIGURE 10.1
Competition of two minima of the underlying Gibbs potential near the coexistence curve.

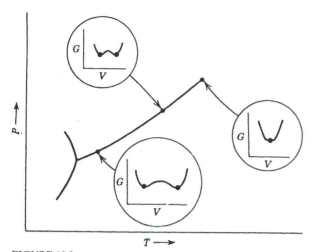

FIGURE 10.2
Coalescence of minima of the underlying Gibbs potential as the critical point is approached.

molar entropy, and so forth. These two states correspond, of course, to the two phases that compete in the first-order phase transition.

Suppose the point P, T on the coexistence curve to be chosen closer to the critical point. As the point approaches the critical T and P the two minima of the underlying Gibbs potential coalesce (Fig. 10.2).

For all points *beyond* the critical point (on the extended or extrapolated coexistence curve) the minimum is single and normal (Fig. 10.3). As the critical point is reached (moving inward toward the physical coexistence curve) the single minimum develops a flat bottom, which in turn develops a "bump" dividing the broadened minimum into two separate minima. The single minimum "bifurcates" at the critical point.

The flattening of the minimum of the Gibbs potential in the region of the critical state implies the absence of a "restoring force" for fluctuations away from the critical state (at least to leading order)—hence the divergent fluctuations.

This classical conception of the development of phase transitions was formulated by Lev Landau,[2] and extended and generalized by Laszlo Tisza,[3] to form the standard classical theory of critical phenomena. The essential idea of that theory is to expand the appropriate underlying thermodynamic potential (conventionally referred to as the "free energy functional") in a power series in $T - T_c$, the deviation of the temperature from its value $T_c(P)$ on the coexistence curve. The qualitative features described here then determine the relative signs of the first several

---

[2]*cf*. L. D. Landau and E. M. Lifshitz, *Statistical Physics*, MIT Press, Cambridge, Massachusetts and London, 1966.

[3]*cf*. L. Tisza, *Generalized Thermodynamics*, MIT Press, Cambridge, Massachusetts and London, 1966 (see particularly papers 3 and 4).

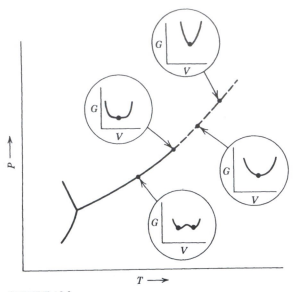

FIGURE 10.3
The classical picture of the development of a first-order phase transition. The dotted curve
is the extrapolated (non-physical) coexistence curve.

coefficients, and these terms in turn permit calculation of the analytic
behavior of the susceptibilities as T_c approaches the critical temperature
T_{cr}. A completely analogous treatment of a simple mechanical analogue
model is given in the Example at the end of this section, and an explicit
thermodynamic calculation will be carried out in Section 10.4. At this
point it is sufficient to recognize that the Landau theory is simple,
straightforward, and deeply rooted in the postulates of macroscopic
thermodynamics; it is based only on those postulates plus the reasonable
assumption of analyticity of the free energy functional. However, a direct
comparison of the theoretical predictions with experimental observations
was long bedeviled by the extreme difficulty of accurately measuring and
controlling temperature in systems that are incipiently unstable, with
gigantic fluctuations.

In 1944 Lars Onsager[4] produced the first rigorous statistical mechanical
solution for a nontrivial model (the "two-dimensional Ising model"), and
it exhibited a type of divergence very different from that expected. The
scientific community was at first loath to accept this disquieting fact,
particularly as the model was two-dimensional (rather than three-dimensional), and furthermore as it was a highly idealized construct bearing
little resemblance to real physical systems. In 1945 E. A. Guggenheim[5]

[4] L. Onsager, *Phys. Rev.* **65**, 117 (1944).
[5] E. A. Guggenheim, *J. Chem. Phys.* **13**, 253 (1945).

observed that the shape of the coexistence curve of fluid systems also cast doubt on classical predictions, but it was not until the early 1960s that precise measurements[6] forced confrontation of the failure of the classical Landau theory and initiated the painful reconstruction[7] that occupied the decades of the 1960s and the 1970s.

Deeply probing insights into the nature of critical fluctuations were developed by a number of theoreticians, including Leo Kadanoff, Michael Fischer, G. S. Rushbrooke, C. Domb, B. Widom, and many others.[8,9] The construction of a powerful analytical theory ("renormalization theory") was accomplished by Kenneth Wilson, a high-energy theorist interested in statistical mechanics as a simpler analogue to similar difficulties that plagued quantum field theory.

The source of the failure of classical Landau theory can be understood relatively easily, although it depends upon statistical mechanical concepts yet to be developed in this text. Nevertheless we shall be able in Section 10.5 to anticipate those results sufficiently to describe the origin of the difficulty in pictorial terms. The correction of the theory by renormalization theory unfortunately lies beyond the scope of this book, and we shall simply describe the general thermodynamic consequences of the Wilson theory. But first we must develop a framework for the description of the analytic form of divergent quantities, and we must review both the classical expectations and the (very different) experimental observations. To all of this the following mechanical analogue is a simple and explicit introduction.

Example

The mechanical analogue of Section 9.1 provides instructive insights into the flattening of the minimum of the thermodynamic potential at the critical point as that minimum bifurcates into two competing minima below T_{cr}. We again consider a length of pipe bent into a semicircle, closed at both ends, standing vertically on a table in the shape of an inverted U, containing an internal piston. On either side of the piston there is 1 mole of a monatomic ideal gas. The metal balls that were inserted in Section 9.1 in order to break the symmetry (and thereby to produce a first-order rather than a second-order transition) are *not* present.

If θ is the angle of the piston with respect to the vertical, $\tilde{R}$ is the radius of curvature of the pipe section, and Mg is the weight of the piston (we neglect gravitational effects on the gas itself), then the potential energy of the piston is

[6]*cf*. P. Heller and G. B. Benedek, *Phys. Rev. Let.* **8**, 428 (1962).

[7]*cf*. H. E. Stanley, *Introduction to Phase Transitions and Critical Phenomena*, Oxford Univ. Press, New York and Oxford, 1971.

[8]*cf*. H. E. Stanley, Ibid.

[9]P. Pfeuty and G. Toulouse, *Introduction to the Renormalization Group and Critical Phenomena*, John Wiley and Sons, NY 1977.

$(Mg\tilde{R})\cos\theta$, and the Helmholtz potential is

$$F = U - TS = (Mg\tilde{R})\cos\theta + F_L + F_R$$

The Helmholtz potentials F_L and F_R of the gases in the left-hand and right-hand sections of the pipe are given by (recall Problem 5.3-1)

$$F_{L,R} = F'(T) - RT\ln\left(\frac{V_{L,R}}{V_0}\right)$$

where $F'(T)$ is a function of T only. The volumes are determined by the position θ of the piston

$$V_L = \left(1 - \frac{2\theta}{\pi}\right)V_0, \qquad V_R = \left(1 + \frac{2\theta}{\pi}\right)V_0$$

where we have taken V_0 as half the total volume of the pipe. It follows then that, for small θ,

$$F(\theta, T) = Mg\tilde{R}\left[1 - \frac{\theta^2}{2} + \frac{\theta^4}{24} + \cdots\right]$$

$$+ 2F'(T) + RT\left[\left(\frac{2\theta}{\pi}\right)^2 + \frac{1}{2}\left(\frac{2\theta}{\pi}\right)^4 + \cdots\right]$$

$$= [Mg\tilde{R} + 2F'(T)] + \left(\frac{4}{\pi^2}RT - \frac{1}{2}Mg\tilde{R}\right)\theta^2$$

$$+ \left(\frac{1}{24}Mg\tilde{R} + \frac{8}{\pi^2}RT\right)\theta^4 + \cdots$$

The coefficient of θ^4 is intrinsically positive, but the coefficient of θ^2 changes sign at a temperature T_{cr}

$$T_{cr} = \frac{\pi^2}{8R}(Mg\tilde{R})$$

For $T > T_{cr}$ there is then only a single minimum; the piston resides at the apex of the pipe and the two gases have equal volumes.

For $T < T_{cr}$ the state $\theta = 0$ is a *maximum* of the Helmholtz potential and there are two symmetric minima at

$$\theta = \pm\sqrt{6}\,\pi\frac{T_{cr} - T}{24T + \pi^2 T_{cr}}$$

For $T = T_{cr}$ the Helmholtz potential has a very flat minimum, arising only from the fourth-order terms. Spontaneous fluctuations thereby experience only weak restoring forces. The "Brownian motion" (fluctuation) of the position of the piston is correspondingly large. Furthermore, even a trivially small force applied to the piston would induce a very large displacement; the "generalized susceptibility" diverges.

Although we have now seen the manner in which this model develops a bifurcating Helmholtz functional at the critical temperature, it may be instructive also to reflect on the manner in which a first-order transition occurs at lower temperatures. For this purpose some additional parameters must be introduced, to bias one minimum of F relative to the other. We might simply tilt the table slightly, thereby inducing a first-order transition from one minimum to the other. Alternatively, and more familiarly, a first-order phase transition can be thermally induced. In Section 9.1 this possibility was built into the model by inclusion of two metal ball-bearings of different coefficients of thermal expansion; a more appealing model would be one in which the two gases are differently nonideal.

Although this example employs a rather artificial system, the fundamental equation mimics that of homogeneous thermodynamic systems, and the analysis given above anticipates many features of the classical Landau theory to be described in Section 10.4.

10-2   DIVERGENCE AND STABILITY

The descriptive picture of the origin of divergences at the critical point, as alluded to in the preceding section, is cast into an illuminating perspective by the stability criteria (equation 8.15 and Problem 8.2-3)

$$\left(\frac{\partial^2 g}{\partial T^2} \right)_P < 0 \qquad \left(\frac{\partial^2 g}{\partial P^2} \right)_T < 0 \qquad (10.1)$$

and

$$\left(\frac{\partial^2 g}{\partial T^2} \right)_P \left(\frac{\partial^2 g}{\partial P^2} \right)_T - \left(\frac{\partial^2 g}{\partial T \, \partial P} \right)^2 > 0 \qquad (10.2)$$

These stability criteria express the concavity requirements of the Gibbs potential. The "flattening" of the Gibbs potential at the critical point corresponds to a failure of these concavity requirements. In fact all three of the stability criteria fail simultaneously, and α, κ_T, and c_p diverge together. Further perspective is provided by a physical, rather than a formal, point of view. Consider a particular point P, T on the coexistence

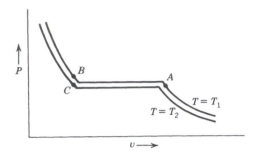

FIGURE 10.4
Schematic isotherms of a two-phase system.

curve of a two-phase system. The isotherms of the system are qualitatively similar to those shown in Fig. 10.4 (recall Fig. 9.12, although the van der Waals equation of state may not be *quantitatively* relevant). In particular, the isotherms have a flat portion in the P–T plane. On this flat portion the system is a mixture of two phases, in accordance with the "lever rule" (Section 9.4). The volume can be increased at constant pressure and temperature, the system responding simply by altering the mole fraction in each of the two coexistent phases. Thus, formally, the isothermal compressibility $\kappa_T = -v^{-1}(\partial v/\partial P)_T$ diverges.

Again considering this same system in the mixed two-phase state, suppose that a small quantity of heat Q $(= T\Delta S)$ is injected. The heat supplies the heat of transition (the heat of vaporization or the heat of melting) and a small quantity of matter transforms from one phase to the other. The temperature remains constant. Thus $c_p = T(\partial s/\partial T)_p$ diverges.

The divergence of κ_T and of c_p exists formally all along the coexistence locus. Across the coexistence locus in the P–T plane both κ_T and c_p are discontinuous, jumping from one finite value to another by passing through an intermediate infinity (in the mixed-phase state), see Fig. 10.5.

As the point of crossing of the coexistence curve is chosen closer to the critical point, classical Landau theory predicts that the "jump" of κ_T should decrease but that the intermediate infinity should remain. This

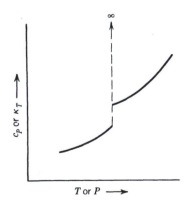

FIGURE 10.5
Discontinuity and divergence of generalized susceptibilities across a coexistence locus. The abscissa can be either T or P, along a line crossing the coexistence locus in the $T - P$ plane.

description is correct except very close to the critical point, in which region nonclassical behavior dominated by the fluctuations intervenes. Nevertheless, the qualitative behavior remains similar—a divergence of κ_T at the critical point, albeit of an altered functional form.

The heat capacity behaves somewhat differently. As we shall see later, Landau theory predicts that as the critical point is approached both the jump in the heat capacity and the intermediate divergence should fade away. In fact the divergence remains, though it is a weaker divergence than that of κ_T.

10-3 ORDER PARAMETERS AND CRITICAL EXPONENTS

Although Landau's classical theory of critical transitions was not quantitatively successful, it did introduce several pivotal concepts. A particularly crucial observation of Landau was that in any phase transition there exists an "order parameter" that can be so defined that it is zero in the high-temperature phase and nonzero in the low-temperature phase. Order parameters for various second-order transitions are listed in Table 10.1. The simplest case, and the prototypical example, is provided by the paramagnetic to ferromagnetic transition (or its electric analogue). An appropriate order parameter is the magnetic moment, which measures the cooperative alignment of the atomic or molecular dipole moments.

Another simple and instructive transition is the binary alloy "order–disorder" transition that occurs, for instance, in copper–zinc (Cu–Zn) alloy. The crystal structure of this material is "body-centered cubic," which can be visualized as being composed of two interpenetrating simple cubic lattices. For convenience we refer to one of the sublattices as the A sublattice and to the other as the B sublattice. At high temperatures the Cu and Zn atoms of the alloy are randomly located, so that any particular lattice point is equally likely to be populated by a zinc or by a copper atom. As the temperature is lowered, a phase transition occurs such that the copper atoms preferentially populate one sublattice and the zinc atoms preferentially populate the other sublattice. Immediately below the transition temperature this preference is very slight, but with decreasing temperature the sublattice segregation increases. At zero temperature one of the sublattices is entirely occupied by copper atoms and the other sublattice is entirely occupied by zinc atoms. An appropriate order parameter is $(N_{Zn}^A - N_{Cu}^A)/N^A$, or the difference between the fraction of A sites occupied by zinc atoms and the fraction occupied by copper atoms. Above the transition temperature the order parameter is zero; it becomes nonzero at the transition temperature; and it becomes either $+1$ or -1 at $T = 0$.

As in the order–disorder transition, the order parameter can always be chosen to have unit magnitude at zero temperature; it is then "normalized." In the ferromagnetic case the normalized order parameter is $I(T)/I(0)$; whereas the extensive parameter is the magnetic moment $I(T)$.

In passing we recall the discussion in Section 3.8 on unconstrainable variables. As was pointed out, it sometimes happens that a formally defined intensive parameter does not have a physical realization. The copper–zinc alloy system is such a case. In contrast to the ferromagnetic case (in which the order parameter is the magnetic moment I and the intensive parameter $\partial U/\partial I$ is the magnetic field B_e), the order parameter for the copper–zinc alloy is $(N_{Zn}^A - N_{Cu}^A)$ but the intensive parameter has no physical reality. Thus the thermodynamic treatment of the Cu–Zn system requires that the intensive parameter always be assigned the value zero. Similarly the intensive parameter conjugate to the order parameter of the superfluid ^{4}He transition must be taken as zero.

Identification of the order parameter, and recognition that various generalized susceptibilities diverge at the critical point, motivates the definition of a set of "critical exponents" that describe the behavior of these quantities in the critical region.

In the thermodynamic context there are four basic critical exponents, defined as follows.

The molar heat capacity (c_v in the fluid case or c_{B_e} in the magnetic case) diverges at the critical point with exponents α above T_{cr} and α' below T_{cr}

$$c_v \text{ or } c_{B_e} \sim (T - T_{cr})^{-\alpha} \qquad (T > T_{cr}) \qquad (10.3)$$

$$c_v \text{ or } c_{B_e} \sim (T_{cr} - T)^{-\alpha'} \qquad (T < T_{cr}) \qquad (10.4)$$

The "generalized susceptibilities", $\kappa_T = -(\partial v/\partial P)_T/v$ in the fluid case or $\chi_T = \mu_0(\partial I/\partial B_e)_T/v$ in the magnetic case, diverge with exponents γ or γ'.

$$\kappa_T \text{ or } \chi_T \sim (T - T_{cr})^{-\gamma} \qquad (T > T_{cr}) \qquad (10.5)$$

$$\kappa_T \text{ or } \chi_T \sim (T_{cr} - T)^{-\gamma'} \qquad (T < T_{cr}) \qquad (10.6)$$

Along the coexistence curve the order parameter varies as $(T_{cr} - T)^\beta$

$$\Delta v \text{ or } I \sim (T_{cr} - T)^\beta \qquad (T < T_{cr}) \qquad (10.7)$$

and, of course, the order parameter vanishes for $T > T_{cr}$. Note that a prime indicates $T < T_{cr}$ for the exponents α' and γ'; whereas β can be defined only for $T < T_{cr}$ so that a prime is superfluous.

Finally, on the critical isotherm (i.e., for $T = T_{cr}$) the order parameter and its corresponding intensive parameter satisfy the relation

$$I \sim B_e^{1/\delta} \quad \text{or} \quad \Delta v \sim (P - P_{cr})^{1/\delta} \tag{10.8}$$

which define the exponent δ.

In addition there are several critical exponents defined in terms of statistical mechanical concepts lying outside the domain of macroscopic thermodynamics. Perhaps the most significant of these additional exponents describes the range of fluctuations, or the size of the correlated regions within the system. The long wavelength fluctuations dominate near the critical point, and the range of the correlated regions diverges.

This onset of long-range correlated behavior is the key to the statistical mechanical (or "renormalization group") solution to the problem. Because large regions are so closely correlated, the details of the particular atomic structure of the specific material become of secondary importance! The atomic structure is so masked by the long-range correlation that large families of materials behave similarly—a phenomenon known as "universality," to which we shall return subsequently.

10-4 CLASSICAL THEORY IN THE CRITICAL REGION: LANDAU THEORY

The classical theory of Landau, which evaluates the critical exponents, provides the standard of expectation to which we can contrast both experimental observations and the results of renormalization group theory.

We consider a system in which the unnormalized order parameter is ϕ. We have in mind, perhaps, the magnetization of a uniaxial crystal (in which the dipoles are equally probably "up" or "down" above the transition temperature), or the binary Cu–Zn alloy. The Gibbs potential G is a function of $T, P, \phi, N_1, N_2, \ldots, N_r$

$$G = G(T, P, \phi, N_1, N_2, \ldots, N_r). \tag{10.9}$$

In the immediate vicinity of the critical point the order parameter is small, suggesting a series expansion in powers of ϕ

$$G = G_0 + G_1\phi + G_2\phi^2 + G_3\phi^3 + \cdots \tag{10.10}$$

where $G_0, G_1, G_2, \ldots$ are functions of $T, P, N_1, \ldots, N_r$. For the magnetic system or binary alloy the symmetry of the problems immediately precludes the odd terms, requiring that the Gibbs potential be even in ϕ; there is no a priori difference between spin up and spin down, or between the A and B sublattices. (This reasoning is a precursor and a prototype of more elaborate symmetry arguments in more complex systems.)

$$G(T, P, \phi, N_1, \ldots, N_r) = G_0 + G_2\phi^2 + G_4\phi^4 + \cdots \qquad (10.11)$$

Each of the expansion coefficients is a function of T, P, and the N_j's; $G_n = G_n(T, P, N_1, \ldots, N_r)$. We now concentrate our attention on the extrapolated coexistence curve—the dotted curve in Fig. 10.3. Along this locus P is a function of T, and all mole numbers are constant, so that each of the expansion coefficients G_n is effectively a function of T only. Correspondingly, G is effectively a function only of T and ϕ.

The shape of $G(T, \phi)$ as a function of ϕ, for small ϕ, is shown in Fig. 10.6 for the four possible combinations of signs of G_2 and G_4.

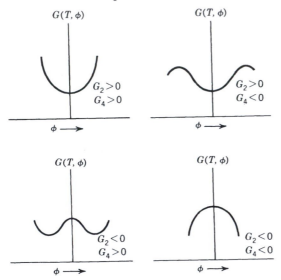

FIGURE 10.6
Possible shapes of $G(T, \phi)$ for various signs of the expansion coefficients.

A point on the extrapolated coexistence curve ("beyond" the critical point) is in the single-phase region of stability where the Gibbs potential has a simple minimum. From this fact we conclude that $G_2(T)$ is positive. Stability to large fluctuations implies also that $G_4(T)$ is positive. As the point of interest approaches and then passes the critical point, along the coexistence curve, the curvature $G_2(T)$ passes through zero and becomes negative (Fig. 10.6). The function $G_4(T)$ normally remains positive. *The critical temperature is viewed simply as the temperature at which G_2 happens to have a zero.*

The change of sign of G_2 at the critical point implies that a series expansion of G_2 in powers of $(T - T_c)$ has the form

$$G_2[T, P(T)] = (T - T_{cr})G_2^0 + \text{terms of order } (T - T_{cr})^2 + \cdots$$

$$(10.12)$$

Now, let the intensive parameter conjugate to ϕ have the value zero. In the magnetic case, in which ϕ is the normalized magnetic moment, this implies that there is no external magnetic field, whereas in the binary alloy the intensive parameter is automatically zero. Then, in either type of case

$$\frac{\partial G}{\partial \phi} = 2(T - T_{cr})G_2^0\phi + 4G_4\phi^3 + \cdots = 0 \qquad (10.13)$$

This equation has different solutions above and below T_{cr}. For $T > T_{cr}$ the only real solution is $\phi = 0$.

$$\phi = 0 \quad (\text{for } T > T_{cr}) \qquad (10.14)$$

Below T_{cr} the solution $\phi = 0$ corresponds to a maximum rather than a minimum value of G (recall Fig. 10.6), but there are two real solutions corresponding to minima

$$\phi = \pm\left[2\frac{G_2^0}{G_4}(T_{cr} - T)\right]^{1/2}, \quad (T \le T_{cr}) \qquad (10.15)$$

This is the basic conclusion of the classical theory of critical points. *The order parameter (magnetic moment, difference in zinc and copper occupation of the A sublattice, etc.) spontaneously becomes nonzero and grows as* $(T_{cr} - T)^{1/2}$ *for temperatures below* T_{cr}. *The critical exponent β, defined in equation 10.7, thereby is evaluated classically to have the value $\frac{1}{2}$.*

$$\beta(\text{classical}) = 1/2 \qquad (10.16)$$

In contrast, experiment indicates that for various ferromagnets or fluids the value of β is in the neighborhood of 0.3 to 0.4.

In equation 10.13 we assumed that the intensive parameter conjugate to ϕ is zero; this was dictated by our interest in the spontaneous value of ϕ below T_{cr}. We now seek the behavior of the "susceptibility" χ_T for temperatures just above T_{cr}, χ_T being defined by

$$\chi_T^{-1} = N\left(\frac{\partial^2 G}{\partial \phi^2}\right)_{T,\phi \to 0} \qquad (10.17)$$

In the magnetic case χ_T^{-1} is equal to $N(\partial B_e/\partial I)_{T,I \to 0}$ so that $\mu_0\chi_T$ is the familiar molar magnetic susceptibility (but in the present context we shall not be concerned with the constant factor μ_0). Then

$$\frac{1}{N}\chi_T^{-1} = 2(T - T_{cr})G_2^0 + 12G_4\phi^2 + \cdots \qquad (10.18)$$

or taking $\phi \rightarrow 0$ according to the definition 10.17,

$$\frac{1}{N}\chi_T^{-1} = 2(T - T_{cr})G_2^0 + \cdots \qquad T \geq T_{cr} \qquad (10.19)$$

This result evaluates the classical value of the exponent γ (equation 10.5) as unity

$$\gamma(\text{classical}) = 1 \qquad (10.20)$$

Again, for ferromagnets and for fluids the measured values of γ are in the region of 1.2 to 1.4.

For $T < T_{cr}$ the order parameter ϕ becomes nonzero. Inserting equation 10.15 for $\phi(T)$ into equation 10.18

$$\frac{1}{N}\chi_T^{-1} = 2(T - T_{cr})G_2^0 + 12G_4 \times \left[\frac{1}{2}\frac{G_2^0}{G_4}\right](T_{cr} - T) + \cdots$$

$$= 4(T_{cr} - T)G_2^0 + \cdots \qquad (10.21)$$

We therefore conclude that the classical value of γ' is unity (recall equation 10.6). Again this does not agree with experiment, which yields values of γ' in the region of 1.0 to 1.2.

The values of the critical exponents that follow from the Landau theory are listed, for convenience, in Table 10.2.

TABLE 10.2
Critical Exponents; Classical Values and Approximate Range of Observed Values

Exponent	Classical value	Approximate range of observed values
α	0	$-0.2 < \alpha < 0.2$
α'	0	$-0.2 < \alpha' < 0.3$
β	$\frac{1}{2}$	$0.3 < \beta < 0.4$
γ	1	$1.2 < \gamma < 1.4$
γ'	1	$1 < \gamma' < 1.2$
δ	3	$4 < \delta < 5$

Example

It is instructive to calculate the classical values of the critical exponents for a system with a given, definite fundamental equation, thereby corroborating the more general Landau analysis. Calculate the critical indices for a system described by the van der Waals equation of state.

Solution

From Example 1 of Section 9.4, the van der Waals equation of state can be written in "reduced variables";

$$\tilde{P} = \frac{8\tilde{T}}{3\tilde{v} - 1} - \frac{3}{\tilde{v}^2}$$

where $\tilde{P} \equiv P/P_{cr}$ and similarly for $\tilde{T}$ and $\tilde{v}$. Then, defining

$$p \equiv \tilde{P} - 1 \qquad \hat{v} \equiv \tilde{v} - 1 \qquad \varepsilon \equiv \tilde{T} - 1$$

and multiplying the van der Waals equation by $(1 + \hat{v})^2$ we obtain[10]

$$2p\left(1 + \tfrac{7}{2}\hat{v} + 4\hat{v}^2 + \tfrac{3}{2}\hat{v}^3\right) = -3v^3 + 8(1 + 2\hat{v} + \hat{v}^2)$$

or

$$p = -\tfrac{3}{2}\hat{v}^3 + \varepsilon\left(4 - 6\hat{v} + 9\hat{v}^2 + \cdots\right) + \cdots$$

If $\varepsilon = 0$ (that is $T = T_{cr}$) then $\hat{v}$ is proportional to $(-p)^{\frac{1}{3}}$, so that the critical exponent δ is identified as $\delta = 3$.

To evaluate γ we calculate

$$\kappa_T^{-1} = -V\left(\frac{\partial P}{\partial V}\right)_T = -\tilde{v}\left(\frac{\partial p}{\partial \hat{v}}\right)_\varepsilon = 6\tilde{v}\varepsilon + \cdots$$

whence $\gamma = \gamma' = 1$.

To calculate β we recall that $\theta(\tilde{v}_g) = \theta(\tilde{v}_\ell)$, where $\theta(\tilde{v})$ is defined by the last equation in Example 2, page 241.

$$\theta(\tilde{v}) \equiv \ln(3\tilde{v} - 1) - (3\tilde{v} - 1)^{-1} + 9/(4\tilde{v}T).$$

$$= \ln(3\hat{v} + 2) - (3\hat{v} + 2)^{-1} + \tfrac{9}{4}(\hat{v} + 1)^{-1}(\varepsilon + 1)^{-1}$$

$$= \ln 2 + \tfrac{7}{4} + \tfrac{9}{16}\hat{v}^3 + \tfrac{9}{4}\varepsilon(1 + \varepsilon + \hat{v} - \hat{v}\varepsilon - \hat{v}^2 + \dots)$$

Then, from $\theta(\tilde{v}_g) = \theta(\tilde{v}_\ell)$ we find

$$\tfrac{1}{4}\left(\hat{v}_g^2 + \hat{v}_g\hat{v}_\ell + \hat{v}_\ell^2\right) + \varepsilon - \varepsilon^2 - \varepsilon(\hat{v}_g + \hat{v}_\ell) = 0$$

Also $p(\hat{v}_g) = p(\hat{v}_\ell)$, which gives

$$\hat{v}_g^2 + \hat{v}_g\hat{v}_\ell + \hat{v}_\ell + 4\varepsilon - 6\varepsilon(\hat{v}_g + \hat{v}_\ell) = 0$$

These latter two equations constitute two equations in the two unknowns $\hat{v}_g$ and $\hat{v}_\ell$. Eliminating $(\hat{v}_g + \hat{v}_\ell)$ we are left with a single equation in $\hat{v}_g - \hat{v}_\ell$; we find

$$\hat{v}_g - \hat{v}_\ell = 4(-\varepsilon)^{\frac{1}{2}} + \cdots$$

which identifies the critical exponent β as $\tfrac{1}{2}$.

The remaining critical exponents are α and α', referring to the heat capacity. The van der Waals equation of state alone does not determine the heat capacity, but we can turn to the "ideal van der Waals fluid" defined in Section 3.5. For that

---

[10] H. Stanley, *Introduction to Phase Transitions and Critical Phenomena*, Oxford Univ. Press, New York and Oxford, 1971. (sect. 5.5).

system the heat capacity c_v is a constant, with no divergence at the critical point, and $\alpha = \alpha' = 0$.

10-5 ROOTS OF THE CRITICAL POINT PROBLEM

The reader may well ask how so simple, direct, and general an argument as that of the preceding section can possibly lead to incorrect results. Does the error lie within the argument itself, or does it lie deeper, at the very foundations of thermodynamics? That puzzlement was shared by thermodynamicists for three decades. Although we cannot enter here into the renormalization theory that solved the problem, it may be helpful at least to identify the source of the difficulty. To do so we return to the most central postulate of thermodynamics—the entropy maximum postulate. In fact that "postulate" is a somewhat over-simplified transcription of the theorems of statistical mechanics. The over-simplification has significant consequences only when fluctuations become dominant—that is, in the critical region.

The crucial theorems of statistical mechanics evaluate the probability of fluctuations in closed composite systems (or in systems in contact with appropriate reservoirs). In particular, for a closed composite system the energy of one of the subsystems fluctuates, and the probability that at any given instant it has a value E is proportional to $\exp(S(E)/k_B)$, where S is the entropy of the composite system. The average energy U is to be obtained from this probability density by a standard averaging process.

Generally the probability density is very "sharp", or narrow. The average energy then is very nearly equal to the most probable energy. The latter is the more easily obtainable from the probability distribution, for it (i.e., *the most probable energy*) *is simply that value of E that maximizes* $\exp(S(E)/k_B)$ *or that maximizes the entropy S.*

The basic postulate of thermodynamics incorrectly identifies the most probable value of the energy as the equilibrium or average value!

Fortunately the probability density of macroscopic systems is almost always extremely narrow. For a narrow probability density the average value and the most probable value coincide, and classical thermodynamics then is a valid theory. However, in the critical region the minimum of the thermodynamic potential becomes very shallow, the probability distribution becomes very broad, and the distinction between average and most probable states can become significant.

To illustrate the consequence of this distinction near the critical point, Fig. 10.7 shows the Gibbs potential schematically as a function of the order parameter ϕ for two temperatures very slightly below T_{cr} (with the intensive parameter equal to zero). Only the positive branch of ϕ is shown, though there is a similar branch for negative ϕ (we assume the system to be in the minimum with $\phi > 0$). For T_1 the potential is shallow and asymmetric, and the probability density for the fluctuating order

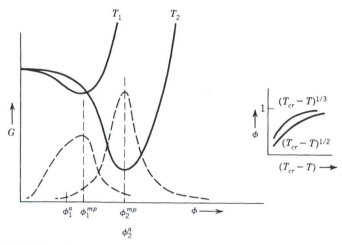

FIGURE 10.7

Probability distributions, average, and most probable values for the fluctuating order parameter. The temperatures are $T_2 \leq T_1 \leq T_c$. The probability distributions are shown as dotted curves. The classical or most probable values are ϕ_1^{mp} and ϕ_2^{mp}, and these coincide with the minima of G. The average or observable values are ϕ_1^a and ϕ_2^a. The rate of change of the average values is more rapid than the rate of change of the most probable values because of the asymmetry of the curves for T_1. This is more consistent with a critical index $\beta \simeq \frac{1}{3}$ rather than $\frac{1}{2}$, as shown in the small figure.

parameter (shown dotted) is correspondingly broad and asymmetric. The *average* value ϕ_1^a of ϕ is shifted to the left of the *most probable* value ϕ_1^{mp}. For a temperature T_2 further removed from the critical temperature the potential well is almost symmetric near its minimum, and the probability density is almost symmetric. The average value ϕ_2^a and the most probable value ϕ_2^{mp} are then almost identical. As the temperature changes from T_1 to T_2 the classically predicted change in the order parameter is $\phi_2^{mp} - \phi_1^{mp}$ whereas the statistical mechanical prediction is $\phi_2^a - \phi_1^a$. Thus we see that classical thermodynamics incorrectly predicts the temperature dependence of the order parameter as the critical temperature is approached, and that this failure is connected with the shallow and asymmetric nature of the minimum of the potential.

To extend the reasoning slightly further, we observe that $\phi_2^{mp} - \phi_1^{mp}$ is smaller than $\phi_2^a - \phi_1^a$ (Fig. 10.7). That is, the classical thermodynamic prediction of the shift in ϕ (for a given temperature change) is smaller than the true shift (i.e., than the shift in the *average* value of ϕ). This is consistent with the classical prediction of $\beta = \frac{1}{2}$ rather than the true value $\beta \simeq \frac{1}{3}$, as indicated in the insert in Fig. 10.7.

This discussion provides, at best, a pictorial insight as to the origin of the failure of classical Landau theory. It gives no hint of the incredible depth and beauty of "renormalization-group theory," about which we later shall have only a few observations to make.

10-6   SCALING AND UNIVERSALITY

As mentioned in the last paragraph of Section 10.3, the dominant effect that emerges in the renormalization group theory is the onset of long-range correlated behavior in the vicinity of the critical point. This occurs because the long wave length excitations are most easily excited. As fluctuations grow the very long wave length fluctuations grow most rapidly, and they dominate the properties in the critical region. Two effects result from the dominance of long range correlated fluctuations.

The first class of effects is described by the term *scaling*. Specifically, the divergence of the susceptibilities and the growth of the order parameter are linked to the divergence of the range of the correlated fluctuations. Rather than reflecting the full atomic complexity of the system, the diverse critical phenomena all scale to the range of the divergent correlations *and thence to each other*. This interrelation among the critical exponents is most economically stated in the "scaling hypothesis," the fundamental result of renormalization–group theory. That result states that the dominant term in the Gibbs potential (or another thermodynamic potential, as appropriate to the critical transition considered) in the region of the critical point, is of the form

$$G_s \sim |T - T_{cr}|^{2-\alpha} f^{\pm}\left(\frac{B_e^{1+1/\delta}}{|T - T_c|^{2-\alpha}} \right), \qquad (T \to T_{cr}) \qquad (10.22)$$

We here use the magnetic notation for convenience, but B_e can be interpreted generally as the intensive parameter conjugate to the order parameter ϕ. The detailed functional form of the Gibbs potential is discontinuous across the coexistence curve, as expected, and this discontinuity in form is indicated by the notation $f^{\pm}$; the function f^+ applies for $T > T_{cr}$ and the (different) function f^- applies for $T < T_{cr}$. Furthermore the Gibbs potential may have additional "regular" terms, the terms written in equation 10.22 being only the dominant part of the Gibbs potential in the limit of approach to the critical point.

The essential content of equation 10.22 is that the quantity $G_s/(T - T_{cr})^{2-\alpha}$ is not a function of both T and B_e separately, but only of the single variable $B_e^{1+1/\delta}/|T - T_{cr}|^{2-\alpha}$. It can equally well be written as a function of the square of this composite variable, or of any other power. We shall later write it as a function of $B_e/(T - T_{cr})^{(2-\alpha)\delta/(1+\delta)}$.

The scaling property expressed in equation 10.22 relates all other critical exponents *by universal relationships* to the two exponents α and δ, as we shall now demonstrate. The procedure is straightforward; we simply evaluate each of the critical exponents from the fundamental equation 10.22.

We first evaluate the critical index α, to corroborate that the symbol α appearing in equation 10.22 does have its expected significance. For this purpose we take $B_e = 0$. The functions $f^\pm(x)$ are assumed to be well behaved in the region of $x = 0$, with $f^\pm(0)$ being finite constants. Then the heat capacity is

$$c_{B_e} \sim \frac{\partial^2 G_s(B_e = 0)}{\partial T^2} \sim (2 - \alpha)(1 - \alpha)|T - T_{cr}|^{-\alpha} f^\pm(0) \qquad (10.23)$$

Hence the critical index for the heat capacity, both above and below T_{cr}, is identified as equal to the parameter α in G_s, whence

$$\alpha' = \alpha \qquad (10.24)$$

Similarly, the equation of state $I = I(T, B_e)$ is obtained from equation 10.22 by differentiation

$$I = -\frac{\partial G_s}{\partial B_e} \sim -|T - T_{cr}|^{2-\alpha} f'^\pm \left(\frac{B_e^{1+1/\delta}}{|T - T_{cr}|^{2-\alpha}} \right) \frac{\partial}{\partial B_e} \left(\frac{B_e^{1+1/\delta}}{|T - T_{cr}|^{2-\alpha}} \right)$$

$$\sim -B_e^{1/\delta} f'^\pm \left(\frac{B_e^{1+1/\delta}}{|T - T_{cr}|^{2-\alpha}} \right) \qquad (10.25)$$

where $f^\pm(x)$ denotes $(d/dx)f^\pm(x)$. Again the functions $f'^\pm(0)$ are assumed finite, and we have therefore corroborated that the symbol δ has its expected significance (as defined in equation 10.8).

To focus on the temperature dependence of I and of χ, in order to evaluate the critical exponents β and γ, it is most convenient to rewrite $f^\pm$ as a function $g^\pm$ of $B_e/(T - T_{cr})^{(2-\alpha)\delta/(1+\delta)}$.

$$G_s \sim |T - T_{cr}|^{2-\alpha} g^\pm \left(\frac{B_e}{|T - T_{cr}|^{(2-\alpha)\delta/(1+\delta)}} \right) \qquad (10.26)$$

Then

$$I = -\frac{\partial G_s}{\partial B_e} \sim |T - T_{cr}|^{(2-\alpha)/(1+\delta)} g'^\pm \left(\frac{B_e}{|T - T_{cr}|^{(2-\alpha)\delta/(1+\delta)}} \right) \qquad (10.27)$$

whence

$$\beta = \frac{2 - \alpha}{1 + \delta} \qquad (10.28)$$

Also

$$\chi = \mu_0 \frac{\partial I}{\partial B_e} \sim |T - T_{cr}|^{(2-\alpha)(1-\delta)/(1+\delta)} g''^{\pm} \left(\frac{B_e}{|T - T_{cr}|^{(2-\alpha)\delta/(1+\delta)}} \right)$$

(10.29)

whence

$$\gamma = \gamma' = (\alpha - 2) \frac{1 - \delta}{1 + \delta}$$

(10.30)

Thus all the critical indices have been evaluated in terms of α and δ. The observed values of the critical indices of various systems are, of course, consistent with these relationships.

As has been stated earlier, there are two primary consequences of the dominance of long range correlated fluctuations. One of these is the scaling of critical properties to the range of the correlations, giving rise to the scaling relations among the critical exponents. The second consequence is that the numerical values of the exponents do not depend on the detailed atomic characteristics of the particular material, but are again determined by very general properties of the divergent fluctuations. Renormalization group theory demonstrates that the numerical values of the exponents of large classes of materials are identical; the values are determined primarily by the *dimensionality of the system* and by the *dimensionality of the order parameter*.

The dimensionality of the system is a fairly self-evident concept. Most thermodynamic systems are three-dimensional. However it is possible to study two-dimensional systems such as monomolecular layers adsorbed on crystalline substrates. Or one-dimensional polymer chains can be studied. An even greater range of dimensions is available to theorists, who can (and do) construct statistical mechanical model systems in four, five, or more dimensions (and even in fractional numbers of dimensions!).

The dimensionality of the order parameter refers to the scalar, vector, or tensorial nature of the order parameter. The order parameter of the binary alloy discussed in Section 10.3 is one-dimensional (scalar). The order parameter of a ferromagnet, which is the magnetic moment, is a vector and is of dimensionality three. The order parameter of a superconductor, or of superfluid ^{4}He, is a complex number; having independent real and imaginary components it is considered as two-dimensional. And again theoretical models can be devised with other dimensionalities of the order parameters.

Systems* with the same spatial dimensionality and with the same dimensionality of their order parameters are said to be in the same "universality class." And systems in the same universality class have the same values of their critical exponents.

PROBLEMS

10.6-1. Show that the following identities hold among the critical indices

$$\alpha + 2\beta + \gamma = 2 \qquad (\text{"Rushbrooke's scaling law"})$$

$$\gamma = \beta(\delta - 1) \qquad (\text{"Widom's scaling law"})$$

10.6-2. Are the classical values of the critical exponents consistent with the scaling relations?

---

*It is assumed that the interatomic forces in the system are not of infinite range.

11

THE NERNST POSTULATE

11-1 NERNST'S POSTULATE, AND THE PRINCIPLE OF THOMSEN AND BERTHELOT

One aspect of classical thermodynamics remains. That is the exploration of the consequences of postulate IV, to the effect that the entropy vanishes at zero temperature.

The postulate as first formulated by Walther Nernst in 1907 was somewhat weaker than our postulate IV, stating only that the entropy *change* in any isothermal process approaches zero as the temperature approaches zero. The statement that we have adopted emerged several decades later through the work of Francis Simon and the formulation of Max Planck; it is nevertheless referred to as the Nernst postulate. It is also frequently called the "third law" of thermodynamics.

Unlike the other postulates of the formalism, the Nernst postulate is not integral to the overall structure of thermodynamic theory. Having developed the theory almost in its entirety, we can now simply append the Nernst postulate. Its implications refer entirely to the low-temperature region, near $T = 0$.

The historical origins of the Nernst theorem are informative; they lie in the "principle of Thomsen and Berthelot"—an empirical (but nonrigorous) rule by which chemists had long predicted the equilibrium state of chemically reactive systems.

Consider a system maintained at constant temperature and pressure (as by contact with the ambient atmosphere), and released from constraints (as by mixture of two previously separated chemical reactants). According to the empirical rule of Thomsen and Berthelot, the equilibrium state to which the system proceeds is such that the accompanying process evolves the greatest efflux of heat, or, in the more usual language, "the process is realized that is most exothermic."

The formal statement of this empirical rule is most conveniently put in terms of the enthalpy. We recall that in isobaric processes the enthalpy acts as a potential for heat, so that the total heat efflux is

$$\text{heat efflux} = H_{\text{initial}} - H_{\text{final}} \tag{11.1}$$

The statement of Thomsen and Berthelot therefore is equivalent to the statement that the equilibrium state is the one that maximizes $H_{\text{initial}} - H_{\text{final}}$, or minimizes H_{final}.

The *proper* criterion of equilibrium at constant temperature and pressure is, of course, the minimization of the Gibbs potential. Why then should these two differing criteria provide similar predictions at low temperatures (and, in fact, sometimes even at or near room temperature)?

In an isothermal process

$$\Delta G = \Delta H - T \Delta S \tag{11.2}$$

so that *at* $T = 0$ the changes in the Gibbs potential and in the enthalpy are equal (ΔS certainly being bounded). But that is not sufficient to explain why they remain approximately equal over some nonnegligible temperature range. However, dividing by T

$$\frac{\Delta H - \Delta G}{T} = \Delta S \tag{11.3}$$

We have seen from equation 11.2 that $\Delta H = \Delta G$ at $T = 0$; hence the left-hand side of equation 11.3 is an indeterminate form as $T \to 0$. The limiting value is obtained by differentiating numerator and denominator separately (L'Hospital's rule), whence

$$\left(\frac{d \Delta H}{dT} \right)_{T=0} - \left(\frac{d \Delta G}{dT} \right)_{T=0} = \lim_{T \to 0} \Delta S \tag{11.4}$$

By *assuming* that

$$\lim_{T \to 0} \Delta S = 0 \tag{11.5}$$

it was ensured by Nernst that ΔH and ΔG have the same initial slope (Fig. 11.1), and that therefore the change in enthalpy is very nearly equal to the change in Gibbs potential over a considerable temperature range.

The Nernst statement, that the *change* in entropy ΔS vanishes in any reversible isothermal process at zero temperature, can be restated: *The $T = 0$ isotherm is also an isentrope* (or "adiabat"). This coincidence of isotherm and isentrope is illustrated in Fig. 11.2.

The Planck restatement assigns a particular value to the entropy: *The $T = 0$ isotherm coincides with the $S = 0$ adiabat.*

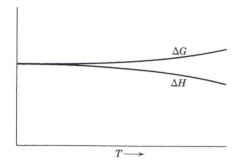

FIGURE 11.1
Illustrating the principle of Thomsen and
Berthelot.

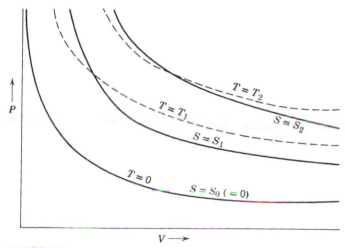

FIGURE 11.2
Isotherms and isentropes ("adiabats") near $T = 0$.

In the thermodynamic context there is no a priori meaning to the
absolute value of the entropy. The Planck restatement has significance
only in its statistical mechanical interpretation, to which we shall turn in
Part II. We have, in fact, chosen the Planck form of the postulate rather
than the Nernst form largely because of the pithiness of its statement
rather than because of any additional thermodynamic content.

The "absolute entropies" tabulated for various gases and other systems
in the reference literature fix the scale of entropy by invoking the Planck
form of the Nernst postulate.

PROBLEMS

11.1-1. Does the two-level system of Problem 5.3-8 satisfy the Nernst postulate?
Prove your assertion.

11-2   HEAT CAPACITIES AND OTHER
DERIVATIVES AT LOW TEMPERATURE

A number of derivatives vanish at zero temperature, for reasons closely associated with the Nernst postulate.

Consider first a change in pressure at $T = 0$. The change in entropy must vanish as $T \to 0$. The immediate consequence is

$$\left(\frac{\partial S}{\partial P} \right)_T = \left(\frac{\partial V}{\partial T} \right)_P \to 0 \qquad (\text{as } T \to 0) \tag{11.6}$$

where we have invoked a familiar Maxwell relation. It follows that the coefficient of thermal expansion α vanishes at zero temperature.

$$\alpha = \frac{1}{V} \left(\frac{\partial V}{\partial T} \right)_P \to 0 \qquad (\text{as } T \to 0) \tag{11.7}$$

Replacing the pressure by the volume in equation 11.6, the vanishing of $(\partial S/\partial V)_T$ implies (again by a Maxwell relation)

$$\left(\frac{\partial P}{\partial T} \right)_v \to 0 \qquad (\text{as } T \to 0) \tag{11.8}$$

The heat capacities are more delicate. If the entropy does not only approach zero at zero temperature, but *if it approaches zero with a bounded derivative* (i.e., if $(\partial s/\partial T)_v$ is not infinite) then

$$c_v = T \left(\frac{\partial s}{\partial T} \right)_v \to 0 \qquad (\text{as } T \to 0) \tag{11.9}$$

and, similarly, *if* $(\partial s/\partial T)_P$ is bounded

$$c_p = T \left(\frac{\partial s}{\partial T} \right)_P \to 0 \qquad (\text{as } T \to 0) \tag{11.10}$$

Referring back to Fig. 11.1 it will be noted that both ΔG and ΔH were drawn with zero slope; whereas equations 11.4 and 11.5 required only that ΔG and ΔH have the *same* slope. The fact that they have zero slope is a consequence of equation 11.10 and of the fact that the temperature derivative of ΔH is just $N \Delta c_p$.

The vanishing of c_v and c_p (and the zero slope of ΔG or ΔH) appears generally to be true. However, whereas the vanishing of α at $T=0$ is a direct consequence of the Nernst postulate, the vanishing of c_v and c_p are observational facts which are suggested by, but not absolutely required by, the Nernst postulate.

Finally, we note that the pressure in equation 11.6 can be replaced by other intensive parameters (such as B_e for the magnetic case) leading to general analogues of equation 11.7, and similarly for equation 11.8.

11-3 THE "UNATTAINABILITY" OF ZERO TEMPERATURE

It is frequently stated that, as a consequence of the Nernst postulate, the absolute zero of temperature can never be reached by any physically realizable process. Temperatures of 10^{-3} K are reasonably standard in cryogenic laboratories; 10^{-7} K has been achieved; and there is no reason to believe that temperatures of 10^{-10} K or less are fundamentally inaccessible. The question of whether the state of *precisely* zero temperature can be realized by any process yet undiscovered may well be an unphysical question, raising profound problems of absolute thermal isolation and of infinitely precise temperature measurability. The theorem that does follow from the Nernst postulate is more modest. It states that *no reversible adiabatic process starting at nonzero temperature can possibly bring a system to zero temperature*. This is, in fact, no more than a simple restatement of the Nernst postulate that the $T = 0$ isotherm is coincident with the $S = 0$ adiabat. As such, the $T = 0$ isotherm cannot be intersected by any *other* adiabat (recall Fig. 11.2).

12

SUMMARY OF PRINCIPLES
FOR GENERAL SYSTEMS

12-1 GENERAL SYSTEMS

Throughout the first eleven chapters the principles of thermodynamics have been so stated that their generalization is evident. The fundamental equation of a simple system is of the form

$$U = U(S, V, N_1, N_2, \ldots, N_r) \tag{12.1}$$

The volume and the mole numbers play symmetric roles throughout, and we can rewrite equation 12.1 in the symmetric form

$$U = U(X_0, X_1, X_2, X_3, \ldots, X_t) \tag{12.2}$$

where X_0 denotes the entropy, X_1 the volume, and the remaining X_j are the mole numbers. For non-simple systems the formalism need merely be re-interpreted, the X_j then representing magnetic, electric, elastic, and other extensive parameters appropriate to the system considered.

For the convenience of the reader we recapitulate briefly the main theorems of the first eleven chapters, using a language appropriate to general systems.

12-2 THE POSTULATES

Postulate I. *There exist particular states (called equilibrium states) that, macroscopically, are characterized completely by the specification of the internal energy U and a set of extensive parameters $X_1, X_2, \ldots, X_t$ later to be specifically enumerated.*

Postulate II. *There exists a function (called the entropy) of the extensive parameters, defined for all equilibrium states, and having the following property. The values assumed by the extensive parameters in the absence of a constraint are those that maximize the entropy over the manifold of constrained equilibrium states.*

Postulate III. *The entropy of a composite system is additive over the constituent subsystems (whence the entropy of each constituent system is a homogeneous first-order function of the extensive parameters). The entropy is continuous and differentiable and is a monotonically increasing function of the energy.*

Postulate IV. *The entropy of any system vanishes in the state for which* $T \equiv (\partial U/\partial S)_{X_1, X_2,...} = 0.$

12-3 THE INTENSIVE PARAMETERS

The differential form of the fundamental equation is

$$dU = T\,dS + \sum_1^t P_k\,dX_k = \sum_0^t P_k\,dX_k \tag{12.3}$$

in which

$$P_k = \frac{\partial U}{\partial X_k} \tag{12.4}$$

The term $T\,dS$ is the flux of heat and $\sum_1^t P_k\,dX_k$ is the work. The intensive parameters are functions of the extensive parameters, the functional relations being the equations of state. Furthermore, the conditions of equilibrium with respect to a transfer of X_k between two subsystems is the equality of the intensive parameters P_k.

The Euler relation, which follows from the homogeneous first-order property, is

$$U = \sum_0^t P_k X_k \tag{12.5}$$

and the Gibbs-Duhem relation is

$$\sum_0^t X_k\,dP_k = 0 \tag{12.6}$$

Similar relations hold in the entropy representation.

12-4   LEGENDRE TRANSFORMS

A partial Legendre transformation can be made by replacing the variables $X_0, X_1, X_2, \ldots, X_s$ by $P_0, P_1, \ldots, P_s$. The Legendre transformed function is

$$U[P_0, P_1, \ldots, P_s] = U - \sum_0^s P_k X_k \tag{12.7}$$

The natural variables of this function are $P_0, \ldots, P_s, X_{s+1}, \ldots, X_t$, and the natural derivatives are

$$\frac{\partial U[P_0, \ldots, P_s]}{\partial P_k} = -X_k, \qquad (k = 0, 1, \ldots, s) \tag{12.8}$$

$$\frac{\partial U[P_0, \ldots, P_s]}{\partial X_k} = P_k \qquad (k = s + 1, \ldots, t) \tag{12.9}$$

and consequently

$$dU[P_0, \ldots, P_s] = \sum_0^s (-X_k)\, dP_k + \sum_{s+1}^t P_k\, dX_k \tag{12.10}$$

The equilibrium values of any unconstrained extensive parameters in a system in contact with reservoirs of constant $P_0, P_1, \ldots, P_s$ minimize $U[P_0, \ldots, P_s]$ at constant $P_0, \ldots, P_s, X_{s+1} \ldots X_t$.

12-5   MAXWELL RELATIONS

The mixed partial derivatives of the potential $U[P_0, \ldots, P_s]$ are equal, whence, from equation 12.10,

$$\frac{\partial X_j}{\partial P_k} = \frac{\partial X_k}{\partial P_j} \qquad (\text{if } j, k \leq s) \tag{12.11}$$

$$\frac{\partial X_j}{\partial X_k} = \frac{-\partial P_k}{\partial P_j} \qquad (\text{if } j \leq s \text{ and } k > s) \tag{12.12}$$

and

$$\frac{\partial P_j}{\partial X_k} = \frac{\partial P_k}{\partial X_j} \qquad (\text{if } j, k > s) \tag{12.13}$$

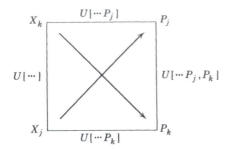

FIGURE 12.1
The general thermodynamic mnemonic diagram. The potential $U[\dots]$ is a general Legendre transform of U. The potential $U[\dots, P_j]$ is $U[\dots] - P_j X_j$. That is, $U[\dots, P_j]$ is transformed with respect to P_j in addition to all the variables of $U[\dots]$. The other functions are similarly defined.

In each of these partial derivatives the variables to be held constant are all those of the set $P_0, \dots, P_s, X_{s+1}, \dots, X_t$, except the variable with respect to which the derivative is taken.

These relations can be read from the mnemonic diagram of Fig. 12.1.

12-6 STABILITY AND PHASE TRANSITIONS

The criteria of stability are the convexity of the thermodynamic potentials with respect to their extensive parameters and concavity with respect to their intensive parameters (at constant mole numbers). Specifically this requires

$$c_p > c_v > 0 \qquad \kappa_T > \kappa_S > 0 \tag{12.14}$$

and analogous relations for more general systems.

If the criteria of stability are not satisfied a system breaks up into two or more phases. The molar Gibbs potential of each component j is then equal in each phase

$$\mu_j^{\mathrm{I}} = \mu_j^{\mathrm{II}} = \mu_j^{\mathrm{III}} \tag{12.15}$$

The dimensionality f of the thermodynamic "space" in which a given number M of phases can exist, for a system with r components, is given by the Gibbs phase rule

$$f = r - M + 2 \tag{12.16}$$

The slope, in the $P–T$ plane, of the coexistence curve of two phases is given by the Clapeyron equation

$$\frac{dP}{dT} = \frac{\Delta s}{\Delta v} = \frac{\ell}{T \Delta v} \tag{12.17}$$

12-7 CRITICAL PHENOMENA

Near a critical point the minimum of the Gibbs potential becomes shallow and possibly asymmetric. Fluctuations diverge, and the most probable values, which are the subject of thermodynamic theory, differ from the average values which are measured by experiment. Thermodynamic behavior near the critical point is governed by a set of "critical exponents." These are interrelated by "scaling relations." The numerical values of the critical exponents are determined by the physical dimensionality and by the dimensionality of the order parameter; these two dimensionalities define "universality classes" of systems with equal critical exponents.

12-8 PROPERTIES AT ZERO TEMPERATURE

For a general system the specific heats vanish at zero temperature.

$$c_{x_1, x_2, \ldots} \equiv T \left(\frac{\partial s}{\partial T} \right)_{x_1, x_2, \ldots} \to 0 \qquad \text{as } T \to 0 \qquad (12.18)$$

and

$$c_{x_1, \ldots, x_{k-1}, P_k, x_{k+1}, \ldots} \to 0 \qquad \text{as } T \to 0 \qquad (12.19)$$

Furthermore, the four following types of derivatives vanish at zero temperature.

$$\left(\frac{\partial s}{\partial x_k} \right)_{T, x_1, \ldots, x_{k-1}, x_{k+1}, \ldots} \to 0 \qquad \text{as } T \to 0 \qquad (12.20)$$

$$\left(\frac{\partial P_k}{\partial T} \right)_{x_1, x_2, \ldots} \to 0 \qquad \text{as } T \to 0 \qquad (12.21)$$

$$\left(\frac{\partial s}{\partial P_k} \right)_{T, x_1, \ldots, x_{k-1}, x_{k+1}, \ldots} \to 0 \qquad \text{as } T \to 0 \qquad (12.22)$$

and

$$\left(\frac{\partial x_k}{\partial T} \right)_{x_1, \ldots, x_{k-1}, P_k, x_{k+1}, \ldots} \to 0 \qquad \text{as } T \to 0 \qquad (12.23)$$

13

PROPERTIES OF MATERIALS

13-1 THE GENERAL IDEAL GAS

A brief survey of the range of physical properties of gases, liquids, and solids logically starts with a recapitulation of the simplest of systems—the ideal gas. All gases approach ideal behavior at sufficiently low density, and all gases deviate strongly from ideality in the vicinity of their critical points.

The essence of ideal gas behavior is that the molecules of the gas do not interact. This single fact implies (by statistical reasoning to be developed in Section 16.10) that

(a) The mechanical equation of state is of the form $PV = NRT$.
(b) For a single-component ideal gas the temperature is a function only of the molar energy (and inversely).
(c) The Helmholtz potential $F(T, V, N_1, N_2, \ldots, N_r)$ of a multicomponent ideal gas is additive over the components ("Gibbs's Theorem"):

$$F(T, V, N_1, \ldots, N_r) = F_1(T, V, N_1) + F_2(T, V, N_2)$$

$$+ \cdots + F_r(T, V, N_r) \qquad (13.1)$$

Considering first a *single-component* ideal gas of molecular species j, property (b) implies

$$U_j = N_j u_j(T) \qquad (13.2)$$

It is generally preferable to express this equation in terms of the heat capacity, which is the quantity most directly observable

$$U_j = N_j u_{j0} + N_j \int_{T_0}^{T} c_{vj}(T') \, dT' \qquad (13.3)$$

where T_0 is some arbitrarily-chosen standard temperature.

The entropy of a single-component ideal gas, like the energy, is determined by $c_{vj}(T)$. Integrating $c_{vj} = N_j^{-1}T(dS_j/dT)_v$, and determining the constant of integration by the equation of state $PV = N_jRT$

$$S_j = N_j s_{j0} + N_j \int_{T_0}^{T} T'^{-1}c_{vj}(T')\, dT' + N_j R \ln\left(\frac{v}{v_0}\frac{N_0}{N_j}\right) \quad (13.4)$$

Finally, the Helmholtz potential of a general multicomponent ideal gas is, by property (c)

$$F(T,V) = \sum_j U_j(T) - T\sum_j S_j(T,V) = U - TS \quad (13.5)$$

Thus *the most general multicomponent ideal gas is completely characterized by the molar heat capacities $c_{vj}(T)$ of its individual constituents* (and by the values of u_{j0}, s_{j0} assigned in some arbitrary reference state).

The first summation in equation 13.5 is the energy of the multicomponent gas, and the second summation is the entropy. The general ideal gas obeys Gibbs's theorem (recall the discussion following equation 3.39).

Similarly, as in equation 3.40, we can rewrite the entropy of the general ideal gas (equation 13.4) in the form (taking $N_0 = N$):

$$S = -\sum_j N_j s_{j0} = \sum_j N_j \int_{T_0}^{T} \frac{1}{T'}c_{vj}(T')\, dT' + \sum_j N_j R \ln\left(\frac{v}{v_0}\frac{N}{N_j}\right)$$

$$= N\int_{T_0}^{T} \frac{1}{T'}\bar{c}_v(T')\, dT' + NR\ln\frac{v}{v_0} - NR\sum_j x_j \ln x_j \quad (13.6)$$

and the last term is again the *entropy of mixing*. We recall that the entropy of mixing is the difference in entropies between that of the mixture of gases and that of a collection of separate gases, each at the same temperature and the same density $N_j/V_j = N/V$ as the original mixture (and hence at the same pressure as the original mixture).

It is left to the reader to show that κ_T, α, and the difference $(c_p - c_v)$ have the same values for a general ideal gas as for a monatomic ideal gas (recall Section 3.8). In particular,

$$\kappa_T = \frac{1}{P}, \qquad \alpha = \frac{1}{T}, \qquad c_p - c_v = R \quad (13.7)$$

The molar heat capacity appearing in equation 13.3 is subject to certain thermostatistical requirements, and these correspond to observational regularities. One such regularity is that the molar heat capacity c_v of real

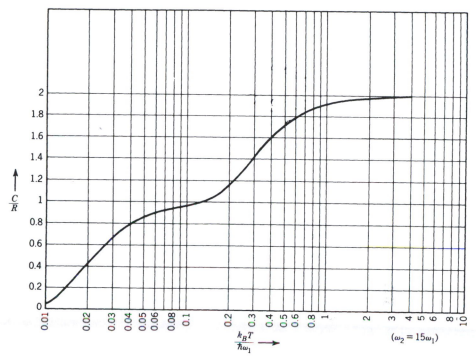

FIGURE 13.1

The molar heat capacity of a system with two vibrational modes, with $\omega_2 = 15\omega_1$.

gases approaches a constant value at high temperatures (but not so high that the molecules ionize or dissociate). *If the classical energy can be written as a sum of quadratic terms* (in some generalized coordinates and momenta), *then the high temperature value of c_v is simply $R/2$ for each such quadratic term.* Thus, for a monatomic ideal gas the energy of each molecule is $(p_x^2 + p_y^2 + p_z^2)/2m$; there are three quadratic terms, and hence $c_v = 3R/2$ at high temperatures. In Section 16.10, we shall explore the thermostatistical basis for this "equipartition value" of c_v at high temperatures.

At zero temperature the heat capacities of all materials in thermodynamic equilibrium vanish, and in particular the heat capacities of gases fall toward zero (until the gases condense). At high temperatures the heat capacities of ideal gases are essentially temperature independent at the "equipartition" value described in the preceding paragraph. In the intermediate temperature region the contribution of each quadratic term in the Hamiltonian tends to appear in a restricted temperature range, so that c_v versus T curves tend to have a roughly steplike form, as seen in Fig. 13.1

The temperatures at which the "steps" occur in the c_v versus T curves, and the "height" of each step, can be understood in descriptive terms (anticipating the statistical mechanical analysis of Chapter 16). The quadratic terms in the energy represent kinetic or potential energies associated with particular *modes of excitation*. Each such mode contributes additively and independently to the heat capacity, and each such mode is responsible for one of the "steps" in the c_v versus T curve.

For a diatomic molecule there is a quadratic term representing the potential energy of stretching of the interatomic bond, and there is another quadratic term representing the kinetic energy of vibration; together the potential and kinetic energies constitute a harmonic oscillator of frequency ω_0.

The contribution of each mode appears as a "step" of height $R/2$ for each quadratic term in the energy (two terms, or $\Delta c_v \simeq R$, for a vibrational mode). The temperature at which the step occurs is such that $k_B T$ is of the order of the energy difference of the low-lying energy levels of the mode ($k_B T \simeq \hbar \omega_0$ for a vibrational mode).

Similar considerations apply to rotational, translational, and other types of modes. A more detailed description of the heat capacity will be developed in Chapter 16.

13-2   CHEMICAL REACTIONS IN IDEAL GASES

The chemical reaction properties of ideal gases is of particular interest. This reflects the fact that in industrial processes many important chemical reactions actually are carried out in the gaseous phase, and the assumption of ideal behavior permits a simple and explicit solution. Furthermore the theory of ideal gas reactions provides the starting point for the theory of more realistic gaseous reaction models.

It follows directly from the fundamental equation of a general ideal gas mixture (as given parametrically in equations 13.3 to 13.5) that the partial molar Gibbs potential of the jth component is of the form

$$\mu_j = RT\left[\phi_j(T) + \ln P + \ln x_j\right] \tag{13.8}$$

The quantity $\phi_j(T)$ is a function of T only, and x_j is the mole fraction of the jth component. The equation of chemical equilibrium is (equation 2.70 or 6.51)

$$\sum_j \nu_j \mu_j = 0 \tag{13.9}$$

whence

$$\sum_j \nu_j \ln x_j = -\sum_j \nu_j \ln P - \sum_j \nu_j \phi_j(T) \tag{13.10}$$

Defining the "equilibrium constant" $K(T)$ for the particular chemical reaction by

$$\ln K(T) \equiv - \sum_j \nu_j \phi_j(T) \tag{13.11}$$

we find the *mass action law*

$$\prod_j x_j^{\nu_j} = P^{-\sum_j \nu_j} K(T) \tag{13.12}$$

The equilibrium constant $K(T)$ can be synthesized from the functions $\phi_j(T)$ by the definition (13.11), and the functions $\phi_j(T)$ are tabulated for common chemical gaseous components. Furthermore the equilibrium constant $K(T)$ is itself tabulated for many common chemical reactions. In either case the equilibrium constant can be considered as known. Thus, given the temperature and pressure of the reaction, the product $\prod x_j^{\nu_j}$ is determined by the mass action law (13.12). Paired with the condition that the sum of the mole fractions is unity, and given the quantities of each atomic constituent in the system, the knowledge of $\prod x_j^{\nu_j}$ determines each of the x_j. We shall illustrate such a determination in an example, but we first note that tabulations of equilibrium constants for simple reactions can be extended to additional reactions by "logarithmic additivity." Certain chemical reactions can be considered as the sum of two other chemical reactions. As an example, consider the reactions

$$2H_2 + O_2 \rightleftharpoons 2H_2O \tag{13.13}$$

and

$$2CO + O_2 \rightleftharpoons 2CO_2 \tag{13.14}$$

Subtracting these two equations in algebraic fashion gives

$$2H_2 - 2CO \rightleftharpoons 2H_2O - 2CO_2 \tag{13.15}$$

or

$$2[H_2 + CO_2 \rightleftharpoons H_2O + CO] \tag{13.16}$$

We now observe that the quantities $\ln K(T)$ of the various reactions can be subtracted in a corresponding fashion.

Consider two reactions

$$0 \rightleftharpoons \sum \nu_j^{(1)} A_j \tag{13.17}$$

and

$$0 \rightleftharpoons \Sigma \nu_j^{(2)} A_j \tag{13.18}$$

and a third reaction obtained by multiplying the first reaction by a constant B_1, the second reaction by B_2, and adding

$$0 \rightleftharpoons \Sigma \nu_j^{(3)} A_j \equiv \Sigma \left(B_1 \nu_j^{(1)} + B_2 \nu_j^{(2)} \right) A_j \tag{13.19}$$

Assume that the equilibrium constant of the first reaction is $K_1(T)$ and that of the second reaction is $K_2(T)$, so that by definition

$$\ln K_1(T) = -\Sigma \nu_j^{(1)} \phi_j(T) \tag{13.20}$$

and

$$\ln K_2(T) = -\Sigma \nu_j^{(2)} \phi_j(T) \tag{13.21}$$

The equilibrium constant for the *resultant reaction* equation 13.19 is defined by an analogous equation, from which it follows that

$$\ln K_3(T) = B_1 \ln K_1(T) + B_2 \ln K_2(T) \tag{13.22}$$

Thus tabulations of equilibrium constants for basic reactions can be extended to additional reactions by the additivity property.

Finally we recall that in the discussion following equation 6.58 it was observed that the heat of reaction is plausibly related to the temperature dependence of the equilibrium constant. We showed there that, in fact,

$$\frac{dH}{d\tilde{N}} = -T \frac{\partial}{\partial T} \left(\Sigma \nu_j \mu_j \right)_{P, N_1, N_2, \dots} \tag{13.23}$$

and, inserting equation 13.8

$$\frac{dH}{d\tilde{N}} = -T \frac{\partial}{\partial T} \left(RT \Sigma \nu_j \phi_j + RT \Sigma \nu_j \ln P + RT \Sigma \nu_j \ln x_j \right) \tag{13.24}$$

$$\frac{dH}{d\tilde{N}} = -\Sigma \nu_j \mu_j - RT^2 \frac{d}{dT} \Sigma \nu_j \phi_j \tag{13.25}$$

Recognizing that $\Sigma_j \nu_j \mu_j$ vanishes at equilibrium and recalling the definition (13.11) of the equilibrium constant, we find the *van't Hoff relation*

$$\frac{dH}{d\tilde{N}} = RT^2 \frac{d}{dT} \ln K(T) \tag{13.26}$$

Thus measurements of the equilibrium constant at various temperatures enable calculation of the heat of reaction without calorimetric methods (the equilibrium constant being measurable by direct determination of the concentrations x_k).

Example

Two moles of H_2O are enclosed in a rigid vessel and heated to a temperature of 2000 K and a pressure of 1 MPa. The equilibrium constant $K(T)$ for the chemical reaction

$$H_2O \rightleftarrows H_2 + \tfrac{1}{2}O_2$$

has the value $K(2000) = 0.0877$ $Pa^{1/2}$. What is the equilibrium composition of the system? What is the composition if the temperature remains constant but the pressure is decreased to 10^4 Pa?

The law of mass action states that

$$\frac{x_{H_2} x_{O_2}^{1/2}}{x_{H_2O}} = P^{-1/2} K(T)$$

The mole numbers of each component are given by

$$N_{H_2O} = 2 - \Delta\tilde{N} \qquad N_{H_2} = \Delta\tilde{N} \qquad N_{O_2} = \tfrac{1}{2}\Delta\tilde{N}$$

so that the sum of the mole numbers is $2 + \Delta\tilde{N}/2$. Consequently

$$x_{H_2O} = \frac{2 - \Delta\tilde{N}}{2 + \tfrac{1}{2}\Delta\tilde{N}} \qquad x_{H_2} = \frac{\Delta\tilde{N}}{2 + \tfrac{1}{2}\Delta\tilde{N}} \qquad x_{O_2} = \frac{\tfrac{1}{2}\Delta\tilde{N}}{2 + \tfrac{1}{2}\Delta\tilde{N}}$$

The law of mass action accordingly becomes

$$\frac{1}{\sqrt{2}} \frac{(\Delta\tilde{N})^{3/2}}{(2 - \Delta\tilde{N})(2 + \tfrac{1}{2}\Delta\tilde{N})^{1/2}} = P^{-1/2} K(T)$$

and with the right-hand side known we can solve numerically for $\Delta\tilde{N}$. We find $\Delta\tilde{N} = 0.005$ for $P = 1$ MPa and $\Delta\tilde{N} = 0.023$ for $P = 10^4$ Pa. Thus, for a pressure of 1 MPa, the mole fractions of the components are

$$x_{H_2O} = 0.9963 \qquad x_{H_2} = 0.0025 \qquad x_{O_2} = 0.0012$$

whereas for a pressure of 10^4 Pa the mole fractions are

$$x_{H_2O} = 0.9828 \qquad x_{H_2} = 0.0114 \qquad x_{O_2} = 0.0057$$

PROBLEMS

13.2-1. How is the equilibrium constant of the reaction in the Example related to that for the same reaction when written with stoichiometric coefficients twice as large? Note this fact with caution!

13.2-2. What are the mole fractions of the constituents in the Example if the pressure is further reduced to 10^3 Pa?

13.2-3. In the Example, what would the final mole fractions be at a pressure of 10^3 Pa if the vessel initially had contained 1 mole of oxygen as well as 2 moles of water?

13.2-4. In an ideal gas reaction an increase in pressure at constant temperature increases the degree of reaction if the sum of the stoichiometric coefficients of the "reactants" is greater in absolute value than the sum of the ν's of the "products," and vice versa. Prove this statement or show it to be false, using the law of mass action. What is the relation of this statement to the Le Chatelier–Braun principle (Sect. 8.5)?

13.2-5. The equilibrium constant of the reaction

$$SO_3 \rightleftharpoons SO_2 + \tfrac{1}{2}O_2$$

has the value 171.9 $Pa^{1/2}$ at $T = 1000$ K. Assuming 1 mole of SO_2 and 2 moles of O_2 are introduced into a vessel and maintained at a pressure of 0.4 MPa, find the number of moles of SO_3 present in equilibrium.

13.2-6. At temperatures above ~ 500 K phosphorus pentachloride dissociates according to the reaction

$$PCl_5 \rightleftharpoons PCl_3 + Cl_2$$

A PCl_5 sample of 1.9×10^{-3} Kg is at a temperature of 593 K and a pressure of 0.314×10^5 Pa. After the reaction has come to equilibrium the system is found to have a volume of 2.4 liters (or 2.4×10^{-3} m^3). Determine the equilibrium constant. What is the "degree of dissociation" (i.e., the degree of reaction ε for this dissociation reaction; recall equation 6.53)?

13.2-7. A system containing 0.02 Kg of CO and 0.02 Kg of O_2 is maintained at a temperature of 3200 K and a pressure of 0.2 MPa. At this temperature the equilibrium constant for the reaction

$$2CO_2 \rightleftharpoons 2CO + O_2$$

is $K = .0424$ MPa. What is the mass of CO_2 at equilibrium?

13.2-8. Apply equation 13.8 to a *single-component* general ideal gas (of species j). Evaluate μ_j for the single-component ideal gas by equation 13.4 (note that by equation 13.3 constant U implies constant T), and in this way obtain an expression for ϕ_j.

13.2-9. An experimenter finds that water vapor is 0.53% dissociated at a temperature of 2000 K and a pressure of 10^5 Pa. Raising the temperature to 2100 K and keeping the pressure constant leads to a dissociation of 0.88%. That is, an initial mole of H_2O remains as 0.9947 moles at 2000 K or as 0.9912 moles at 2100 K after the reaction comes to completion. Calculate the heat of reaction of the dissociation of water at $P = 10^5$ Pa and $T \simeq 2050$ K.

Answer:
$$\Delta H \simeq 2.7 \times 10^5 \text{ J/mole}$$

13-3 SMALL DEVIATIONS FROM "IDEALITY"—THE VIRIAL EXPANSION

Although all gases behave "ideally" at sufficiently large molar volume, they exhibit more complicated behavior as the molar volume v is decreased. To describe at least the initial deviations from ideal gas behavior the mechanical equation of state can be expanded in inverse powers of v

$$\frac{P}{T} = \frac{R}{v}\left(1 + \frac{B(T)}{v} + \frac{C(T)}{v^2} + \cdots\right) \qquad (13.27)$$

This expansion is called a "virial expansion"; $B(T)$ is called the "second virial coefficient," $C(T)$ is the "third virial coefficient," and so forth. The forms of these functions depend on the form of the intermolecular forces in the gas. The second virial coefficient is shown in Fig. 13.2 as a function of temperature for several simple gases.

Corresponding to the virial expansion of the mechanical equation of state, in inverse powers of v, the molar Helmholtz potential can be similarly expanded

$$f = f_{ideal} + RT\left[\frac{B(T)}{v} + \frac{C(T)}{2v^2} + \frac{D(T)}{3v^3} + \cdots\right] \qquad (13.28)$$

The equality of the coefficients $B(T)$, $C(T),\ldots$ in these expansions follows, of course, from $P = -\partial f/\partial v$.

All thermodynamic quantities thereby are expressible in virial-type expansions, in inverse powers of v. The molar heat capacity c_v, for instance, is

$$c_v = c_{v,ideal} + RT\left[\frac{1}{v}\frac{d^2(BT)}{dT^2} + \frac{1}{2v^2}\frac{d^2(CT)}{dT^2} + \frac{1}{3v^3}\frac{d^2(DT)}{dT^2} + \cdots\right] \qquad (13.29)$$

and the molar energy is

$$u = u_{ideal} + RT^2\left[\frac{1}{v}\frac{dB}{dT} + \frac{1}{2v^2}\frac{dC}{dT} + \frac{1}{3v^3}\frac{dD}{dT} + \cdots\right] \qquad (13.30)$$

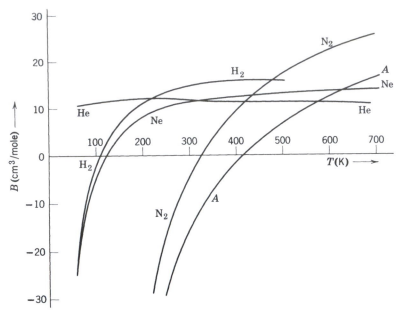

FIGURE 13.2
Second virial coefficient as a function of temperature for several gases. Measurements by Holborn and Otto. Data from *Statistical Thermodynamics*, by R. H. Fowler and E. A. Guggenheim, Cambridge University Press, 1939.

PROBLEMS

13.3-1. In a thermostatistical model in which each atom is treated as a small hard sphere of volume τ, the leading virial coefficients are

$$B = 4N_A\tau \qquad C = 10N_A^2\tau^2 \qquad D = 18.36N_A^3\tau^3$$

Using the value of B determined from Fig. 13.2, find the approximate radius of a He atom. Given Fig. 13.2, what would be a reasonable (though fairly crude) guess as to the value of the third virial coefficient of He?

13.3-2. Expand the mechanical equation of state of a van der Waals gas (equation 3.41) in a virial expansion, and express the virial coefficients in terms of the van der Waals constants a and b.

13.3-3. Show that the second virial coefficient of gaseous nitrogen (Fig. 13.2) can be fit reasonably by an equation of the form

$$B = B_0 - \frac{B_1}{T}$$

and find the values of B_0 and B_1. Assume that all higher virial coefficients can be neglected. Also take the molar heat capacity c_v of the *noninteracting* gas to be $5R/2$.
(*a*) Explain why c_v (noninteracting) reasonably can be taken as $5R/2$.
(*b*) Evaluate the values of B_0 and B_1 from Fig. 13.2.

(*c*) What is the value of $c_v(T, v)$ for N_2, to second order in a virial expansion?

13.3-4. The simplest analytic form suggested by the qualitative shape of $B(T)$ of H_2 and Ne in Fig. 13.2 is $B(T) = B_0 - B_1/T$ (as in Problem 13.3-3). With this assumption calculate $c_p(T, v)$ for H_2 and Ne.

13.3-5. A "porous plug" experiment is carried out by installing a porous plug in a plastic pipe. To the left of the plug the gas is maintained at a pressure slightly higher than atmospheric by a movable piston. To the right of the plug there is a freely sliding piston, and the right-hand end of the pipe is open to the atmosphere. What is the fractional difference of velocities of the pistons?

(*a*) Express the answer in terms of atmospheric pressure P_a, the driving pressure P_h, and c_p, α, κ_T, and v (assuming that the pressure difference is small enough that no distinction need be made between the values of the latter quantities on the two sides of the plug).

(*b*) Evaluate this result for an ideal gas, and express the deviation from this result in terms of the second virial coefficient, carrying results only to first order in $B(T)$ or its derivatives (the heat capacity c_p is to be left as an unspecified quantity in the solution).

13-4 THE "LAW OF CORRESPONDING STATES" FOR GASES

A complete virial expansion can describe the properties of any gas with high precision, but only at the cost of introducing an infinite number of expansion constants. In contrast the van der Waals equation of state captures the essential features of fluid behavior, including the phase transition, with only *two* adjustable constants. The question arises as to whether the virial coefficients of real gases are indeed independent, or whether there exists some general relationships among them. Alternatively stated, does there exist a more or less universal form of the equation of state of fluids, involving some finite (or even small) number of independent constants?

In the equation of state of any fluid there is one unique point—the critical point, characterized by T_{cr}, P_{cr}, and v_{cr}. A dimensionless equation of state would, then, be most naturally expressed in terms of the "reduced temperature" T/T_{cr}, the "reduced pressure" P/P_{cr}, and the "reduced molar volume" v/v_{cr}.

It might be expected that the three parameters T_{cr}, P_{cr}, and v_{cr} are themselves independent. But evaluation of the dimensionless ratio $P_c v_{cr}/RT_c$ for various gases reveals a remarkable regularity, as shown in Table 13.1. The ratio is strikingly constant (with small deviations to lower values for a few polar fluids such as water or ammonia). *The dimensionless constant $P_{cr}v_{cr}/RT_{cr}$ has a value on the order of 0.27 for all "normal" fluids. Of the three parameters that characterize the critical point, only two are independent* (in the semiquantitative sense of this section).

TABLE 13.1
Critical Constants and the Ratio $P_{cr}v_{cr}/RT_{cr}$ of Various Fluids*

Substance	Molecular Weight	$T_{cr}(K)$	$P_{cr}(10^6 \ Pa)$	$v_{cr}(10^{-3} \ m^3)$	$P_{cr}v_{cr}/RT_{cr}$
H_2	2.016	33.3	1.30	0.0649	0.30
He	4.003	5.3	0.23	0.0578	0.30
CH_4	16.043	191.1	4.64	0.0993	0.29
NH_3	17.03	405.5	11.28	0.0724	0.24
H_2O	18.015	647.3	22.09	0.0568	0.23
Ne	20.183	44.5	2.73	0.0417	0.31
N_2	28.013	126.2	3.39	0.0899	0.29
C_2H_6	30.070	305.5	4.88	0.1480	0.28
O_2	31.999	154.8	5.08	0.0780	0.31
C_2H_8	44.097	370	4.26	0.1998	0.28
C_2H_5OH	46.07	516	6.38	0.1673	0.25
SO_2	64.063	430.7	7.88	0.1217	0.27
C_6H_6	78.115	562	4.92	0.2603	0.27
Kr	83.80	209.4	5.50	0.0924	0.29
CCl_4	153.82	556.4	4.56	0.2759	0.27

* Abstracted from K. A. Kobe and R. E. Lynn, Jr., *Chem. Rev.* **52**, 117 (1953).

Proceeding further, then, one can plot v/v_{cr} as a function of P/P_{cr} and T/T_{cr} for a variety of fluids. Again there is a remarkable similarity among all such "reduced equations of state".

There exists, at least semiquantitatively, a universal equation of state containing no arbitrary constants if expressed in the reduced variables v/v_{cr}, P/P_{cr}, and T/T_{cr}. This empirical fact is known as the "Law of Corresponding States".

The universal reduced equation of state can be represented in a convenient two-dimensional form, as in Fig. 13.3 from Sonntag and Van Wylen.[1] The dependent variable, the ordinate in the figure, is the dimensionless quantity Pv/RT, or $0.27 \ (P/P_{cr}) \ (v/v_{cr})/(T/T_{cr})$. The independent variables are P/P_{cr} and T/T_{cr}. The reduced pressure P/P_{cr} is the abscissa in the graph. In order to avoid a third dimension the reduced temperature scale is superimposed as a set of constant reduced temperature loci in the plane.

To find v/v_{cr} at a given value of P/P_{cr} and T/T_{cr} one reads P/P_{cr} on the abscissa and locates the appropriate T/T_{cr} curve. These values determine a point, of which one can read the ordinate. The ordinate is $\approx 0.27(P/P_{cr})(v/v_{cr})/(T/T_{cr})$, so that v/v_{cr} is thereby evaluated.

The existence of such an approximate universal equation of state is given a rational basis by statistical mechanical models. The force between

[1] R. E. Sonntag and G. J. Van Wylen, *Introduction to Thermodynamics, Classical and Statistical*, 2nd ed. (Wiley, New York, 1982).

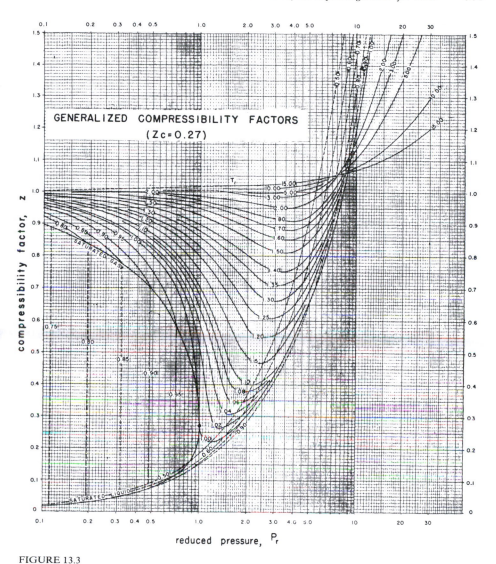

FIGURE 13.3

"Generalized" or universal equation of state of gases in terms of reduced variables. From R. E. Sonntag and G. van Wylen, *Introduction to Thermodynamics, Classical and Statistical*, 2nd edition, 1982, John Wiley & Sons, New York.

molecules is generally repulsive at small distance (where the molecules physically overlap) and attractive at larger distances. The long-range attraction in nonpolar molecules is due to the polarization of one molecule by the instantaneous fluctuating dipole moment of the other; such a "van der Waals force" falls as the sixth power of the distance. Thus the force between two molecules can be parametrized by the radii of the molecules (describing the short-range repulsion) and the strength of the long-range attractive force. It is this two-parameter characterization of the intermolecular forces that underlies the two-parameter equation of state.

13-5 DILUTE SOLUTIONS: OSMOTIC PRESSURE AND VAPOR PRESSURE

Whereas the "law of corresponding states" applies most accurately in the gaseous region of the fluid state (with increasing validity as the density decreases below that at the critical point), the liquid region is less subject to a generalized treatment. There is, however, a very useful general regularity that applies to dilute solutions of arbitrary density. That regularity consists of the carry-over of the "entropy of mixing" terms (recall equation 13.6) from ideal gas mixtures to general fluid mixtures.

Consider a single-component fluid system, for which the chemical potential is $\mu_1^0(P, T)$. Then let a second component (the solute) be added, in small concentration. The Gibbs potential of the dilute solution can be written in the general form

$$G(T, P, N_1, N_2) \simeq N_1 \mu_1^0(P, T) + N_2 \psi(P, T) + N_1 RT \ln \frac{N_1}{N_1 + N_2}$$

$$+ N_2 RT \ln \frac{N_2}{N_1 + N_2} \tag{13.31}$$

where ψ is an unspecified function of P and T and where the latter two terms are suggested by the entropy of mixing terms (equation 13.6) of an ideal gas. From a statistical mechanical perspective ψ represents the effect of the interaction energy between the two types of molecules; whereas the entropy of mixing terms arise purely from combinational considerations (to be developed in Chapters 15 *et seq.*). For our present purposes, however, equation 13.31 is to be viewed as an empirical thermodynamic approximation.

In the region of validity (i.e., small concentrations, or $N_2 \ll N_1$) we can expand the third term to first order in N_2/N_1 and we can neglect N_2 relative to N_1 in the denominator of the logarithm in the last term, obtaining

$$G(T, P, N_1, N_2) \simeq N_1 \mu_1^0(P, T) + N_2 \psi(P, T) - N_2 RT + N_2 RT \ln \frac{N_2}{N_1}$$

$$\tag{13.32}$$

It follows that the partial molar Gibbs potentials of solvent and solute are, respectively

$$\mu_1(P,T,x) = \frac{\partial G}{\partial N_1} = \mu_1^0(P,T) - xRT \tag{13.33}$$

where x is the mole fraction of solute ($\simeq N_2/N_1$), and

$$\mu_2(P,T,x) = \frac{\partial G}{\partial N_2} = \psi(P,T) + RT \ln x \tag{13.34}$$

It is of interest to examine some simple consequences of these results. Consider first the case of the osmotic pressure difference across a semipermeable membrane. Suppose the membrane to be permeable to a liquid (water, for instance). A small amount of solute (such as sugar) is introduced on one side of the membrane. Assume that the pressure on the pure solvent side of the membrane is maintained constant ($= P$), but that the pressure on the solute side can alter (as by a change in height of the liquid in a vertical tube). Then the condition of equilibrium with respect to diffusion of the solvent across the membrane is

$$\mu_1(P,T,0) = \mu_1(P',T,X) \tag{13.35}$$

where P' is the as yet unknown pressure on the solute side of the membrane. Then, by equation 13.33

$$\mu_1(P,T,0) = \mu_1(P',T,0) - xRT \tag{13.36}$$

where we have altered the notation slightly to write $\mu_1(P,T,0)$ for $\mu_1^0(P,T)$. Then, expanding $\mu_1(P',T,0)$ around the pressure P

$$\mu_1(P',T,0) = \mu_1(P,T,0) + \frac{\partial \mu_1(P,T,0)}{\partial P} \times (P' - P)$$

$$= \mu_1(P,T,0) + (P' - P)v \tag{13.37}$$

or, from equation 13.36

$$(P' - P)v = xRT \tag{13.38}$$

Multiplying by N_1 we find the *van't Hoff relation* for osmotic pressure in dilute solutions

$$V\Delta P = N_2 RT \tag{13.39}$$

Another interesting effect in liquids is the reduction in the vapor pressure (recall Sections 9.1 to 9.3) by the addition of a low concentration of nonvolatile solute. In the absence of the solute

$$\mu^{liq}(P,T) = \mu^{gas}(P,T) \tag{13.40}$$

But with the addition of the solute, as in 13.36

$$\mu^{liq}(P',T) - xRT = \mu^{gas}(P',T) \tag{13.41}$$

Expanding the first term around the original pressure P

$$\mu^{liq}(P',T) = \mu^{liq}(P,T) + v^{liq}(P,T) \times (P' - P) \tag{13.42}$$

and similarly for the gaseous phase, whence we find

$$P' - P = -\frac{xRT}{v_g - v_\ell} \tag{13.43}$$

Thus the addition of a solute decreases the vapor pressure.

If we make the further approximation that $v_g \gg v_\ell$ and that $v_g = RT/P$ (the ideal gas equation) we obtain

$$\frac{\Delta P}{P} \simeq -x \tag{13.44}$$

which is known as *Raoult's Law*.

PROBLEMS

13.5-1. Assuming the latent heat of vaporization of a fluid to be constant over the temperature range of interest, and assuming that the density of the vapor can be neglected relative to that of the liquid, plot the vapor pressure (i.e., the liquid–gas coexistence curve) as a function of the dimensionless temperature RT/ℓ. Plot the corresponding graphs for five and ten percent dissolved solute.

13.5-2. One hundred grams of a particular solute are dissolved in one liter of water. The vapor pressure of the water is decreased by roughly 6%. Is the solute more likely to be sugar ($C_{12}H_{22}O_{11}$), table salt (NaCl), or potassium iodide (KI)? Simple ionic solutions double their effective Raoult concentration!

13.5-3. If 20 grams of sugar ($C_{12}H_{22}O_{11}$) are dissolved in 250 cm^3 of water, what is the change in the boiling temperature at atmospheric pressure?

13-6 SOLID SYSTEMS

The heat capacities and various other properties of a wide variety of solid systems show marked similarities, as we shall see in specific detail in Section 16.6 (where we shall carry out an explicit statistical mechanical calculation of the thermal equation of state of a solid). Accordingly, we defer further description of the properties of solids, other than to stress that the *thermal* properties of solids are not qualitatively different than those of liquids; it is the *thermomechanical* properties of solids that introduce new elements in the theory.

Whereas the mechanical state of a fluid is adequately characterized by the volume, a solid system can be characterized by a set of *elastic strain components*. These describe both the shape and the angular dilatations ("twists") of the system. The corresponding intensive parameters are the *elastic stress components*. These conjugate variables follow the structure of the general thermodynamic formalism. For specific details the reader is referred to the monograph by Duane C. Wallace[2], or to references cited therein. However, it is important to stress that *conventional thermodynamic theory, in which the volume is the single mechanical parameter, fully applies to solids*. The more detailed analysis in terms of elastic strains gives additional information, but it does not invalidate the results obtained by the less specific, conventional form of thermodynamics.

In the full theory the extensive parameters include both the volume (the "fully symmetric strain") and various other strain components. The conjugate intensive parameters are the stress components, including the pressure (the "fully symmetric stress component"). If the walls of the system impose no stress components other than the pressure, then these stress components vanish and the formalism reduces to the familiar form in which the volume is the only explicit mechanical parameter. Inversely, in the more general case the additional strain components can be appended to the simple theory in a manner fully analogous to the addition of any generalized extensive parameter.

---

[2] Duane C. Wallace, *Thermodynamics of Crystals* (Wiley, New York, 1972).

14

IRREVERSIBLE THERMODYNAMICS

14-1 GENERAL REMARKS

As useful as the characterization of equilibrium states by thermostatic theory has proven to be, it must be conceded that our primary interest is frequently in processes rather than in states. In biology, particularly, it is the life process that captures our imagination, rather than the eventual equilibrium state to which each organism inevitably proceeds. Thermostatics does provide two methods that permit us to infer some limited information about processes, but each of these methods is indirect and each yields only the most meager return. First, by studying the initial and terminal equilibrium states it is possible to bracket a process and thence to determine the effect of the process in its totality. Second, if some process occurs *extremely* slowly, we may compare it with an idealized, nonphysical, quasistatic process. But neither of these methods confronts the central problem of *rates* of real physical processes.

The extension of thermodynamics that has reference to the rates of physical processes is the theory of *irreversible thermodynamics*.

Irreversible thermodynamics is based on the postulates of equilibrium thermostatics plus the additional postulate of *time reversal symmetry of physical laws*. This additional postulate states that *the laws of physics remain unchanged if the time t is everywhere replaced by its negative* $-t$, *and if simultaneously the magnetic field* B_e *is replaced by its negative* $-B_e$ (and, if the process of interest is one involving the transmutation of fundamental particles, that the charge and "parity" of the particles also be reversed in sign). For macroscopic processes the parenthetical restriction has no observable consequences, and we shall henceforth refer to time reversal symmetry in its simpler form.

The thermodynamic theory of irreversible processes is based on the Onsager Reciprocity Theorem, formulated by Lars Onsager[1] in brilliant

---

[1] Lars Onsager, *Physical Review* **37**, 405 (1931); **38**, 2265 (1931).

pioneering papers published in 1931, but not widely recognized for almost 20 years thereafter. Powerful statistical mechanical theorems also exist; the "fluctuation-dissipation theorem",[2] the "Kubo relations," and the formalism of "linear response theory" based on the foregoing theorems[3]. We review only the thermodynamic theory, rooted in the Onsager theorem.

14-2   AFFINITIES AND FLUXES

Preparatory to our discussion of the Onsager theorem, we define certain quantities that appropriately describe irreversible processes. Basically we require two types of parameters: one to describe the "force" that drives a process and one to describe the response to this force.

The processes of most general interest occur in continuous systems, such as the flow of energy in a bar with a continuous temperature gradient. However, to suggest the proper way to choose parameters in such continuous systems, we first consider the relatively simple case of a discrete system. A typical process in a discrete system would be the flow of energy from one homogeneous subsystem to another through an infinitely thin diathermal partition.

Consider a composite system composed of two subsystems. An extensive parameter has values X_k and X_k' in the two subsystems, and the closure condition requires that

$$X_k + X_k' = X_k^\circ, \qquad \text{a constant} \tag{14.1}$$

If X_k and X_k' are unconstrained, their equilibrium values are determined by the vanishing of the quantity

$$\mathscr{F}_k \equiv \left(\frac{\partial S^\circ}{\partial X_k} \right)_{X_k^\circ} = \left(\frac{\partial (S + S')}{\partial X_k} \right)_{X_k^\circ} = \frac{\partial S}{\partial X_k} - \frac{\partial S'}{\partial X_k'} = F_k - F_k' \tag{14.2}$$

Thus, if $\mathscr{F}_k$ is zero the system is in equilibrium, but if $\mathscr{F}_k$ is nonzero an irreversible process occurs, taking the system toward the equilibrium state. The quantity $\mathscr{F}_k$, which is the difference in the entropy-representation intensive parameters, acts as a "generalized force" which "drives" the process. Such generalized forces are called *affinities*.

[2] H. Callen and T. Welton, *Phys. Rev.* **83**, 34 (1951).
[3] c.f. R. Kubo, *Lectures in Theoretical Physics*, vol. 1 (Interscience, New York, 1959, p. 120–203.)

For definiteness, consider two systems separated by a diathermal wall, and let X_k be the energy U. Then the affinity is

$$\mathscr{F}_k = \frac{1}{T} - \frac{1}{T'} \tag{14.3}$$

No heat flows across the diathermal wall if the difference in inverse temperatures vanishes. But a nonzero difference in inverse temperature, acting as a generalized force, drives a flow of heat between the subsystems.

Similarly, if X_k is the volume the affinity $\mathscr{F}_k$ is $[P/T - (P'/T')]$, and if X_k is a mole number the associated affinity is $[\mu'_k/T' - (\mu_k/T)]$.

We characterize the response to the applied force by the rate of change of the extensive parameter X_k. The *flux* J_k is then defined by

$$J_k \equiv \frac{dX_k}{dt} \tag{14.4}$$

Therefore, the flux vanishes if the affinity vanishes, and a nonzero affinity leads to a nonzero flux. It is the relationship between fluxes and affinities that characterizes the rates of irreversible processes.

The identification of the affinities in a particular type of system is frequently rendered more convenient by considering the rate of production of entropy. Differentiating the entropy $S(X_0, X_1, \ldots)$ with respect to the time, we have

$$\frac{dS}{dt} = \sum_k \frac{\partial S}{\partial X_k} \frac{dX_k}{dt} \tag{14.5}$$

or

$$\dot{S} = \sum_k \mathscr{F}_k J_k \tag{14.6}$$

Thus *the rate of production of entropy is the sum of products of each flux with its associated affinity.*

The entropy production equation is particularly useful in extending the definition of affinities to continuous systems rather than to discrete systems. If heat flows from one homogeneous subsystem to another, through an infinitely thin diathermal partition, the generalized force is the difference $[1/T - (1/T')]$; but if heat flows along a metal rod, in which the temperature varies in a continuous fashion, it is difficult to apply our previous definition of the affinity. Nevertheless we can compute the rate of production of entropy, and thereby we can identify the affinity.

With the foregoing considerations to guide us, we now turn our attention to continuous systems. We consider a three-dimensional system in which energy and matter flow, driven by appropriate forces. We choose the components of the vector current densities of energy and matter as fluxes. Thus, associated with the energy U we have the three energy fluxes J_{ox}, J_{oy}, J_{oz}. These quantities are the x, y, and z components of the vector current density $\mathbf{J}_o$. By definition the magnitude of $\mathbf{J}_o$ is the amount of energy that flows across the unit area in unit time, and the direction of $\mathbf{J}_o$ is the direction of this energy flow. Similarly, the current density $\mathbf{J}_k$ may describe the flow of a particular chemical component per unit area and per unit time; the components J_{kx}, J_{ky}, and J_{kz} are fluxes.

In order to identify the affinities, we now seek to write the rate of production of entropy in a form analogous to equation 14.6. One problem that immediately arises is that of defining entropy in a nonequilibrium system. This problem is solved in a formal manner as follows.

To any infinitesimal region we associate a local entropy $S(X_0, X_1, \ldots)$, where, *by definition, the functional dependence of S on the local extensive parameters $X_0, X_1, \ldots$ is taken to be identical to the dependence in equilibrium.* That is, we merely adopt the equilibrium fundamental equation to associate a local entropy with the local parameters $X_0, X_1, \ldots$. Then

$$dS = \sum_k F_k \, dX_k \tag{14.7}$$

or, taking all quantities per unit volume,[4]

$$ds = \sum_k F_k \, dx_k \tag{14.8}$$

The summation in this equation omits the term for volume and consequently has one less term than that in equation 14.7.

Again, the local intensive parameter F_k is taken to be the same function of the local extensive parameters as it would be in equilibrium. It is because of this convention, incidentally, that we can speak of the temperature varying continuously in a bar, despite the fact that thermostatics implies the existence of temperature only in equilibrium systems.

Equation 14.7 immediately suggests a reasonable definition of the *entropy current density* $\mathbf{J}_S$

$$\mathbf{J}_S = \Sigma F_k \mathbf{J}_k \tag{14.9}$$

in which $\mathbf{J}_k$ is the current density of the extensive parameter X_k. The magnitude of the entropy flux $\mathbf{J}_S$ is the entropy transported through unit area per unit time.

---

[4]It should be noted that in the remainder of this chapter we use lowercase letters to indicate extensive parameters *per unit volume* rather than *per mole*.

The rate of local production of entropy is equal to the entropy leaving the region, plus the rate of increase of entropy within the region. If $\dot{s}$ denotes the rate of production of entropy per unit volume and $\partial s / \partial t$ denotes the increase in entropy per unit volume, then

$$\dot{s} = \frac{\partial s}{\partial t} + \nabla \cdot \mathbf{J}_S \tag{14.10}$$

If the extensive parameters of interest are *conserved*, as are the energy and (in the absence of chemical reactions) the mole numbers, the equations of continuity for these parameters become

$$0 = \frac{\partial x_k}{\partial t} + \nabla \cdot \mathbf{J}_k \tag{14.11}$$

We are now prepared to compute $\dot{s}$ explicitly and thence to identify the affinities in continuous systems.

The first term in equation 14.10 is easily computed from equation 14.8.

$$\frac{\partial s}{\partial t} = \sum_k F_k \frac{\partial x_k}{\partial t} \tag{14.12}$$

The second term in equation 14.10 is computed by taking the divergence of equation 14.9

$$\nabla \cdot \mathbf{J}_S = \nabla \cdot \left(\sum_k F_k \mathbf{J}_k \right) = \sum_k \nabla F_k \cdot \mathbf{J}_k + \sum_k F_k \nabla \cdot \mathbf{J}_k \tag{14.13}$$

Thus equation 14.10 becomes

$$\dot{s} = \sum_k F_k \frac{\partial x_k}{\partial t} + \sum_k \nabla F \cdot \mathbf{J}_k + \sum_k F_k \nabla \cdot \mathbf{J}_k \tag{14.14}$$

Finally, by equation 14.11, we observe that the first and third terms cancel, giving

$$\dot{s} = \sum_k \nabla F_k \cdot \mathbf{J}_k \tag{14.15}$$

Although the affinity is defined as the difference in the entropy-representation intensive parameters for discrete systems, it is the gradient of the entropy-representation intensive parameters in continuous systems.

If $\mathbf{J}_{oz}$ denotes the z component of the energy current density, the associated affinity $\mathscr{F}_{oz}$ is $\nabla_z(1/T)$, the z component of the gradient of the inverse temperature. And if $\mathbf{J}_k$ denotes the kth mole number current density (the number of moles of the kth component flowing through unit area per second), the affinity associated with J_{kz} is $\mathscr{F}_{kz} = -\nabla_z(\mu_k/T)$.

14-3 PURELY RESISTIVE AND LINEAR SYSTEMS

For certain systems the fluxes at a given instant depend only on the values of the affinities at that instant. Such systems are referred to as "purely resistive."

For other than purely resistive systems the fluxes may depend upon the values of the affinities at previous times as well as upon the instantaneous values. In the electrical case a "resistor" is a purely resistive system, whereas a circuit containing an inductance or a capacitance is not purely resistive. A non-purely-resistive system has a "memory."

Although it might appear that the restriction to purely resistive systems is very severe, it is found in practice that a very large fraction of the systems of interest, other than electrical systems, are purely resistive. Extensions to non-purely-resistive systems do exist, based on the fluctuation–dissipation theorem or Kubo formula referred to in Section 14.1.

For a purely resistive system, by definition, each local flux depends only upon the instantaneous local affinities and upon the local intensive parameters. That is, dropping the indices denoting vector components

$$J_k = J_k\left(\mathscr{F}_0, \mathscr{F}_1, \ldots, \mathscr{F}_j, \ldots ; F_0, F_1, \ldots, F_j, \ldots\right) \qquad (14.16)$$

Thus, the local mole number current density of the kth component depends on the gradient of the inverse temperature, on the gradients of μ_j/T for each component, and upon the local temperature, pressure, and so forth. It should be noted that we do not assume that each flux depends only on its own affinity but rather that each flux depends on *all* affinities. It is true that each flux tends to depend most strongly on its own associated affinity, but the dependence of a flux on other affinities as well is the source of some of the most interesting phenomena in the field of irreversibility.

Each flux J_k is known to vanish as the affinities vanish, so we can expand J_k in powers of the affinities with no constant term

$$J_k = \sum_j L_{jk}\mathscr{F}_j + \frac{1}{2!}\sum_i\sum_j L_{ijk}\mathscr{F}_i\mathscr{F}_j + \cdots \qquad (14.17)$$

where

$$L_{jk} = \left(\frac{\partial J_k}{\partial \mathscr{F}_j}\right)_0 \qquad (14.18)$$

and

$$L_{ijk} = \left(\frac{\partial^2 J_k}{\partial \mathscr{F}_i \, \partial \mathscr{F}_j} \right)_0 \tag{14.19}$$

The functions L_{jk} are called *kinetic coefficients. They are functions of the local intensive parameters*

$$L_{jk} = L_{jk}(F_0, F_1, \dots) \tag{14.20}$$

The functions L_{ijk} are called *second-order kinetic coefficients*, and they are also functions of the local intensive parameters. Third-order and higher-order kinetic coefficients are similarly defined.

For the purposes of the Onsager theorem, which we are about to enunciate, it is convenient to adopt a notation that exhibits the functional dependence of the kinetic coefficients on an externally applied magnetic field $\mathbf{B}_e$, suppressing the dependence on the other intensive parameters

$$L_{jk} = L_{jk}(\mathbf{B}_e) \tag{14.21}$$

The Onsager theorem states that

$$L_{jk}(\mathbf{B}_e) = L_{kj}(-\mathbf{B}_e) \tag{14.22}$$

That is, the value of the kinetic coefficient L_{jk} measured in an external magnetic field $\mathbf{B}_e$ is identical to the value of L_{kj} measured in the reversed magnetic field $-\mathbf{B}_e$.

The Onsager theorem states a symmetry between the linear effect of the jth affinity on the kth flux and the linear effect of the kth affinity on the jth flux when these effects are measured in opposite magnetic fields.

A situation of great practical interest arises if the affinities are so small that all quadratic and higher-order terms in equation 14.17 can be neglected. A process that can be adequately described by the truncated approximate equations

$$J_k = \sum_j L_{jk} \mathscr{F}_j \tag{14.23}$$

is called a *linear* purely resistive process. For the analysis of such processes the Onsager theorem is a particularly powerful tool.

It is perhaps surprising that so many physical processes of interest are linear. But the affinities that we commonly encounter in the laboratory are quite small in the sense of equation 14.17, and we therefore recognize that we generally deal with systems that deviate only slightly from equilibrium.

Phenomenologically, it is found that the flow of energy in a thermally conducting body is proportional to the gradient of the temperature. Denoting the energy current density by $\mathbf{J}_o$, we find that experiment yields the linear law

$$\mathbf{J}_o = -\kappa \nabla T \qquad (14.24)$$

in which κ is the thermal conductivity of the body. We can rewrite this in the more appropriate form

$$\mathbf{J}_{oz} = \kappa T^2 \nabla_z \left(\frac{1}{T} \right) \qquad (14.25)$$

and similarly for x and y components, and we see that (κT^2) is the kinetic coefficient. The absence of higher-order terms, such as $[\nabla(1/T)]^2$ and $[\nabla(1/T)]^3$, in the phenomenological law shows that commonly employed temperature gradients are small in the sense of equation 14.17.

Ohm's law of electrical conduction and Fick's law of diffusion are other linear phenomenological laws which demonstrate that for the common values of the affinities in these processes higher-order terms are negligible. On the other hand, both the linear region and the nonlinear region can be realized easily in chemical systems, depending upon the deviations of the molar concentrations from their equilibrium values. Although the class of linear processes is sufficiently common to merit special attention, it is by no means all inclusive, and the Onsager theorem is *not* restricted to this special class of systems.

14-4 THE THEORETICAL BASIS OF THE ONSAGER RECIPROCITY

The Onsager reciprocity theorem has been stated but not proved in the preceding sections. Before turning to applications in the following sections we indicate the relationship of the theorem to the underlying principle of time reversal symmetry of physical laws.

From the purely thermodynamic point of view, the extensive parameters of a system in contact with a reservoir are constants. In fact, if an extensive parameter (such as the energy) is permitted to flow to and from a reservoir, it does so in continual spontaneous fluctuations. These fluctuations tend to be very rapid, and macroscopic observations average over the fluctuations (as discussed in some detail in Chapter 1). Occasionally a large fluctuation occurs, depleting the energy of the system by a non-negligible amount. If the system were to be decoupled from the reservoir before this rare large fluctuation were to decay, we would then associate a lower temperature to the system. But if the system were *not* decoupled, the fluctuation would decay by the spontaneous flow of energy from the reservoir to the system.

Onsager connected the theory of macroscopic processes to thermodynamic theory by the assumption that *the decay of a spontaneous fluctuation is identical to the macroscopic process of flow of energy or other analogous quantity between the reservoir and the system of depleted energy.*

We consider a system in equilibrium with a pair of reservoirs corresponding to the extensive parameters X_j and X_k. Let the instantaneous values of these parameters be denoted by $\hat{X}_j$ and $\hat{X}_k$, and let $\delta\hat{X}_j$ denote the deviation of $\hat{X}_j$ from its average value. Thus $\delta\hat{X}_j$ describes a fluctuation, and the average value of $\delta\hat{X}_j$ is zero. Nevertheless the average value of $(\delta\hat{X}_j)^2$, denoted by $\langle(\delta\hat{X}_j)^2\rangle$, is not zero. Nor is the correlation moment $\langle\delta\hat{X}_j\delta\hat{X}_k\rangle$. A very slight extension of the thermodynamic formalism, invoking only very general features of statistical mechanics, permits exact evaluation of the correlation moments of the fluctuations (as we shall see in Chapter 19).

More general than the correlation moment $\langle\delta\hat{X}_j\delta\hat{X}_k\rangle$ is the *delayed correlation moment* $\langle\delta\hat{X}_j\delta\hat{X}_k(\tau)\rangle$, which is the average product of the deviations $\delta\hat{X}_j$ and $\delta\hat{X}_k$, with the latter being observed a time τ after the former. It is this delayed correlation moment upon which Onsager focused attention.

The delayed correlation moment is subject to certain symmetries that follow from the time reversal symmetry of physical laws. In particular, assuming no magnetic field to be present, the delayed correlation moment must be unchanged under the replacement of τ by $-\tau$

$$\langle\delta\hat{X}_j\delta\hat{X}_k(\tau)\rangle = \langle\delta\hat{X}_j\delta\hat{X}_k(-\tau)\rangle \tag{14.26}$$

or, since only the relative times in the two factors are significant,

$$\langle\delta\hat{X}_j\delta\hat{X}_k(\tau)\rangle = \langle\delta\hat{X}_j(\tau)\delta\hat{X}_k\rangle \tag{14.27}$$

If we now subtract $\langle\delta\hat{X}_j\delta\hat{X}_k\rangle$ from each side of the equation and divide by τ, we find

$$\left\langle\delta\hat{X}_j\frac{\delta\hat{X}_k(\tau)-\delta\hat{X}_k}{\tau}\right\rangle = \left\langle\frac{\delta\hat{X}_j(\tau)-\delta\hat{X}_j}{\tau}\delta\hat{X}_k\right\rangle \tag{14.28}$$

In the limit as $\tau \to 0$ we can write the foregoing equation in terms of time derivatives.

$$\langle\delta\hat{X}_j\,\delta\dot{\hat{X}}_k\rangle = \langle\delta\dot{\hat{X}}_j\,\delta\hat{X}_k\rangle \tag{14.29}$$

Now we *assume* that the decay of a fluctuation $\delta \hat{X}_k$ is governed by the *same* linear dynamical laws as are macroscopic processes

$$\delta \hat{X}_k = \sum_i L_{ik} \delta \hat{\mathscr{F}}_i \tag{14.30}$$

Inserting these equations in equation 14.29 gives

$$\sum_i L_{ik} \langle \delta \hat{X}_j \delta \hat{\mathscr{F}}_i \rangle = \sum_i L_{ij} \langle \delta \hat{\mathscr{F}}_i \delta \hat{X}_k \rangle \tag{14.31}$$

The theory of fluctuations reveals (Chapter 19) the plausible result that in the absence of a magnetic field the fluctuation of each affinity is associated only with the fluctuation of its own extensive parameter; *there are no cross-correlation terms of the form* $\langle \delta \hat{X}_j \delta \hat{\mathscr{F}}_i \rangle$ *with* $i \neq j$. Furthermore it will be shown that the "diagonal" correlation function (with $i = j$) has the value $-k_B$ (though the specific value is not of importance for our present purposes)

$$\langle \delta \hat{X}_j \delta \hat{\mathscr{F}}_i \rangle = \begin{cases} -k_B & \text{if } i = j \\ 0 & \text{if } i \neq j \end{cases} \quad \mathbf{B}_e = 0 \tag{14.32}$$

It follows that in the absence of a magnetic field $L_{ij} = L_{ji}$, which is the Onsager reciprocity theorem (equation 14.22).

In the presence of a magnetic field the proof follows in similar fashion, depending upon a similar symmetry in the correlation functions of the spontaneous fluctuations.

Despite this fundamental basis in fluctuation theory, the applications of the Onsager theory are purely macroscopic, expressed in terms of phenomenological dynamical equations. This thermodynamic emphasis of application has motivated interjection of the subject prior to the statistical mechanical chapters to follow. Accordingly we turn to thermoelectric effects as an illustrative application of the Onsager theorem.

14-5 THERMOELECTRIC EFFECTS

Thermoelectric effects are phenomena associated with the simultaneous flow of electric current and "heat current" in a system. Relationships among various such phenomena were proposed in 1854 by Lord Kelvin on the basis of empirical observations. Kelvin also presented a heuristic argument leading to the relations, carefully pointing out, however, that the argument was not only unjustified but that it could be made to yield incorrect relations as well as correct ones. Unfortunately the argument continually resurfaces with renewed claims of rigor—of which the reader of the thermodynamic literature should be forewarned.

To analyze the thermoelectric effects in terms of the Onsager reciprocity we focus attention on a conductor in which both electric current and heat current flow in one dimension, and we describe the electric current as being carried by electrons. Then if s is the local entropy density

$$ds = \frac{1}{T} du - \sum_k \left(\frac{\mu_k}{T} \right) dn_k \qquad (14.33)$$

in which u is the local energy density, μ is the electrochemical potential (per particle) of the electrons, n is the number of electrons per unit volume, and in which the sum refers to other "components." These other components are the various types of atomic nuclei that together with the electrons constitute the solid. It will be noted that we have taken n as the number of electrons rather than the number of moles of electrons, and μ is accordingly the electrochemical potential per particle rather than per mole. In this regard we deviate from the more usual parameters merely by multiplication and division by Avogadro's number, respectively.

Just as equation 14.7 led to equation 14.9, equation 14.33 now leads to

$$\mathbf{J}_S = \frac{1}{T}\mathbf{J}_U - \frac{\mu}{T}\mathbf{J}_N \qquad (14.34)$$

in which $\mathbf{J}_S$, $\mathbf{J}_U$, and $\mathbf{J}_N$ are current densities of entropy, energy, and number of electrons, respectively. The other components in equation 14.33 are assumed immobile and consequently do not contribute flux terms to equation 14.34.

Repeating the logic leading to equation 14.15, we find

$$\dot{s} = \nabla \frac{1}{T} \cdot \mathbf{J}_U - \nabla \frac{\mu}{T} \cdot \mathbf{J}_N \qquad (14.35)$$

Thus if the components of $\mathbf{J}_U$ and $-\mathbf{J}_N$ are taken as fluxes, the associated affinities are the components of $\nabla(1/T)$ and $\nabla\mu/T$. Assuming for simplicity that all flows and forces are parallel to the x-direction, and omitting the subscript x, the linear dynamical laws become

$$-J_N = L'_{11}\nabla \frac{\mu}{T} + L'_{12}\nabla \frac{1}{T} \qquad (14.36)$$

$$J_U = L'_{21}\nabla \frac{\mu}{T} + L'_{22}\nabla \frac{1}{T} \qquad (14.37)$$

and the Onsager theorem gives the relation

$$L'_{12}(\mathbf{B}_e) = L'_{21}(-\mathbf{B}_e) \tag{14.38}$$

Before drawing physical conclusions from equation 14.38 we recast the dynamical equations into an equivalent but instructive form. Although $\mathbf{J}_U$ is a current density of total internal energy, we generally prefer to discuss the current density of heat. In analogy with the relation $dQ = T\,dS$ we therefore define a heat current density $\mathbf{J}_Q$ by the relation

$$\mathbf{J}_Q = T\mathbf{J}_S \tag{14.39}$$

or, by equation 14.34,

$$\mathbf{J}_Q = \mathbf{J}_U - \mu\mathbf{J}_N \tag{14.40}$$

In a very rough intuitive way we can look on μ as the potential energy per particle and on $\mu\mathbf{J}_N$ as a current density of potential energy; subtraction of the potential energy current density from the total energy current density yields the heat current density as a sort of kinetic energy current density. At any rate, eliminating $\mathbf{J}_U$ in favor of $\mathbf{J}_Q$ from equation 14.34 gives

$$\dot{s} = \nabla\frac{1}{T}\cdot\mathbf{J}_Q - \frac{1}{T}\nabla\mu\cdot\mathbf{J}_N \tag{14.41}$$

It follows from this equation that if the components of $\mathbf{J}_Q$ and of $-\mathbf{J}_N$ are chosen as fluxes the associated affinities are the corresponding components of $\nabla(1/T)$ and of $(1/T)\nabla\mu$, respectively. The dynamical equations can then be written, in the one-dimensional case, as

$$-J_N = L_{11}\frac{1}{T}\nabla\mu + L_{12}\nabla\frac{1}{T} \tag{14.42}$$

$$J_Q = L_{21}\frac{1}{T}\nabla\mu + L_{22}\nabla\frac{1}{T} \tag{14.43}$$

and the Onsager relation is

$$L_{12}(\mathbf{B}_e) = L_{21}(-\mathbf{B}_e) \tag{14.44}$$

The reader should verify that the dynamical equations 14.42 and 14.43 can also be obtained by direct substitution of equation 14.40 into the previous pair of dynamical equations 14.36 and 14.37 without recourse to the entropy production equation 14.41.

The significance of the heat current can be exhibited in another manner. We consider, for a moment, a steady-state flow. Then both $\mathbf{J}_U$ and $\mathbf{J}_N$ are divergenceless and taking the divergence of equation 14.40 gives

$$\nabla \cdot \mathbf{J}_Q = -\nabla\mu \cdot \mathbf{J}_N \quad \text{(in the steady state)} \tag{14.45}$$

which states that in the steady state the rate of increase in heat current is equal to the rate of decrease in the potential energy current. Furthermore, the insertion of this equation into equation 14.41 gives

$$\dot{s} = \nabla \frac{1}{T} \cdot \mathbf{J}_Q + \frac{1}{T}\nabla \cdot \mathbf{J}_Q \tag{14.46}$$

which can be interpreted as stating that the production of entropy is due to two causes: The first term is the production of entropy due to the flow of heat from high to low temperature, and the second term is the increase in entropy due to the appearance of heat current.

We now accept the dynamical equations 14.42 and 14.43 and the symmetry condition (equation 14.44) as the basic equations with which to study the flow of heat and electric current in a system.

14-6 THE CONDUCTIVITIES

We consider a system in which an electric current and a heat current flow parallel to the x-axis in a steady state, with no applied magnetic field. Then omitting the subscript x

$$-J_N = L_{11}\frac{1}{T}\nabla\mu + L_{12}\nabla\frac{1}{T} \tag{14.47}$$

$$J_Q = L_{12}\frac{1}{T}\nabla\mu + L_{22}\nabla\frac{1}{T} \tag{14.48}$$

where the Onsager theorem has reduced to the simple symmetry

$$L_{12} = L_{21} \tag{14.49}$$

The three kinetic coefficients appearing in the dynamical equations can be related to more familiar quantities, such as conductivities. In developing this connection we first comment briefly on the nature of the electrochemical potential μ of the electrons. We can consider μ as being composed of two parts, a chemical portion μ_c and an electrical portion μ_e

$$\mu = \mu_c + \mu_e \tag{14.50}$$

If the charge on an electron is e, then μ_e is simply $e\phi$, where ϕ is the ordinary electrostatic potential. The chemical potential μ_c is a function of the temperature and of the electron concentration. Restating these facts in terms of gradients, the electrochemical potential *per unit charge* is $(1/e)\mu$; its gradient $(1/e)\nabla\mu$ is the sum of the *electric field* $(1/e)\nabla\mu_e$, plus an effective driving force $(1/e)\nabla\mu_c$ arising from a concentration gradient.

The *electric conductivity* σ is defined as the electric current density (eJ_N) per unit potential gradient $(1/e)\nabla\mu$ in an isothermal system. It is easily seen that $(1/e)\nabla\mu$ is actually the emf, for in a homogeneous isothermal system $\nabla\mu_c = 0$ and $\nabla\mu = \nabla\mu_e$. Thus, by definition

$$\sigma \equiv -eJ_N \bigg/ \frac{1}{e}\nabla\mu \qquad \text{for } \nabla T = 0 \qquad (14.51)$$

whence equation 14.47 gives

$$\sigma = e^2 L_{11}/T \qquad (14.52)$$

Similarly the *heat conductivity* κ is defined as the heat current density per unit temperature gradient for zero electric current

$$\kappa \equiv -J_Q/\nabla T \qquad \text{for } J_N = 0 \qquad (14.53)$$

Solving the two kinetic equations simultaneously, we find

$$\kappa = \frac{D}{T^2 L_{11}} \qquad (14.54)$$

where D denotes the determinant of the kinetic coefficients

$$D \equiv L_{11}L_{22} - L_{12}^2 \qquad (14.55)$$

14-7 THE SEEBECK EFFECT AND THE THERMOELECTRIC POWER

The Seebeck effect refers to the production of an electromotive force in a thermocouple under conditions of zero electric current.

Consider a thermocouple with junctions at temperatures T_1 and T_2 $(T_2 > T_1)$, as indicated in Fig. 14.1. A voltmeter is inserted in one arm of the thermocouple at a point at which the temperature is T'. This voltmeter is such that it allows no passage of electric current but offers no resistance to the flow of heat. We designate the two materials composing the

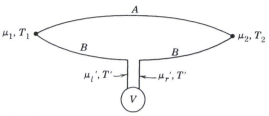

FIGURE 14.1

thermocouple by A and B. With $J_N = 0$, we obtain from the kinetic equations, for either conductor

$$\nabla\mu = \frac{L_{12}}{TL_{11}}\nabla T \tag{14.56}$$

Thus

$$\mu_2 - \mu_1 = \int_1^2 \frac{L_{12}^A}{TL_{11}^A}\,dT \tag{14.57}$$

$$\mu_2 - \mu_r' = \int_r^2 \frac{L_{12}^B}{TL_{11}^B}\,dT \tag{14.58}$$

$$\mu_l' - \mu_1 = \int_1^l \frac{L_{12}^B}{TL_{11}^B}\,dT \tag{14.59}$$

Eliminating μ_1 and μ_2 from these equations

$$\mu_r' - \mu_l' = \int_1^2 \left(\frac{L_{12}^A}{TL_{11}^A} - \frac{L_{12}^B}{TL_{11}^B} \right) dT \tag{14.60}$$

But, because there is no temperature difference across the voltmeter, the voltage is simply

$$V = \frac{1}{e}(\mu_r' - \mu_l') = \int_1^2 \left(\frac{L_{12}^A}{eTL_{11}^A} - \frac{L_{12}^B}{eTL_{11}^B} \right) dT \tag{14.61}$$

The *thermoelectric power* of the thermocouple, ε_{AB}, is defined as the change in voltage per unit change in temperature difference. The sign of ε_{AB} is chosen as positive if the voltage increment is such as to drive the current from A to B at the hot junction. Then

$$\varepsilon_{AB} = \frac{\partial V}{\partial T_2} = \left(\frac{-L_{12}^B}{eTL_{11}^B}\right) - \left(\frac{-L_{12}^A}{eTL_{11}^A}\right) \qquad (14.62)$$

Defining the *absolute thermoelectric power* of a single medium by the relation

$$\varepsilon_A \equiv \frac{-L_{12}^A}{eTL_{11}^A} \qquad (14.63)$$

the thermoelectric power of the thermocouple is

$$\varepsilon_{AB} = \varepsilon_B - \varepsilon_A \qquad (14.64)$$

If we accept the electric conductivity σ, the heat conductivity κ, and the absolute thermoelectric power ε as the three physically significant dynamical properties of a medium, we can eliminate the three kinetic coefficients in favor of these quantities and rewrite the kinetic equations in the following form

$$-J_N = \left(\frac{T\sigma}{e^2}\right)\frac{1}{T}\nabla\mu - \left(\frac{T^2\sigma\varepsilon}{e}\right)\nabla\frac{1}{T} \qquad (14.65)$$

$$J_Q = -\left(\frac{T^2\sigma\varepsilon}{e}\right)\frac{1}{T}\nabla\mu + (T^3\sigma\varepsilon^2 + T^2\kappa)\nabla\frac{1}{T} \qquad (14.66)$$

An interesting insight to the physical meaning of the absolute thermoelectric power can be obtained by eliminating $(1/T)\nabla\mu$ between the two foregoing dynamical equations and writing J_Q in terms of J_N and $\nabla(1/T)$

$$J_Q = T\varepsilon e J_N + T^2\kappa\nabla\frac{1}{T} \qquad (14.67)$$

or, recalling that $J_S = J_Q/T$

$$J_S = \varepsilon e J_N + T\kappa\nabla\frac{1}{T} \qquad (14.68)$$

According to this equation, each electron involved in the electric current carries with it an entropy of εe. This flow of entropy is in addition to the entropy current $T\kappa\nabla(1/T)$, which is independent of the electronic current. The thermoelectric power can be looked on as the entropy transported per coulomb by the electron flow.

14-8 THE PELTIER EFFECT

The Peltier effect refers to the evolution of heat accompanying the flow of an electric current across an isothermal junction of two materials.

FIGURE 14.2

Consider an isothermal junction of two conductors A and B and an electric current $e\mathbf{J}_N$ to flow as indicated in Fig. 14.2. Then the total energy current will be discontinuous across the junction, and the energy difference appears as *Peltier heat* at the junction. We have $J_U = J_Q + \mu J_N$, and since both μ and J_N are continuous across the junction it follows that the discontinuity in J_U is equal to the discontinuity in J_Q

$$J_U^A - J_U^B = J_Q^A - J_Q^B \tag{14.69}$$

Because of the isothermal condition, the dynamical equations 14.65 and 14.66 give, in either conductor

$$J_Q = T\varepsilon(eJ_N) \tag{14.70}$$

whence

$$J_Q^B - J_Q^A = T(\varepsilon_B - \varepsilon_A)(eJ_N) \tag{14.71}$$

The *Peltier coefficient* π_{AB} is defined as the heat that must be supplied to the junction when unit electric current passes from conductor A to conductor B. Thus

$$\pi_{AB} \equiv \left(J_Q^B - J_Q^A\right)/eJ_N = T(\varepsilon_B - \varepsilon_A) \tag{14.72}$$

Equation 14.72, which relates the Peltier coefficient to the absolute thermoelectric powers, is one of the relations presented on empirical evidence by Kelvin in 1854. It is called the *second Kelvin relation*.

The method by which we have derived equation 14.72 is typical of all applications of the Onsager relations, so that it may be appropriate to review the procedure. We first write the linear dynamical equations, reducing the number of kinetic coefficients appearing therein by invoking the Onsager relations. We then proceed to analyze various effects, expressing each in terms of the kinetic coefficients. When we have analyzed as many effects as there are kinetic coefficients, we rewrite the dynamical equations in terms of those effects rather than in terms of the kinetic coefficients (as in equations 14.65 and 14.66). Thereafter every additional effect analyzed on the basis of the dynamical equations results in a relation analogous to equation 14.72 and expresses this new effect in terms of the coefficients in the dynamical equation.

14-9   THE THOMSON EFFECT

The Thomson effect refers to the evolution of heat as an electric current traverses a temperature gradient in a material.

Consider a conductor carrying a heat current but no electric current. A temperature distribution governed by the temperature dependence of the kinetic coefficients will be set up. Let the conductor now be placed in contact at each point with a heat reservoir of the same temperature as that point, so that there is no heat interchange between conductor and reservoirs. Now let an electric current pass through the conductor. An interchange of heat will take place between conductor and reservoirs. This heat exchange consists of two parts—the *Joule heat* and the *Thomson heat*.

As the electric current passes along the conductor, any change in total energy flow must be supplied by an energy interchange with the reservoirs. Thus we must compute $\nabla \cdot \mathbf{J}_U$

$$\nabla \cdot \mathbf{J}_U = \nabla \cdot (\mathbf{J}_Q + \mu \mathbf{J}_N) = \nabla \cdot \mathbf{J}_Q + \nabla \mu \cdot \mathbf{J}_N \qquad (14.73)$$

which can be expressed in terms of $\mathbf{J}_N$ and $\nabla(1/T)$ by using equations 14.67 and 14.68

$$\nabla \cdot \mathbf{J}_U = \nabla \cdot \left(T \varepsilon e \mathbf{J}_N + T^2 \kappa \nabla \frac{1}{T} \right) + \left(-\frac{e^2}{\sigma} \mathbf{J}_N + T^2 e \varepsilon \nabla \frac{1}{T} \right) \cdot \mathbf{J}_N$$

$$(14.74)$$

or

$$\nabla \cdot \mathbf{J}_U = T \nabla \varepsilon \cdot (e \mathbf{J}_N) + \nabla \cdot \left(T^2 \kappa \nabla \frac{1}{T} \right) - \frac{e^2}{\sigma} \mathbf{J}_N \qquad (14.75)$$

However the temperature distribution is that which is determined by the steady state with no electric current, and we know that $\nabla \cdot \mathbf{J}_U$ vanishes in that state. By putting $\mathbf{J}_N = 0$ and $\nabla \cdot \mathbf{J}_U = 0$ in equation 14.75 we conclude that the temperature distribution is such as to make the second term vanish, and consequently

$$\nabla \cdot \mathbf{J}_U = T \nabla \varepsilon \cdot (e \mathbf{J}_N) - \frac{1}{\sigma} (e \mathbf{J}_N)^2 \qquad (14.76)$$

Furthermore, noting that the thermoelectric power is a function of the local temperature, we write

$$\nabla \varepsilon = \frac{d\varepsilon}{dT} \nabla T \qquad (14.77)$$

and

$$\nabla \cdot \mathbf{J}_U = T \frac{d\varepsilon}{dT} \nabla T \cdot (e\mathbf{J}_N) - \frac{1}{\sigma}(e\mathbf{J}_N)^2 \qquad (14.78)$$

The second term is the Joule heat, produced by the flow of electric current even in the absence of a temperature gradient. The first term represents the Thomson heat, absorbed from the heat reservoirs when the current $e\mathbf{J}_N$ traverses the temperature gradient ∇T. The Thomson coefficient τ is defined as the Thomson heat absorbed per unit electric current and per unit temperature gradient

$$\tau \equiv \frac{\text{Thomson heat}}{\nabla T \cdot (e\mathbf{J}_N)} = T\frac{d\varepsilon}{dT} \qquad (14.79)$$

Thus the coefficient of the Thomson effect is related to the temperature derivative of the thermoelectric power.

Equations 14.72 and 14.79 imply the "first Kelvin relation"

$$\frac{d\pi_{AB}}{dT} + \tau_A - \tau_B = \varepsilon_A - \varepsilon_B \qquad (14.80)$$

which was obtained by Kelvin on the basis of energy conservation alone.

Various other thermoelectric effects can be defined, and each can be expressed in terms of the three independent coefficients L_{11}, L_{12}, and L_{22}, or in terms of α, κ, and ε.

In the presence of an orthogonal magnetic field the number of "thermomagnetic" effects becomes quite large. If the field is in the z-direction an x-directed electric current produces a y-directed gradient of the electrochemical potential; this is the "Hall effect." Similarly an x-directed thermal gradient produces a y-directed gradient of the electrochemical potential; the Nernst effect. The method of analysis[5] is identical to that of the thermoelectric effects, with the addition of the field dependence (equation 14.22) of the Onsager reciprocity theorem.

[5] H. Callen, *Phys. Rev.* **73**, 1349 (1948).

PART **II**

STATISTICAL MECHANICS

15

STATISTICAL MECHANICS IN THE ENTROPY REPRESENTATION: THE MICROCANONICAL FORMALISM

15-1 PHYSICAL SIGNIFICANCE OF THE ENTROPY FOR CLOSED SYSTEMS

Thermodynamics constitutes a powerful formalism of great generality, erected on a basis of a very few, very simple hypotheses. The central concept introduced through those hypotheses is the entropy. It enters the formulation abstractly as the variational function in a mathematical extremum principle determining equilibrium states. In the resultant formalism, however, the entropy is one of a set of extensive parameters, together with the energy, volume, mole numbers and magnetic moment. As these latter quantities each have clear and fundamental physical interpretations it would be strange indeed if the entropy alone were to be exempt from physical interpretation.

The subject of statistical mechanics provides the physical interpretation of the entropy, and it accordingly provides a heuristic justification for the extremum principle of thermodynamics. For some simple systems, for which we have tractable models, this interpretation also permits explicit calculation of the entropy and thence of the fundamental equation.

We focus first on a closed system of given volume and given number of particles. For definiteness we may think of a fluid, but this is in no way necessary. The parameters U, V, and N are the only constraints on the system. Quantum mechanics tells us that, if the system is macroscopic, there may exist many discrete quantum states consistent with the specified values of U, V, and N. The system may be in any of these permissible states.

Naively we might expect that the system, finding itself in a particular quantum state, would remain forever in that state. Such, in fact, is the lore of elementary quantum mechanics; the "quantum numbers" that specify a particular quantum state are ostensibly "constants of the motion." This naive fiction, relatively harmless to the understanding of atomic systems (to which quantum mechanics is most commonly applied) is flagrantly misleading when applied to macroscopic systems.

The apparent paradox is seated in the assumption of *isolation* of a physical system. *No physical system is, or ever can be, truly isolated*. There exist weak, long-range, random gravitational, electromagnetic and other forces that permeate all physical space. These forces not only couple spatially separated material systems, but the force fields *themselves* constitute physical systems in direct interaction with the system of interest. The very vacuum is now understood to be a complex fluctuating entity in which occur continual elaborate processes of creation and reabsorption of electrons, positrons, neutrinos, and a myriad of other esoteric subatomic entities. All of these events can couple with the system of interest.

For a simple system such as a hydrogen atom in space the very weak interactions to which we have alluded seldom induce transitions between quantum states. This is so because the quantum states of the hydrogen atom are widely spaced in energy, and the weak random fields in space cannot easily transfer such large energy differences to or from the atom. Even so, such interactions occasionally do occur. An excited atom may "spontaneously" emit a photon, decaying to a lower energy state. Quantum field theory reveals that such ostensibly "spontaneous" transitions actually are *induced* by the interactions between the excited atom and the modes of the vacuum. The quantum states of atoms are *not* infinitely long lived, precisely because of their interaction with the random modes of the vacuum.

For a macroscopic system the energy differences between successive quantum states become minute. In a macroscopic assembly of atoms each energy eigenstate of a single atom "splits" into some 10^{23} energy eigenstates of the assembly, so that the average energy difference between successive states is decreased by a factor of $\sim 10^{-23}$. The slightest random field, or the weakest coupling to vacuum fluctuations, is then sufficient to buffet the system chaotically from quantum state to quantum state.

A realistic view of a macroscopic system is one in which the system makes enormously rapid random transitions among its quantum states. A macroscopic measurement senses only an average of the properties of myriads of quantum states.

All "statistical mechanicians" agree with the preceding paragraph, but not all would agree on the *dominant* mechanism for inducing transitions. Various mechanisms compete and others may well dominate in some or even in all systems. No matter—it is sufficient that *any* mechanism exists, and it is only the conclusion of rapid, random transitions that is needed to validate statistical mechanical theory.

Because the transitions are induced by purely random processes, it is reasonable to suppose that *a macroscopic system samples every permissible quantum state with equal probability*—a permissible quantum state being one consistent with the external constraints.

The assumption of equal probability of all permissible microstates is the fundamental postulate of statistical mechanics. Its justification will be examined more deeply in Part III, but for now we adopt it on two bases; its a priori reasonableness, and the success of the theory that flows from it.

Suppose now that some external constraint is removed—such as the opening of a valve permitting the system to expand into a larger volume. From the microphysical point of view the removal of the constraint activates the possibility of many microstates that previously had been precluded. Transitions occur into these newly available states. After some time the system will have lost all distinction between the original and the newly available states, and the system will thenceforth make random transitions that sample the *augmented* set of states with equal probability. *The number of microstates among which the system undergoes transitions, and which thereby share uniform probability of occupation, increases to the maximum permitted by the imposed constraints.*

This statement is strikingly reminiscent of the entropy postulate of thermodynamics, according to which the entropy increases to the maximum permitted by the imposed constraints. It suggests that the entropy can be identified with the number of microstates consistent with the imposed macroscopic constraints.

One difficulty arises: The entropy is additive (extensive), whereas the number of microstates is multiplicative. The number of microstates available to two systems is the *product* of the numbers available to each (the number of "microstates" of two dice is $6 \times 6 = 36$). To interpret the entropy, then, we require an *additive* quantity that measures the number of microstates available to a system. The (unique!) answer is to *identify the entropy with the logarithm of the number of available microstates* (the logarithm of a product being the sum of the logarithms). Thus

$$S = k_B \ln \Omega \tag{15.1}$$

where Ω is the number of microstates consistent with the macroscopic constraints. The constant prefactor merely determines the scale of S; it is chosen to obtain agreement with the Kelvin scale of temperature, defined by $T^{-1} = \partial S/\partial U$. We shall see that this agreement is achieved by taking the constant to be Boltzmann's constant $k_B = R/N_A = 1.3807 \times 10^{-23} \mathrm{J/K}$.

With the definition 15.1 the basis of statistical mechanics is established.

Just as the thermodynamic postulates were elaborated through the formalism of Legendre transformations, so this single additional postulate will be rendered more powerful by an analogous structure of mathematical formalism. Nevertheless this single postulate is dramatic in its brevity, simplicity, and completeness. The statistical mechanical formalism that derives directly from it is one in which we "simply" calculate the logarithm of the number of states available to the system, thereby obtaining S as a function of the constraints U, V, and N. That is, it is statistical mechanics in the entropy representation, or, in the parlance of the field, it is statistical mechanics in the *microcanonical formalism*.

In the following sections of this chapter we treat a number of systems by this microcanonical formalism as examples of its logical completeness.

As in thermodynamics, the entropy representation is not always the most convenient representation. For statistical mechanical calculations it is frequently so inconvenient that it is analytically intractable. The Legendre transformed representations are usually far preferable, and we shall turn to them in the next chapter. Nevertheless the microcanonical formulation establishes the clear and basic logical foundation of statistical mechanics.

PROBLEMS

15.1-1. A system is composed of two harmonic oscillators each of natural frequency ω_0 and each having permissible energies $(n + \frac{1}{2})\hbar\omega_0$, where n is any non-negative integer. The total energy of the system is $E' = n'\hbar\omega_0$, where n' is a positive integer. How many microstates are available to the system? What is the entropy of the system?

A second system is also composed of two harmonic oscillators, each of natural frequency $2\omega_0$. The total energy of this system is $E'' = n''\hbar\omega_0$, where n'' is an even integer. How many microstates are available to this system? What is the entropy of this system?

What is the entropy of the system composed of the two preceding subsystems (separated and enclosed by a totally restrictive wall)? Express the entropy as a function of E' and E''.

Answer:

$$S_{\text{tot}} = k_B \ln\left(\frac{E'E''}{2\hbar^2\omega_0^2}\right)$$

15.1-2. A system is composed of two harmonic oscillators of natural frequencies ω_0 and $2\omega_0$, respectively. If the system has total energy $E = (n + \frac{1}{2})\hbar\omega_0$, where n is an odd integer, what is the entropy of the system?

If a composite system is composed of two non-interacting subsystems of the type just described, having energies E_1 and E_2, what is the entropy of the composite system?

15-2 THE EINSTEIN MODEL
OF A CRYSTALLINE SOLID

With a identification of the meaning of the entropy we proceed to calculate the fundamental equation of macroscopic systems. We first apply the method to Einstein's simplified model of a nonmetallic crystalline solid.

It is well to pause immediately and to comment on so early an introduction of a specific model system. In the eleven chapters of this book devoted to thermodynamic theory there were few references to specific model systems, and those occasional references were kept carefully distinct from the logical flow of the general theory. In statistical mechanics we almost immediately introduce a model system, and this will be followed by a considerable number of others. The difference is partially a matter of convention. To some extent it reflects the simplicity of the general formalism of statistical mechanics, which merely adds the logical interpretation of the entropy to the formalism of thermodynamics; the interest therefore shifts to applications of that formalism, which underlies the various material sciences (such as solid state physics, the theory of liquids, polymer physics, and the like). But, most important, it reflects the fact that counting the number of states available to physical systems requires computational skills and experience that can be developed only by explicit application to concrete problems.

To account for the thermal properties of crystals, Albert Einstein, in 1907, introduced a highly idealized model focusing only on the vibrational modes of the crystal. Electronic excitations, nuclear modes, and various other types of excitations were ignored. Nevertheless, for temperatures that are neither very close to absolute zero nor very high, the model is at least qualitatively successful.

Einstein's model consists of the assumption that each of the $\tilde{N}$ atoms in the crystal can be considered to be bound to its equilibrium position by a harmonic force. Each atom is free to vibrate around its equilibrium position in any of the three coordinate directions, with a natural frequency ω_0.

More realistically (recall Section 1.2) the atoms of crystals are harmonically bound to their neighboring atoms rather than to fixed points. Accordingly the vibrational modes are strongly coupled, giving rise to $3\tilde{N}$ collective normal modes. The frequencies are distributed from zero (for very long wave length modes) to some maximum frequency (for the modes of minimum permissible wave length, comparable to the interatomic distance). There are far more high frequency modes than low frequency modes, with the consequence that the frequencies tend to cluster mainly in a narrow range of frequencies, to which the Einstein frequency ω_0 is a rough approximation.

In the Einstein model, then, a crystal of $\tilde{N}$ atoms is replaced by $3\tilde{N}$ harmonic oscillators, all with the same natural frequency ω_0.

For the present purposes it is convenient to choose the zero of energy so that the energy of a harmonic oscillator of natural frequency ω_0 can take only the discrete values $n\hbar\omega_0$, with $n = 0, 1, 2, 3, \ldots$. Here $\hbar = h/2\pi = 1.055 \times 10^{-34}$ J-s., h being Planck's constant.

In the language of quantum mechanics, each oscillator can be "occupied by an integral number of energy quanta," each of energy $\hbar\omega_0$.

The number of possible states of the system, and hence the entropy, can now be computed easily. If the energy of the system is U it can be considered as constituting $U/\hbar\omega_0$ quanta. These quanta are to be distributed among $3\tilde{N}$ vibrational modes. The number of ways of distributing the $U/\hbar\omega_0$ quanta among the $3\tilde{N}$ modes is the number of states Ω available to the system.

The problem is isomorphic to the calculation of the number of ways of placing $U/\hbar\omega_0$ identical (indistinguishable) marbles in $3\tilde{N}$ numbered (distinguishable) boxes.

FIGURE 15.1

Illustrating the combinatorial problem of distributing $U/\hbar\omega_0$ indistinguishable objects ("marbles") in $3\tilde{N}$ distinguishable "boxes."

The combinatorial problem can be visualized as follows. Suppose we have $U/\hbar\omega_0$ marbles and $3\tilde{N} - 1$ match sticks. We lay these out in a linear array, in any order. One such array is shown in Fig. 15.1. The interpretation of this array is that three quanta (marbles) are assigned to the first mode, two quanta to the second, none to the third, and so forth, and two quanta are assigned to the last mode (the $3\tilde{N}$-th). Thus the number of ways of distributing the $U/\hbar\omega_0$ quanta among the $3\tilde{N}$ modes is the number of permutations of $(3\tilde{N} - 1 + U/\hbar\omega_0)$ objects, of which $U/\hbar\omega_0$ are identical (marbles or quanta), and $3\tilde{N} - 1$ are identical (match sticks). That is

$$\Omega = \frac{(3\tilde{N} - 1 + U/\hbar\omega_0)!}{(3\tilde{N} - 1)!(U/\hbar\omega_0)!} \simeq \frac{(3\tilde{N} + U/\hbar\omega_0)!}{(3\tilde{N})!(U/\hbar\omega_0)!} \tag{15.2}$$

This completes the calculation, for the entropy is simply the logarithm of this quantity (multiplied by k_B). To simplify the result we employ the Stirling approximation for the logarithm of the factorial of a large number

$$\ln(M!) \simeq M \ln M - M + \cdots \qquad \text{(if } M \gg 1) \tag{15.3}$$

whence the molar entropy is

$$s = 3R \ln\left(1 + \frac{u}{u_0}\right) + 3R\frac{u}{u_0}\ln\left(1 + \frac{u_0}{u}\right) \tag{15.4}$$

where

$$u_0 \equiv 3N_A\hbar\omega_0 \tag{15.5}$$

This is the fundamental equation of the system.

It will be left to the problems to show that the fundamental equation implies reasonable thermal behavior. The molar heat capacity is zero at zero temperature, rises rapidly with increasing temperature, and approaches a constant value $(3R)$ at high temperature, in qualitative agreement with experiment. The rate of increase of the heat capacity is not quantitatively correct because of the naiveté of the model of the vibrational modes. This will be improved subsequently in the "Debye model" (Section 16.7), in which the vibrational modes are treated more realistically.

The heat capacity of the Einstein model is plotted in Fig. 15.2. The molar heat capacity c_v is zero at $T = 0$, and it asymptotes to $3R$ at high temperature. The rise in c_v occurs in the region $k_BT \simeq \frac{1}{3}\hbar\omega_0$ (in particular $c_v/3R = \frac{1}{2}$ and the point of maximum slope both occur near $k_BT/\hbar\omega_0 \simeq \frac{1}{3}$). At low temperature c_v rises exponentially, whereas experimentally the heat capacity rises approximately as T^3.

The mechanical implications of the model—the pressure–volume relationship and compressibility—are completely unreasonable. The entropy, according to equation 15.5, is independent of the volume, whence the pressure $T\partial S/\partial V$ is identically zero! Such a nonphysical result is, of course, a reflection of the naive omission of volume dependent effects from the model.

Certain consequences of the model give important general insights. Consider the thermal equation of state

$$\frac{1}{T} = \frac{\partial S}{\partial U} = \frac{k_B}{\hbar\omega_0}\ln\left(1 + \frac{3N}{U}N_A\hbar\omega_0\right) \tag{15.6}$$

Now, noting that there are $3NN_A$ oscillators in the system

$$\text{mean energy per oscillator} = \frac{U}{3NN_A} = \frac{\hbar\omega_0}{e^{\hbar\omega_0/k_BT} - 1} \tag{15.7}$$

The quantity $\hbar\omega_0/k_B$ is called the "Einstein temperature" of the crystal, and it generally is of the same order of magnitude as the melting

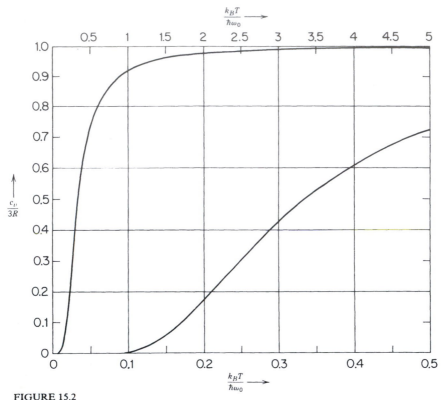

FIGURE 15.2
Heat capacity of the Einstein model, or of a single harmonic oscillator. The upper curve refers to the upper scale of $k_B T/\hbar\omega_0$, and the lower curve to the lower (expanded) scale. The ordinate can be interpreted as the heat capacity of one harmonic oscillator in units of k_B, or as the molar heat capacity in units of $3R$.

temperature of the solid. Thus below the melting temperature, the mean energy of an oscillator is less than, or of the order of, $\hbar\omega_0$. Alternatively stated, the solid melts before the Einstein oscillators attain quantum numbers appreciably greater than unity.

PROBLEMS

15.2-1. Calculate the molar heat capacity of the Einstein model by equation 15.7. Show that the molar heat capacity approaches $3R$ at high temperatures. Show that the temperature dependence of the molar heat capacity is exponential near zero temperature, and calculate the leading exponential term.

15.2-2. Obtain an equation for the mean quantum number $\bar{n}$ of an Einstein oscillator as a function of the temperature. Calculate $\bar{n}$ for $k_B T/\hbar\omega_0 = 0, 1, 2, 3, 4, 10, 50, 100$ (ignore the physical reality of melting of the crystal!).

15.2-3. Assume that the Einstein frequency ω_0 for a particular crystal depends upon the molar volume:

$$\omega_0 = \omega_0^0 - A \ln\left(\frac{v}{v_0}\right)$$

a) Calculate the isothermal compressibility of this crystal.
b) Calculate the heat transfer if a crystal (of one mole) is compressed at constant temperature from v_i to v_f.

15-3   THE TWO-STATE SYSTEM

Another model that illustrates the principles of statistical mechanics in a simple and transparent fashion is the "two-state model." In this model each "atom" can be either in its "ground state" (with energy zero) or in its "excited state" (with energy ε).

To avoid conflict with certain general theorems about energy spectra we assume that each atom has additional states, but all of such high energy as to exceed the total energy of the system under consideration. Such states are then inaccessible to the system and need not be considered further in the calculation.

If U is the energy of the system then U/ε atoms are in the excited state and $(\tilde{N} - U/\varepsilon)$ atoms are in the ground state. The number of ways of choosing U/ε atoms from the total number $\tilde{N}$ is

$$\Omega = \frac{\tilde{N}!}{(U/\varepsilon)!(\tilde{N} - U/\varepsilon)!} \tag{15.8}$$

The entropy is therefore

$$S = k_B \ln \Omega = k_B \ln(\tilde{N}!) - k_B \ln\left(\frac{U}{\varepsilon}!\right) - k_B \ln\left[\left(\tilde{N} - \frac{U}{\varepsilon}\right)!\right] \tag{15.9}$$

or, invoking Stirling's approximation (equation 15.3)

$$S = \left(\frac{U}{\varepsilon} - \tilde{N}\right) k_B \ln\left(1 - \frac{U}{\tilde{N}\varepsilon}\right) - \frac{U}{\varepsilon} k_B \ln\frac{U}{\tilde{N}\varepsilon} \tag{15.10}$$

Again, because of the artificiality of the model, the fundamental equation is independent of the volume. The thermal equation of state is easily calculated to be

$$\frac{1}{T} = \frac{k_B}{\varepsilon} \ln\left(\frac{\tilde{N}\varepsilon}{U} - 1\right) \tag{15.11}$$

Recalling that the calculation is subject to the condition $U < \tilde{N}\varepsilon$, we observe that the temperature is a properly positive number. Solving for the energy

$$U = \frac{\tilde{N}\varepsilon}{1 + e^{\varepsilon/k_BT}} \tag{15.12}$$

The energy approaches $\tilde{N}\varepsilon/2$ as the temperature approaches infinity in this model (although we must recall that additional states of high energy would alter the high temperature properties). At infinite temperature half the atoms are excited and half are in their ground state.

The molar heat capacity is

$$c = \frac{du}{dT} = N_A \frac{\varepsilon^2}{k_BT^2} \frac{e^{\varepsilon/k_BT}}{(1 + e^{\varepsilon/2k_BT})^2} = N_A \frac{\varepsilon^2}{k_BT^2}(e^{\varepsilon/2k_BT} + e^{-\varepsilon/2k_BT})^{-2}$$

$$\tag{15.13}$$

A graph of this temperature dependence is shown in Fig 15.3. The molar heat capacity is zero both at very low temperatures and at very high

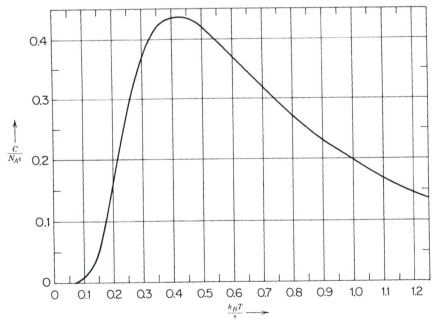

FIGURE 15.3
Heat capacity of the two-state model; the "Schottky hump."

temperatures, peaking in the region of $k_B T = .42\varepsilon$. This behavior is known as a "Schottky hump." Such a maximum, when observed in empirical data, is taken as an indication of a pair of low lying energy states, with all other energy states lying at considerably higher energies. This is an example of the way in which thermal properties can reveal information about the atomic structure of materials.

PROBLEMS

15.3-1. In the two-state model system of this section suppose the excited state energy ε of an "atom" depends on its average distance from its neighboring atoms so that

$$\varepsilon = \frac{a}{\tilde{v}^\gamma}. \qquad \tilde{v} \equiv \frac{V}{N}$$

where a and γ are positive constants. This assumption, applied to a somewhat more sophisticated model of a solid, was introduced by Gruneisen, and γ is the "Gruneisen parameter." Calculate the pressure P as a function of $\tilde{v}$ and T.

Answer:
$$P = \frac{\gamma a}{\tilde{v}^{\gamma+1}} (e^{a/k_B T \tilde{v}^\gamma} + 1)^{-1}$$

15-4   A POLYMER MODEL—THE RUBBER BAND REVISITED

There exists another model of appealing simplicity that is euphemistically referred to as a "polymer model." Its connection with a real polymer is tenuous, but that connection is perhaps close enough to serve the pedagogical purpose of providing some sense of physical reality while again illustrating the basic algorithm of statistical mechanics. And in particular the model provides an insight to the behavior of a "rubber band," as discussed on purely phenomenological grounds in Section 3.7. As we saw in that section the extensive parameter of interest, which replaces the volume, is the *length*; the corresponding intensive parameter, analogous to the pressure, is the *tension*. We are interested in the equation of state relating tension to length and temperature.

The "rubber band" can be visualized as a bundle of long chain polymers. Each polymer chain is considered to be composed of N monomer units each of length a, and we focus our attention on one particular polymer chain in the bundle. One end of the polymer chain is fixed at a point that is taken as the origin of coordinates. The other end of the chain is subject to an externally applied tension $\mathcal{T}$, parallel to the positive x-axis (Fig. 15.4).

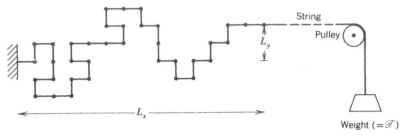

FIGURE 15.4

"Polymer" model. The string should be much longer than shown, so that the end of th
polymer is free to move in the y-direction, and the applied tension $\mathscr{T}$ is directed along tl
x-direction.

In the polymer model each monomer unit of the chain is permitted to
lie either parallel or antiparallel to the x-axis, and zero energy is associ·
ated with these two orientations. Each monomer unit has the additiona
possibility of lying *perpendicular* to the x-axis, in the $+y$ or $-y$ directions *only*. Such a "perpendicular" monomer unit presumably suffer'
interference with other polymer chains in the bundle; we represent thi.
interference by assigning a positive energy ε to such a perpendiculai
monomer.

A somewhat more reasonable model of the polymer might permit the
perpendicular monomers to lie along the $\pm z$ directions as well as alon;
the $\pm y$ directions, and, more importantly, would account for the interference of a chain doubling back on itself. Such models complicate th·
analysis without adding to the pedagogic clarity or qualitative content of
the result.

We calculate the entropy S of one polymer chain as a function of the
energy $U \equiv U'\varepsilon$, of the coordinates L_x and L_y of the end of the polyme·
chain, and of the number $\tilde{N}$ of monomer units in the chain.

Let N_x^+ and N_x^- be the numbers of monomers along the $+x$ and $-x$
directions respectively, and similarly for N_y^+ and N_y^-. Then

$$N_x^+ + N_x^- + N_y^+ + N_y^- = \tilde{N}$$

$$N_x^+ - N_x^- = \frac{L_x}{a} \equiv L_x'$$

$$N_y^+ - N_y^- = \frac{L_y}{a} \equiv L_y'$$

$$N_y^+ + N_y^- = \frac{U}{\varepsilon} \equiv U' \tag{15.14}$$

from which we find

$$N_x^+ = \tfrac{1}{2}\left(\tilde{N} - U' + L_x'\right)$$

$$N_x^- = \tfrac{1}{2}\left(\tilde{N} - U' - L_x'\right)$$

$$N_y^+ = \tfrac{1}{2}\left(U' + L_y'\right)$$

$$N_y^- = \tfrac{1}{2}\left(U' - L_y'\right) \tag{15.15}$$

The number of configurations of the polymer consistent with given coordinates L_x and L_y of its terminus, and with given energy U, is

$$\Omega\left(U, L_x, L_y, \tilde{N}\right) = \frac{\tilde{N}!}{N_x^+! N_x^-! N_y^+! N_y^-!} \tag{15.16}$$

The entropy is, then, using the Stirling approximation (equation 15.3)

$$S = k_B \ln \Omega = \tilde{N} k_B \ln \tilde{N} - N_x^+ k_B \ln N_x^+ - N_x^- k_B \ln N_x^-$$

$$- N_y^+ k_B \ln N_y^+ - N_y^- k_B \ln N_y^- \tag{15.17}$$

or

$$S = \tilde{N} k_B \ln \tilde{N} - \tfrac{1}{2}\left(\tilde{N} - U' + L_x'\right) k_B \ln\left[\tfrac{1}{2}\left(\tilde{N} - U' + L_x'\right)\right]$$

$$- \tfrac{1}{2}\left(\tilde{N} - U' - L_x'\right) k_B \ln\left[\tfrac{1}{2}\left(\tilde{N} - U' - L_x'\right)\right]$$

$$- \tfrac{1}{2}\left(U' + L_y'\right) k_B \ln\left[\tfrac{1}{2}\left(U' + L_y'\right)\right]$$

$$- \tfrac{1}{2}\left(U' - L_y'\right) k_B \ln\left[\tfrac{1}{2}\left(U' - L_y'\right)\right] \tag{15.18}$$

With the statistical mechanical phase of the calculation completed, the thermodynamic formalism comes into play. The y-component of the tension $\mathcal{T}_y$ is conjugate to the extensive coordinate L_y (see Problem 15.4-1). Setting $\mathcal{T}_y = 0$ gives

$$-\frac{\mathcal{T}_y}{T} = \frac{\partial S}{\partial L_y} = \frac{k_B}{2a} \ln \frac{U' - L_y'}{U' + L_y'} = 0 \tag{15.19}$$

from which we conclude (as expected) that

$$L_y = L_y' = 0 \tag{15.20}$$

Similarly

$$-\frac{\mathcal{T}_x}{T} = \frac{\partial S}{\partial L_x} = \frac{k_B}{2a} \ln \frac{\tilde{N} - U' - L'_x}{\tilde{N} - U' + L'_x} \tag{15.21}$$

and

$$\frac{1}{T} = \frac{\partial S}{\partial U} = \frac{k_B}{2\varepsilon} \ln(\tilde{N} + L'_x - U') + \frac{k_B}{2\varepsilon} \ln(\tilde{N} - L'_x - U') - \frac{k_B}{\varepsilon} \ln U' \tag{15.22}$$

or

$$e^{2\varepsilon/k_B T} = \frac{(\tilde{N} - U')^2 - L'^2_x}{U'^2} \tag{15.23}$$

This is the "thermal equation of state." The "mechanical equation of state" (15.21) can be written in an analogous exponential form

$$e^{-2\mathcal{T}_x a/k_B T} = \frac{\tilde{N} - U' - L'_x}{\tilde{N} - U' + L'_x} \tag{15.24}$$

The two preceding equations are the equations of state in the entropy representation, and accordingly they involve the energy U'. That is not generally convenient. We proceed, then, to eliminate U' between the two equations. With some algebra we find (see Problem 15.4-2) that

$$\frac{L'_x}{\tilde{N}} = \frac{\sinh(\mathcal{T}_x a/k_B T)}{\cosh(\mathcal{T}_x a/k_B T) + e^{-\varepsilon/k_B T}} \tag{15.25}$$

For small $\mathcal{T}_x a$ (relative to $k_B T$) the equation can be expanded to first order

$$L_x = \frac{\mathcal{T}_x \tilde{N} a^2}{k_B T} \frac{1}{1 + e^{-\varepsilon/k_B T}} + \cdots \tag{15.26}$$

The modulus of elasticity of the rubber band (the analogue of the compressibility $-1/V(\partial V/\partial P)_T$) is, for small $\mathcal{T}_x$

$$\frac{1}{\tilde{N}} \left(\frac{\partial L_x}{\partial \mathcal{T}_x} \right)_T = \frac{a^2}{k_B T} (1 + e^{-\varepsilon/k_B T})^{-1} \tag{15.27}$$

The fact that this elastic modulus *decreases* with increasing temperature (or that the "stiffness" increases) is in dramatic contrast to the behavior of a spring or of a stretched wire. The behavior of the polymer is sometimes compared to the behavior of a snake; if we grasp a snake by the head and tail and attempt to stretch it straight the resistance is attributable to the writhing activity of the snake. The snake, in its writhing, assumes all possible configurations, and more configurations are accessible if the two ends are not greatly distant from each other. At low temperatures the rubber band is like a torpid snake. At high temperatures the number of configurations available, and the rate of transitions among them, is greater, resulting in a greater contractive tension. It is the *entropy* of the snake and of the rubber band that is responsible for the tendency of the ends to draw together!

The behavior described is qualitatively similar to that of the simple phenomenological model of Section 3.7. But compared to a truly realistic model of a rubber band, both models are extremely naive.

PROBLEMS

15.4-1. Is the sign correct in equation 15.19? Explain.

15.4-2. Eliminate U/ε between equations 15.23 and 15.24 and show that the formal solution is equation 15.25 with a $\pm$ sign before the second term in the denominator. Consider the qualitative dependence of $L_x/\tilde{N}a$ on ε, and show that physical reasoning rejects the negative sign in the denominator, thus validating equation 15.25.

15.4-3. A rubber band consisting of n polymer chains is stretched from zero length to its full extension ($L = \tilde{N}a$) at constant temperature T. Does the energy of the system increase or decrease? Calculate the work done on the system and the heat transfer to the system.

15.4-4. Calculate the heat capacity at constant length for a "rubber band" consisting of n polymer chains. Express the answer in terms of T and L_x.

15.4-5. Calculate the "coefficient of longitudinal thermal expansion" defined by

$$\kappa'_T \equiv \frac{1}{L_x}\left(\frac{\partial L_x}{\partial T}\right)_{\mathcal{F}_x}$$

Express κ'_T as a function of T and sketch the qualitative behavior. Compare this with the behavior of a metallic wire and discuss the result.

15-5 COUNTING TECHNIQUES AND THEIR CIRCUMVENTION; HIGH DIMENSIONALITY

To repeat, the basic algorithm of statistical mechanics consists of counting the number of states consistent with the constraints imposed; the entropy is then the product of Boltzmann's constant and the logarithm of the permissible number of states.

Unfortunately counting problems tend to require difficult and sophisticated techniques of combinatorial mathematics (if they can be done at all!) In fact only a few highly artificial, idealized models permit explicit solution of the counting problem, even with the full armamentarium of combinatorial theory. If statistical mechanics is to be a useful and practical science it is necessary that the difficulties of the counting problem somehow be circumvented. One method of simplifying the counting problem is developed in this section. It is based on certain rather startling properties of systems of "high dimensionality"—a concept to be defined shortly. The method is admittedly more important for the insights it provides to the behavior of complex systems than for the aid it provides in practical calculations. More general and powerful methods of circumventing the counting problem are based on a transfer from thermodynamics to statistical mechanics of the technique of Legendre transformations. That transfer will be developed in the following chapters.

For now we turn our attention to the simplifying effects of *high dimensionality*, a concept that can best be introduced in terms of an explicit model. We choose the simplest model with which we are already familiar—the Einstein model.

Recall that the Einstein solid is a collection of $\tilde{N}$ atoms, each of which is to be associated with three harmonic oscillators (corresponding to the oscillations of the atom along the x, y, and z axes). A quantum state of the system is specified by the $3\tilde{N}$ quantum numbers $n_1, n_2, n_3, \ldots, n_{3\tilde{N}}$, and the energy of the system is

$$U(n_1, n_2, \ldots, n_{3\tilde{N}}) = \sum_{j=1}^{3\tilde{N}} n_j \hbar \omega_0 \tag{15.28}$$

Each such state can be represented by a "point," with coordinates $n_1, n_2, n_3, \ldots, n_{3\tilde{N}}$, in a $3\tilde{N}$-dimensional "state space." Only points with positive integral coordinates are permissible, corresponding to the discreteness or "quantization" of states in quantum mechanics. It is to be stressed that a single point represents the quantum state of the *entire crystal*.

The locus of states with a given energy U is a "diagonal" hyperplane with intercepts $U/\hbar\omega_0$ on each of the $3\tilde{N}$ coordinate axes (Fig. 15.5). All states lying "inside" the plane (i.e., closer to the origin) have energies less than U, and all states lying outside the plane, further from the origin, have energies greater than U.

The first critical observation which is called to our attention by Fig. 15.5 is that an arbitrary "diagonal plane," corresponding to an arbitrary energy U, will generally pass through *none* of the discrete coordinate points in the space! That is, an arbitrarily selected number U generally

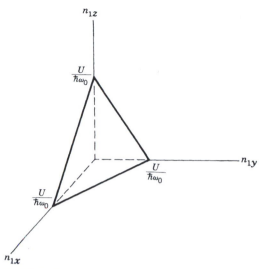

FIGURE 15.5

Quantum state space for the Einstein solid. The three-dimensional state space shown is for an Einstein solid composed of a single atom. Each additional atom would increase the dimensionality of the space by three. The hyperplane U has intercepts $U/\hbar\omega_0$ on all axes. There is one state for each unit of hypervolume, and (neglecting surface corrections) the number of states with energy less than U is equal to the volume inside the diagonal hyperplane U.

cannot be represented in the form of equation 15.28, such a decomposition being possible only if $U/\hbar\omega_0$ is an integer.

More generally, if we inquire as to the number of quantum states of a system with an *arbitrarily chosen and mathematically precise* energy, we almost always find zero. But such a question is unphysical. As we have stressed previously the random interactions of every system with its environment make the energy slightly imprecise. Furthermore we never know (and cannot measure) the energy of any system with absolute precision.

The entropy is not the logarithm of the number of quantum states that lie on the diagonal hyperplane U of Fig. 15.5, but rather it is the logarithm of the number of quantum states that lie in the close vicinity of the diagonal hyperplane.

This consideration leads us to study the number of states between two hyperplanes: U and $U - \Delta$. The energy separation Δ is determined by the imprecision of the energy of the macroscopic system. That imprecision may be thought of as a consequence either of environmental interactions or of imprecision in the preparation (measurement) of the system.

The remarkable consequence of high dimensionality is that *the volume between the two planes ($U - \Delta$ and U), and hence the entropy, is essentially independent of the separation Δ of the planes!*

This result is (at first) so startlingly counter-intuitive, and so fundamental, that it warrants careful analysis and discussion. We shall first corroborate the assertion on the basis of the geometrical representation of the states of the Einstein solid. Then we shall reexamine the geometrical representation to obtain a heuristic understanding of the general geometrical basis of the effect.

The number of states $\tilde{\Omega}(U)$ with energies less than (or equal to) a given value U is equal to the hypervolume lying "inside" the diagonal hyperplane U. This hypervolume is (see problem 15.5-1)

$$\tilde{\Omega}(U) = (\text{number of states with energies less than } U)$$

$$= \frac{1}{(3\tilde{N})!}\left(\frac{U}{\hbar\omega_0}\right)^{3\tilde{N}} \tag{15.29}$$

The fact that this result is proportional to $U^{3\tilde{N}}$, where $3\tilde{N}$ is the dimensionality of the "state space," is the critical feature of this result. The precise form of the coefficient in equation 15.29 will prove to be of only secondary importance.

By subtraction we find the number of states with energies between $U - \Delta$ and U to be

$$\tilde{\Omega}(U) - \tilde{\Omega}(U - \Delta) = \frac{1}{(3\tilde{N})!}\left(\frac{U}{\hbar\omega_0}\right)^{3\tilde{N}} - \frac{1}{(3\tilde{N})!}\left(\frac{U - \Delta}{\hbar\omega_0}\right)^{3\tilde{N}}$$

or

$$\tilde{\Omega}(U) - \tilde{\Omega}(U - \Delta) = \tilde{\Omega}(U)\left[1 - \left(1 - \frac{\Delta}{U}\right)^{3\tilde{N}}\right] \tag{15.30}$$

But $(1 - \Delta/U)$ is less than unity; raising this quantity to an exponent $3\tilde{N} \simeq 10^{23}$ results in a totally negligible quantity (see Problem 15.5-2), so that

$$\Omega(U) = \tilde{\Omega}(U) - \tilde{\Omega}(U - \Delta) \simeq \tilde{\Omega}(U) \tag{15.31}$$

That is, the number $\Omega(U)$ of states with energies between $U - \Delta$ and U is essentially equal to the total number $\tilde{\Omega}(U)$ of states with energies less than U—and this result is essentially independent of Δ!

Thus having corroborated the assertion for our particular model, let us reexamine the geometry to discern the more general geometrical roots of this strange, but enormously useful, result.

The physical volume in Fig. 15.5 can be looked at as one eighth of a regular octahedron (but only the portion of the octahedron in the physical octant of the space has physical meaning). With higher dimensionality the regular polyhedron would become more nearly "spherical." The dimensionless energy $U/\hbar\omega_0$ is analogous to the "radius" of the figure, being the distance from the origin to any of the corners of the polyhedron. This viewpoint makes evident the fact (equation 15.29) that the volume is proportional to the radius raised to a power equal to the dimensionality of the space (r^2 in two dimensions, r^3 in three, etc.). The volume between two concentric polyhedra, with a difference in radii of dr, is $dV = (\partial V/\partial r)\,dr$. The ratio of the volume of this "shell" to the total volume is

$$\frac{dV}{V} = \frac{\partial V}{\partial r}\frac{dr}{V} \tag{15.32}$$

or, if $V = A_n r^n$

$$\frac{dV}{V} = n\frac{dr}{r} \tag{15.33}$$

If we take $n = 10^{23}$ we find $dV/V \simeq 0.1$ only if $dr/r \simeq 10^{-24}$. For dr/r greater than $\sim 10^{-24}$ the equation fails, telling us that the use of differentials is no longer valid. The failure of the differential analysis is evidence that dV/V already becomes on the order of unity for values of dr/r as small as $dr/r \simeq 10^{-23}$.

In an imaginary world of high dimensionality there would be an automatic and perpetual potato famine, for the skin of a potato would occupy essentially its entire volume!

In the real world in which three-dimensional statistical mechanicians calculate entropies as volumes in many-dimensional state spaces, the properties of high dimensionality are a blessing. We need not calculate the number of states "in the vicinity of the system energy U"—it is quite as satisfactory, and frequently easier, to calculate the number of states with energies less than or equal to the energy of the physical system.

Returning to the Einstein solid, we can calculate the fundamental equation using the result 15.29 for $\tilde{\Omega}(U)$, the number of states with energies less than U; the entropy is $S = k_B \ln \tilde{\Omega}(U)$, and it is easily corroborated that this gives the same result as was obtained in equation 15.4.

The two methods that we have used to solve the Einstein model of a solid should be clearly distinguished. In Section 15.2 we assumed that $U/\hbar\omega_0$ was an integer, and we counted the number of ways of distributing quanta among the modes. This was a combinatorial problem, albeit a simple and tractable one because of the extreme simplicity of the model. The second method, in this section, involved no combinatorial calculation whatsoever. Instead we defined a volume in an abstract "state space" and the entropy was related to the total volume *inside* the bounding surface defined by the energy U. The combinatorial approach is not easily transferable to more complicated systems—the method of hypervolumes is general and is usually more tractable. However the last method is not applicable at very low temperature where only a few states are occupied, and where the occupied volume in state space shrinks toward zero.

PROBLEMS

15.5-1. To establish equation 15.29 let Ω_n be the hypervolume subtended by the diagonal hyperplane in n dimensions. Draw appropriate figures for $n = 1, 2,$ and 3 and show that if L is the intercept on each of the coordinate axes

$$\Omega_1 = L$$

$$\Omega_2 = \Omega_1 \int_0^L \left(1 - \frac{x}{L}\right) dx = \frac{L^2}{2!}$$

$$\Omega_3 = \Omega_2 \int_0^L \left(1 - \frac{x}{L}\right)^2 dx = \frac{L^3}{3!}$$

and by mathematical induction

$$\Omega_n = \Omega_{n-1} \int_0^L \left(1 - \frac{x}{L}\right)^{n-1} dx = \frac{L^n}{n!}$$

15.5-2. Recalling that

$$\lim_{x \to 0} (1 + x)^{1/x} = e \qquad (\equiv 2.718\ldots)$$

show that

$$\left(1 - \frac{\Delta}{U}\right) \simeq e^{-\Delta/U} \qquad \text{for } \frac{\Delta}{U} \ll 1$$

With this approximation discuss the accuracy of equation 15.31 for a range of reasonable values of Δ/U (ranging perhaps from 10^{-3} to 10^{-10}).

With what precision Δ/U would the energy have to be known in order that corrections to equation 15.31 might become significant? Assume a system with $\tilde{N} \simeq 10^{23}$.

15.5-3. Calculate the fraction of the hypervolume between the radii $0.9r$ and r for hyperspheres in 1, 2, 3, 4, and 5 dimensions. Similarly for 10, 30, and 50 dimensions.

16

THE CANONICAL FORMALISM: STATISTICAL MECHANICS IN HELMHOLTZ REPRESENTATION

16-1 THE PROBABILITY DISTRIBUTION

The microcanonical formalism of the preceding chapter is simple in principle, but it is computationally feasible only for a few highly idealized models. The combinatorial calculation of the number of ways that a given amount of energy can be distributed in arbitrarily sized "boxes" is generally beyond our mathematical capabilities. The solution is to remove the limitation on the amount of energy available—to consider a system in contact with a thermal reservoir rather than an isolated system. The statistical mechanics of a system in contact with a thermal reservoir may be viewed as statistical mechanics "in Helmholtz representation"; or, in the parlance of the field, "in canonical formalism."

States of all energies, from zero to arbitrarily large energies, are available to a system in contact with a thermal reservoir. But, in contrast to the state probabilities in a closed system, each state does *not* have the same probability. That is, the system does not spend the same fraction of time in each state. The key to the canonical formalism is the determination of the probability distribution of the system among its microstates. And this problem is solved by the realization that the system plus the reservoir constitute a *closed* system, to which the principle of equal probability of microstates again applies.

A simple analogy is instructive. Consider a set of three dice, one of which is red (the remaining two being white). The three dice have been "thrown" many thousands of times. Whenever the sum of the numbers on the three dice has been 12 (and only then), the number on the red die has been recorded. In what fraction of these recorded throws has the red die shown a one, a two, ..., a six?

The result, left to the reader, is that the red die has shown a one in $\frac{2}{25}$ of the throws, a two in $\frac{3}{25}, \ldots$, a five in $\frac{5}{25}$, and a six in $\frac{5}{25}$ of the recorded throws. The probability of a (red) six, in this restricted set of throws, is $\frac{1}{5}$.

The red die is the analogue of our system of interest, the white dice correspond to the reservoir, the numbers shown correspond to the energies of the respective systems, and the restriction to throws in which the sum is 12 corresponds to the constancy of the total energy (of system plus reservoir).

The probability f_j of the subsystem being in state j is equal to the fraction of the total number of states (of system-plus-reservoir) in which the subsystem is in the state j (with energy E_j):

$$f_j = \frac{\Omega_{res}(E_{tot} - E_j)}{\Omega_{tot}(E_{tot})} \tag{16.1}$$

Here E_{tot} is the total energy of the system-plus-reservoir, and Ω_{tot} is the total number of states of the system-plus-reservoir. The quantity in the numerator, $\Omega_{res}(E_{tot} - E_j)$ is the number of states available to the reservoir when the subsystem is in the state j (leaving energy $E_{tot} - E_j$ in the reservoir).

This is the seminal relation in the canonical formalism, but it can be re-expressed in a far more convenient form. The denominator is related to the entropy of the composite system by equation 15.1. The numerator is similarly related to the entropy of the reservoir, so that

$$f_j = \frac{\exp\left\{ k_B^{-1} S_{res}(E_{tot} - E_j) \right\}}{\exp\left\{ k_B^{-1} S_{tot}(E_{tot}) \right\}} \tag{16.2}$$

If U is the average value of the energy of the subsystem, then the additivity of the entropy implies

$$S_{tot}(E_{tot}) = S(U) + S_{res}(E_{tot} - U) \tag{16.3}$$

Furthermore, expanding $S_{res}(E_{tot} - E_j)$ around the equilibrium point $E_{tot} - U$;

$$S_{res}(E_{tot} - E_j) = S_{res}(E_{tot} - U + U - E_j)$$

$$= S_{res}(E_{tot} - U) + (U - E_j)/T \tag{16.4}$$

No additional terms in the expansion appear (this being the very definition of a reservoir). Inserting these latter two equations in the expression for f_j,

$$f_j = e^{(1/k_B T)\{U - TS(U)\}} e^{-(1/k_B T)E_j} \tag{16.5}$$

The quantity $1/k_BT$ appears so pervasively throughout the theory that it is standard practice to adopt the notation

$$\beta \equiv 1/(k_BT) \tag{16.6}$$

Furthermore $U - TS(U)$ is the Helmholtz potential of the system, so that we finally achieve the fundamental result for the probability f_j of the subsystem being in the state j

$$f_j = e^{\beta F}e^{-\beta E_j} \tag{16.7}$$

Of course the Helmholtz potential is not known; it is in fact our task to compute it. The key to its evaluation is the observation that $e^{\beta F}$ plays the role of a state-independent normalization factor in equation 16.7.

$$\sum_j f_j = e^{\beta F}\sum_j e^{-\beta E_j} = 1 \tag{16.8}$$

or

$$e^{-\beta F} = Z \tag{16.9}$$

where Z, the "canonical partition sum," is defined by

$$Z \equiv \sum_j e^{-\beta E_j} \tag{16.10}$$

We have now formulated a complete algorithm for the calculation of a fundamental relation in the canonical formalism. Given a list of all states j of the system, and their energies E_j, we calculate the partition sum (16.10). The partition sum is thus obtained as a function of temperature (or β) and of the parameters $(V, N_1, N_2, \ldots)$ that determine the energy levels. Equation 16.9 in turn determines the Helmholtz potential as a function also of T, V, N_1, N_r. This is the sought for fundamental relation.

The entire algorithm is summarized in the relation

$$-\beta F = \ln \sum_j e^{-\beta E_j} \equiv \ln Z$$

which should be committed to memory.

A corroboration of the consistency of the formalism follows from recalling that f_j is the probability of occupation of the jth state, which (from equations 16.7, 16.9 and 16.10) can be written in the very useful form

$$f_j = e^{-\beta E_j}\Big/\sum_i e^{-\beta E_i} \tag{16.11}$$

The average energy is then expected to be

$$U = \sum_j E_j f_j = \sum_j E_j e^{-\beta E_j} / \sum_i e^{-\beta E_i} \tag{16.12}$$

or

$$U = -(d/d\beta) \ln Z \tag{16.13}$$

Insertion of equation 16.9, expressing Z in terms of F, and recalling that $\beta = 1/k_B T$ reduces this equation to the familiar thermodynamic relation $U = F + TS = F - T(\partial F/\partial T)$ and thereby confirms its validity. Equations 16.12 and 16.13 are very useful in statistical mechanics, but it must be stressed that these equations do not constitute a fundamental relation. The fundamental relation is given by equations 16.9 and 16.10, giving F (rather than U) as a function of β, V, N.

A final observation on units and on formal structure is revealing. The quantity β is, of course, merely the reciprocal temperature in "natural units." The canonical formalism then gives the quantity βF in terms of β, V, and N. That is, F/T is given as a function of $1/T$, V, and N. *This is a fundamental equation in the representation $S[1/T]$* (recall Section 5.4). Just as the microcanonical formalism is naturally expressed in entropy representation, the canonical formalism is naturally expressed in $S[\beta]$ representation. The generalized canonical representations to be discussed in Chapter 17 will similarly all be expressed most naturally in terms of Massieu functions. Nevertheless we shall conform to universal usage and refer to the canonical formalism as being based on the Helmholtz potential. No formal difficulties arise from this slight "misrepresentation."

PROBLEMS

16.1-1. Show that equation 16.13 is equivalent to $U = F + TS$.

16.1-2. From the canonical algorithm expressed by equations 16.9 and 16.10, express the pressure in terms of a derivative of the partition sum. Further, express the pressure in terms of the derivatives $\partial E_j/\partial V$ (and of T and the E_j). Can you give a heuristic interpretation of this equation?

16.1-3. Show that $S/k_B = \beta^2 \partial F/\partial \beta$ and thereby express S in terms of Z and its derivatives (with respect to β).

16.1-4. Show that $c_v = -\beta(\partial s/\partial \beta)_v$ and thereby express c_v in terms of the partition sum and its derivatives (with respect to β).

Answer:

$$c_v = N^{-1} k_B \beta^2 \frac{\partial^2 \ln Z}{\partial \beta^2}$$

16-2 ADDITIVE ENERGIES AND FACTORIZABILITY OF THE PARTITION SUM

To illustrate the remarkable simplicity of the canonical formalism we recall the two-state system of Section 15.3. In that model $\tilde{N}$ distinguishable "atoms" each were presumed to have two permissible states, of energies 0 and ε. Had we attributed even only three states to each atom the problem would have become so difficult as to be insoluble by the microcanonical formalism, at least for general values of the excitation energies. By the canonical formalism it is simple indeed!

We consider a system composed of $\tilde{N}$ distinguishable "elements," an element being an independent (noninteracting) excitation mode of the system. If the system is composed of noninteracting material constituents, such as the molecules of an ideal gas, the "elements" refer to the excitations of the individual molecules. In strongly interacting systems the elements may be wavelike collective excitations such as vibrational modes or electromagnetic modes. *The identifying characteristic of an "element" is that the energy of the system is a sum over the energies of the elements, which are independent and noninteracting.*

Each element can exist in a set of *orbital states* (we henceforth use the term *orbital state* to distinguish the states of an element from the states of the collective system). The energy of the ith element in its jth orbital state is ε_{ij}. Each of the elements need not be the same, either in the energies or the number of its possible orbital states. *The total energy of the system is the sum of the single-element energies, and each element is permitted to occupy any one of its orbital states independently of the orbital states of the other elements.* Then the partition sum is

$$Z = \sum_{j,\,j',\,j'',\,\ldots} e^{-\beta(\varepsilon_{1j} + \varepsilon_{2j'} + \varepsilon_{3j''} + \,\cdots\,)} \tag{16.14}$$

$$= \sum_{j,\,j',\,j'',\,\ldots} e^{-\beta\varepsilon_{1j}} e^{-\beta\varepsilon_{2j'}} e^{-\beta\varepsilon_{3j''}} \cdots \tag{16.15}$$

$$= \sum_{j} e^{-\beta\varepsilon_{1j}} \sum_{j'} e^{-\beta\varepsilon_{2j'}} \sum_{j''} e^{-\beta\varepsilon_{3j''}} \cdots \tag{16.16}$$

$$= z_1 z_2 z_3 \cdots \tag{16.17}$$

where z_i, the "partition sum of the ith element," is

$$z_i = \sum_{j} e^{-\beta\varepsilon_{ij}} \tag{16.18}$$

The partition sum factors. Furthermore the Helmholtz potential is additive over elements

$$-\beta F = \ln Z = \ln z_1 + \ln z_2 + \cdots \tag{16.19}$$

This result is so remarkably simple, powerful, and useful that we emphasize again that it applies to any system in which (a) the energy is additive over elements and (b) each element is permitted to occupy any of its orbital states independently of the orbital state of any other element.

The "two-state model" of Section 15.3 satisfies the above criteria, whence

$$Z = z^{\tilde{N}} = \left(1 + e^{-\beta\varepsilon}\right)^{\tilde{N}} \tag{16.20}$$

and

$$F = -\tilde{N}k_B T \ln\left(1 + e^{-\beta\varepsilon}\right) \tag{16.21}$$

It is left to the reader to demonstrate that this solution is equivalent to that found in Section 15.3. If the number of orbitals had been three rather than two, the partition sum per particle z would merely have contained three terms and the Helmholtz potential would have contained an additional term in the argument of the logarithm.

The Einstein model of a crystal (Section 15.2) similarly yields to the simplicity of the canonical formalism. Here the "elements" are the vibrational modes, and the partition sum per mode is

$$z = 1 + e^{-\beta\hbar\omega_0} + e^{-2\beta\hbar\omega_0} + \cdots = \sum_{n=0}^{\infty} e^{-n\beta\hbar\omega_0} \tag{16.22}$$

This "geometric series" sums directly to

$$z = \frac{1}{1 - e^{-\beta\hbar\omega_0}} \tag{16.23}$$

There are $3\tilde{N}$ vibrational modes so that the fundamental equation of the Einstein model, in the canonical formalism, is

$$F = -\beta^{-1} \ln z^{3\tilde{N}} = 3\tilde{N}k_B T \ln\left(1 - e^{-\beta\hbar\omega_0}\right) \tag{16.24}$$

Clearly Einstein's drastic assumption that all modes of vibration of the crystal have the same frequency is no longer necessary in this formalism. A more physically reasonable approximation, due to P. Debye, will be discussed in Section 16.7.

PROBLEMS

16.2-1. Consider a system of three particles, each different. The first particle has two orbital states, of energies ε_{11} and ε_{12}. The second particle has permissible energies ε_{21} and ε_{22}, and the third particle has permissible energies ε_{31} and ε_{32}. Write the partition sum explicitly in the form of equation 16.14, and by explicit algebra, factor it in the form of equation 16.17.

16.2-2. Show that for the two-level system the Helmholtz potential calculated in equation 16.21 is equivalent to the fundamental equation found in Section 15.3.

16.2-3. Is the energy additive over the particles of a gas if the particles are uncharged mass points (with negligible gravitational interaction)? Is the partition sum factorizable if half the particles carry a positive electric charge and half carry a negative electric charge? Is the partition sum factorizable if the particles are "fermions" obeying the Pauli exclusion principle (such as neutrinos)?

16.2-4. Calculate the heat capacity per mode from the fundamental equation 16.24.

16.2-5. Calculate the energy per mode from equation 16.24. What is the leading term in $U(T)$ in the regions of $T \simeq 0$ and of T large?

16.2-6. A binary alloy is composed of $\tilde{N}_A$ atoms of type A and of $\tilde{N}_B$ atoms of type B. Each A-type atom can exist in its ground state or in an excited state of energy ε (all other states are of such high energy that they can be neglected at the temperatures of interest). Each B-type atom similarly can exist in its ground state of energy zero or in an excited state of energy 2ε. The system is in equilibrium at temperature T.
a) Calculate the Helmholtz potential of the system.
b) Calculate the heat capacity of the system.

16.2-7. A paramagnetic salt is composed of 1 mole of noninteracting ions, each with a magnetic moment of one Bohr magneton ($\mu_B = 9.274 \times 10^{-24}$ joules/tesla). A magnetic field B_e is applied along a particular direction; the permissible states of the ionic moments are either parallel or antiparallel to this direction.
a) Assuming the system is maintained at a temperature $T = 4$ K and B_e is increased from 1 Tesla to 10 Tesla, what is the magnitude of the heat transfer from the thermal reservoir?
b) If the system is now thermally isolated and the applied magnetic field B_e is decreased from 10 Tesla to 1 Tesla, what is the final temperature of the system? (This process is referred to as cooling by adiabatic demagnetization.)

16-3 INTERNAL MODES IN A GAS

The excitations of the molecules of a gas include the three translational modes of the molecules as a whole, vibrational modes, rotational modes, electronic modes, and modes of excitation of the nucleus. For simplicity we initially assume that each of these modes is independent, later returning to reexamine this assumption. Then the partition sum factors with respect to the various modes

$$Z = Z_{trans} Z_{vib} Z_{rot} Z_{elect} Z_{nuc} \qquad (16.25)$$

and, further, with respect to the molecules

$$Z_{vib} = z_{vib}^{\tilde{N}}, \qquad Z_{rot} = z_{rot}^{\tilde{N}} \qquad (16.26)$$

and similarly for Z_{elect} and Z_{nuc}.

The "ideality" or "nonideality" of the gas is a property primarily of the translational partition sum. The translational modes in any case warrant a separate and careful treatment, which we postpone to Section 16.10. We now simply assume that any intermolecular collisions do not couple to the internal modes (rotation, vibration, etc.).

The $\tilde{N}$ identical vibrational modes of a given type (one centered on each molecule) are formally identical to the vibrational modes of the Einstein model of a crystal; that is, they are just simple harmonic oscillators. For a mode of frequency ω_0

$$Z_{vib} = z_{vib}^{\tilde{N}} = \left(1 - e^{-\beta\hbar\omega_0}\right)^{-\tilde{N}} \qquad (16.27)$$

and the contribution of this vibrational mode to the Helmholtz potential is as given in equation 16.24 (with $3\tilde{N}$ replaced by $\tilde{N}$). The contribution of a vibrational mode to the heat capacity of the gas is then as shown in Fig. 15.2 (the ordinate being c/R rather than $c/3R$). As described in Section 13.1, the heat capacity "rises in a roughly steplike fashion" in the vicinity of $k_B T \simeq \hbar\omega_0$, and it asymptotes to $c = R$. Figure 13.1 was plotted as the sum of contributions from two vibrational modes, with $\omega_2 = 15\omega_1$.

The characteristic vibrational temperature $\hbar\omega_0/k_B$ ranges from several thousand kelvin for molecules containing very light elements ($\simeq 6300$ K for H_2) to several hundred kelvin for molecules containing heavier elements ($\simeq 309$ K for Br_2).

To consider the rotational modes of a gas we focus particularly on heteronuclear diatomic molecules (such as HCl), which require two angular coordinates to specify their orientation. The rotational energy of such heteronuclear diatomic molecules is quantized, with energy eigenvalues given by

$$\varepsilon_\ell = \ell(\ell + 1)\varepsilon \qquad \ell = 0, 1, 2, \dots \qquad (16.28)$$

Each energy level is $(2\ell + 1)$-fold degenerate. The energy unit ε is equal to $\frac{1}{2}\hbar^2/$(moment of inertia), or approximately 2×10^{-21} J for the HCl molecule. The characteristic separation between levels is of the order of ε, which corresponds to a temperature $\varepsilon/k_B \approx 15$ K for HCl—larger for lighter molecules and smaller for heavier molecules.

The rotational partition sum per molecule is

$$z_{\text{rot}} = \sum_{\ell=0}^{\infty} (2\ell + 1) e^{-\beta\ell(\ell+1)\varepsilon} \tag{16.29}$$

If $k_B T \gg \varepsilon$ the sum can be replaced by an integral. Then, noting that $2\ell + 1$ is the derivative of $\ell(\ell + 1)$, and writing x for the quantity $\ell(\ell + 1)$,

$$z_{\text{rot}} \simeq \int_0^{\infty} e^{-\beta\varepsilon x} \, dx = \frac{1}{\beta\varepsilon} = \frac{k_B T}{\varepsilon} \tag{16.30}$$

If $k_B T$ is less than or of the order of ε it may be practical to calculate several terms of the series explicitly, to some ℓ' such that $\ell'(\ell' + 1) \gg k_B T$, and to integrate over the remaining range (from ℓ' to infinity); see Problem 16.3-2.

It is left to the reader to show that for $k_B T \gg \varepsilon$ the average energy is $k_B T$.

The case of homonuclear diatomic molecules, such as O_2 or H_2, is subject to quantum mechanical symmetry conditions into which we shall not enter. Only the even terms in the partition sum, or only the odd terms, are permitted (depending upon detailed characteristics of the atoms). At high temperatures this restriction merely halves the rotational partition sum per molecule.

The nuclear and electronic contributions can be computed in similar fashion, but generally only the lowest energy levels of each contribute. Then z_{nuc} is simply the "degeneracy" (multiplicity) of the lowest energy configuration. Each of these factors simply contributes $\tilde{N}k_B T \ln$ (multiplicity) to the Helmholtz potential.

It is of interest to return to the assumption that the various modes are independent. This assumption is generally a good (but *not* a rigorous) approximation. Thus the vibrations of a diatomic molecule change the instantaneous interatomic distance and thereby change the instantaneous moment of inertia of rotation. It is only because the vibrations generally are very fast relative to the rotations that the rotations sense only the *average* interatomic distance, and thereby become effectively independent of the vibrations.

PROBLEMS

16.3-1. Calculate the average rotational energy per molecule and the rotational heat capacity per molecule for heteronuclear diatomic molecules in the region $k_B T \gg \varepsilon$.

16.3-2. Calculate the rotational contribution to the Helmholtz potential per molecule by evaluating the first two terms of equation 16.29 explicitly and by integrating over the remaining terms. For this purpose note that the leading terms in the Euler–McLaurin sum formula are

$$\sum_{j=0}^{\infty} f(j) \simeq \int_0^{\infty} f(\theta)\, d\theta + \frac{1}{2} f(0) - \frac{1}{12} f'(0) + \cdots$$

where f' denotes the derivative of $f(\theta)$.

16.3-3. A particular heteronuclear diatomic gas has one vibrational mode, of frequency ω, and its characteristic rotational energy parameter is ε (equation 16.28). Assume no intermolecular forces, so that the gas is ideal. Calculate its full fundamental equation in the temperature region in which $T \gg \varepsilon/k_B$ but $T \simeq \hbar\omega/k_B$.

16-4 PROBABILITIES IN FACTORIZABLE SYSTEMS

We may inquire as to the *physical* significance of the factor z associated with a single element in the partition sum of a factorizable macroscopic system. Following equation 16.17 we referred to z as the "partition sum per element." And in equation 16.19 we saw that $-k_B T \ln z$ is the additive contribution of that element to the Helmholtz potential. It is easily shown (Problem 16.4-1) that the probability of occupation by the ith element of its jth orbital state, in a factorizable system, is

$$f_j^i = e^{-\beta \varepsilon_{ij}} / z_i \tag{16.31}$$

In all these respects the statistical mechanics of the single element is closely analogous to that of a macroscopic system.

The polymer model of Section 15.4 is particularly instructive. Consider a polymer chain with a weight suspended as shown in Fig. 15.4. The magnitude of the weight is equal to the tension $\mathscr{T}$ applied to the chain. The length of the chain is (equation 15.14)

$$L_x = (N_x^+ - N_x^-) a \tag{16.32}$$

and the total energy (of chain plus weight) in a given configuration is

$$E = (N_y^+ + N_y^-)\varepsilon - \mathscr{T} L_x = (N_y^+ + N_y^-)\varepsilon + (N_x^- - N_x^+) a \mathscr{T} \tag{16.33}$$

The term $-\mathcal{T}L_x$ is the potential energy of the suspended weight (the potential energy being the weight $\mathcal{T}$ multiplied by the height, and the height being taken as zero when $L_x = 0$). According to equation 16.33 we can associate an energy $a\mathcal{T}$ with every monomer unit along $-x$, an energy $-a\mathcal{T}$ with every monomer unit along $+x$, and an energy ε with every monomer unit along either $+y$ or $-y$. The partition sum factors and the partition sum per monomer unit is

$$z = e^{-\beta a\mathcal{T}} + e^{+\beta a\mathcal{T}} + e^{-\beta \varepsilon} + e^{-\beta \varepsilon} \qquad (16.34)$$

The Helmholtz potential is given by

$$-\beta F = \tilde{N} \ln z \qquad (16.35)$$

Furthermore the probability that a monomer unit is along $-x$ is

$$p_{-x} = e^{-\beta a\mathcal{T}}/z \qquad (16.36)$$

and the probability that it is along $+x$ is

$$p_{+x} = e^{\beta a\mathcal{T}}/z \qquad (16.37)$$

Consequently the mean length of the chain is

$$\langle L_x \rangle = \tilde{N}(p_{+x} - p_{-x})a \qquad (16.38)$$

$$= 2\tilde{N}a \sinh(\beta a\mathcal{T})/z \qquad (16.39)$$

It is left to the reader to calculate the mean energy U from the fundamental equations (16.34 and 16.35) and to show that both the energy and the length agree with the results of Section 15.4.

PROBLEMS

16.4-1. The probability that the ith element is in its jth orbital state is the sum of the probabilities of all microstates of the system in which the ith element is in its jth orbital state. Use this fact to show that for a "factorizable system" the probability of the ith element being in its jth orbital state is as given in equation 16.31.

16.4-2. Demonstrate the equivalence of the fundamental equations found in this section and in Section 15.4.

16-5   STATISTICAL MECHANICS OF SMALL SYSTEMS: ENSEMBLES

The preceding sections have demonstrated a far reaching similarity between the statistical mechanics of a macroscopic system and that of an individual "element" of a factorizable system. The partition sum per element has the same structure as the full partition sum, and it is subject to the same probability interpretation. The logarithm of the partition sum of an element is an additive contribution to the total Helmholtz potential. Does this imply that we can simply apply the statistical mechanics to each element? We can indeed, *when the elements satisfy the factorizability criteria* of Section 16.2.

A further conclusion can be drawn from the preceding observations. *We can apply the canonical formalism to small (nonmacroscopic) systems in diathermal contact with a thermal reservoir.*

Suppose that we are given such a small system. We can imagine it to be replicated many times over, with each replica put into diathermal contact with the reservoir and hence (indirectly) with all other replicas. The "ensemble" of replicas then constitutes a thermodynamic system to which statistical mechanics and thermodynamics apply. Nevertheless no property of the individual element is influenced by its replicas, from which it is "shielded" by the intermediate thermal reservoir. Application of statistical mechanics to the individual element is isomorphic to its application to the full ensemble.

Statistical mechanics is fully valid when applied to a single element in diathermal contact with a thermal reservoir. In contrast, *thermodynamics, with its emphasis on extensivity of potentials, applies only to an ensemble of elements, or to macroscopic systems.*

Example
An atom has energy levels of energies, $0, \varepsilon_1, \varepsilon_2, \varepsilon_3, \ldots$ with degeneracies of $1, 2, 2, 1, \ldots$ The atom is in equilibrium with electromagnetic radiation which acts as a thermal reservoir at temperature T. The temperature is such that $e^{-\beta \varepsilon_j}$ is negligible (with respect to unity) for all energies ε_j with $j \geq 4$. Calculate the mean energy and the mean square deviation of the energy from its average value.

Solution
The partition sum is

$$z = 1 + 2e^{-\beta \varepsilon_1} + 2e^{-\beta \varepsilon_2} + e^{-\beta \varepsilon_3}$$

The mean energy is

$$\langle \varepsilon \rangle = \left(2\varepsilon_1 e^{-\beta \varepsilon_1} + 2\varepsilon_2 e^{-\beta \varepsilon_2} + \varepsilon_3 e^{-\beta \varepsilon_3} \right)/z$$

and the mean squared energy is

$$\langle \varepsilon^2 \rangle = \left(2\varepsilon_1^2 e^{-\beta \varepsilon_1} + 2\varepsilon_2^2 e^{-\beta \varepsilon_2} + \varepsilon_3^2 e^{-\beta \varepsilon_3} \right)/z$$

The mean square deviation is $\langle \varepsilon^2 \rangle - \langle \varepsilon \rangle^2$. For such a small system the mean square deviation may be very large. Only for macroscopic systems are the fluctuations negligible relative to average or observed values.

It should be noted that an energy level with a two-fold degeneracy implies *two* states that have the same energy. The partition sum is over states, not over "levels."

PROBLEMS

16.5-1. The energies of the orbital states of a given molecule are such that $\varepsilon_0 = 0$, $\varepsilon_1/k_B = 200$ K, $\varepsilon_2/k_B = 300$ K, $\varepsilon_3/k_B = 400$ K and all other orbital states have very high energy. Calculate the dispersion $\sigma \equiv \sqrt{\langle \varepsilon^2 \rangle - \langle \varepsilon \rangle^2}$ of the energy if the molecule is in equilibrium at $T = 300$ K. What is the probability of occupation of each orbital state?

16.5-2. A hydrogen atom in equilibrium with a radiation field at temperature T can be in its ground orbital level (the "1-*s*" level, which is two-fold spin degenerate), or it can be in its first excited energy level (eight-fold degenerate). Neglect the probability of higher energy states. What is the probability that the atom will be in an "orbital *p*-state"?

16.5-3. A small system has two normal modes of vibration, with natural frequencies ω_1 and $\omega_2 = 2\omega_1$. What is the probability that, at temperature T, the system has an energy less than $5\omega_1/2$? The zero of energy is taken as its value at $T = 0$.

Answer:
$$(1 + x)(1 + x^2)(1 + x + 2x^2) \qquad \text{where } x \equiv \exp(-\beta\hbar\omega_1)$$

16.5-4. DNA, the genetic molecule deoxyribonucleic acid, exists as a twisted pair of polymer molecules, each with $\tilde{N}$ monomer units. The two polymer molecules are cross-linked by $\tilde{N}$ "base pairs." It requires energy ε to unlink each base pair, and a base pair can be unlinked only if it has a neighboring base pair that is already unlinked (or if it is at the end of the molecule). Find the probability that n pairs are unlinked at temperature T if

a) one end of the molecule is prevented from unlinking, so that the molecule "unwinds" from one end only.

b) the molecule can unwind from both ends.

Reference: C. Kittel, *Amer. J. Phys.* **37**, 917 (1969).

16.5-5. Calculate the probability that a harmonic oscillator of natural frequency ω_0 is in a state of odd quantum number ($n = 1, 3, 5, \ldots$) at temperature T. To what values do you expect this probability to reduce in the limits of zero and infinite temperature? Show that your result conforms to these limiting values. Find the dominant behavior of the probability P_{odd} *near* $T = 0$ and in the high temperature region.

16.5-6. A small system has two energy levels, of energies 0 and ε, and of degeneracies g_0 and g_1. Find the entropy of this system at temperature T. Calculate the energy and the heat capacity of the system at temperature T. What is the dominant behavior of the heat capacity at very low and at very high temperature? Sketch the heat capacity. How would this sketch be affected by an increase in the ratio g_1/g_0? Explain this effect qualitatively.

16.5-7. Two simple harmonic oscillators, each of natural frequency ω, are coupled in such a way that there is no interaction between them if the oscillators have different quantum numbers, whereas their combined energy is $(2n + 1)\hbar\omega + \Delta$ if the oscillators have the same quantum number n. The system is in thermal equilibrium at temperature T. Find the probability that the two oscillators have identical quantum numbers. Find and interpret the zero-temperature limit of your result, for all values of Δ.

16-6 DENSITY OF STATES AND DENSITY OF ORBITAL STATES

We return to large systems, and we shall shortly demonstrate several applications of the canonical formalism to crystals and to electromagnetic radiation. These applications, and a wide class of other applications, call on the concept of a "density of states function." Because this concept lies outside statistical mechanics proper, and because we shall find it so pervasively useful, it is convenient to discuss it briefly in advance.

In the canonical formalism we repeatedly are called upon to compute sums of the form

$$\text{"sum"} = \sum_j (\cdots) e^{-\beta E_j} \tag{16.40}$$

The sum is over all states j of the system, and E_j is the energy of the jth state. If the quantity in the parenthesis is unity, the "sum" is the partition sum Z. If the parenthetical quantity is the energy then the "sum" divided by Z is the average energy U (equation 16.12). And similar situations hold for other dynamical variables.

For macroscopic systems the energies E_j are generally (but not always) closely spaced, in the sense that $\beta(E_{j+1} - E_j) \ll 1$. Under these circumstances the sum can be replaced by an integral

$$\text{"sum"} \simeq \int_{E_{\min}}^{\infty} (\cdots) e^{-\beta E} D(E)\, dE \tag{16.41}$$

where $E_{\min}$ is the energy of the ground state of the system (the minimum possible energy) and $D(E)$ is the "density of states" function defined by

$$\text{number of states in interval } dE = D(E)\, dE \tag{16.42}$$

In many systems the energy eigenstates are combinations of orbital (single-element) states, the partition sum factors, and analogues of equations 16.41 and 16.42 can be applied to single elements. The quantity analogous to $D(E)$ is then a "density of orbital states"; we shall designate it also by $D(E)$.

Further, the orbital states are very commonly normal modes that are wavelike in character. This is true of the vibrational modes of a crystal and of the electromagnetic modes of a cavity containing electromagnetic radiation. From the viewpoint of quantum mechanics it is even the case for the translational modes of a gas, the waves being the quantum mechanical wave functions of the molecules. The density of orbital states function is then subject to certain general considerations, which we briefly review.

Consider a system in a cubic "box" of linear dimension L (the results are independent of this arbitrary but convenient choice of shape). A standing wave parallel to an edge must have a wavelength λ such that an integral number of half wavelengths "fit" in the length L. That is, the wave vector $k \equiv 2\pi/\lambda$ must be of the form $n\pi/L$. For a wave of general orientation, in three dimensions, we have similar restrictions on each of the three components of $\mathbf{k}$

$$\mathbf{k} = \left(\frac{\pi}{L}\right)(n_1, n_2, n_3) = \left(\frac{\pi}{V^{1/3}}\right)(n_1, n_2, n_3) \qquad (16.43)$$

$$n_1, n_2, n_3 = \text{integers}$$

We consider only isotropic media, for which the frequency is a function only of the amplitude k of $\mathbf{k}$

$$\omega = \omega(k), \text{ or inversely, } k = k(\omega) \qquad (16.44)$$

Then the number of orbital states with frequency less than ω is the number of sets of positive integers for which

$$\left(n_1^2 + n_2^2 + n_3^2\right)^{1/2} \leq V^{1/3}\frac{k(\omega)}{\pi} \qquad (16.45)$$

We can think of $(n_1^2 + n_2^2 + n_3^2)^{\frac{1}{2}}$ as the radius in an abstract space in which n_1, n_2, and n_3 are integral distances along the three coordinate axes. The number of such integral lattice points with radii less than $V^{\frac{1}{3}}k(\omega)/\pi$ is the volume inside this radius. Only one octant of this spherical volume is physically acceptable, because n_1, n_2, and n_3 in equation 16.43 must be positive. Thus the number of orbital states with frequency less than ω is

$$\text{number of orbital states with frequency} \leq \omega = \left(\frac{1}{8}\right)\left(\frac{4\pi}{3}\right)\left[V^{1/3}\frac{k(\omega)}{\pi}\right]^3$$

$$(16.46)$$

Differentiating we find the number of orbital states $D'(\omega)\,d\omega$ in the interval $d\omega$

$$D'(\omega)\,d\omega = \frac{V}{6\pi^2}\frac{dk^3(\omega)}{d\omega}\,d\omega = \frac{V}{2\pi^2}k^2(\omega)\frac{dk(\omega)}{d\omega}\,d\omega \quad (16.47)$$

The quantity $D'(\omega)\,d\omega$ then is analogous to $D(E)\,dE$ in the "sum" (equation 16.41); see Problem 16.6-1.

This is the general result we require. Because various models of interest correspond to various functional relations $\omega(k)$, we shall be able to convert sums to integrals simply by evaluating the "density of orbital states" function $D'(\omega)$ by equation 16.41. So prepared, we proceed to several applications of the canonical formalism.

PROBLEMS

16.6-1. Show that the number of orbital states in the energy interval $d\varepsilon = \hbar\,d\omega$ is $D(\varepsilon) = D'(\omega)/\hbar$, where $D'(\omega)\,d\omega$ is the number of orbital states in the frequency interval $d\omega$.

16.6-2. For the particles of a gas $\varepsilon = p^2/2m = (\hbar^2/2m)k^2$, or $\omega = \varepsilon/\hbar = \hbar k^2/2m$. Find the density of orbital states function $D'(\omega)$.

Answer:

$$D'(\omega) = \frac{V}{2\pi^2}k^2\bigg/\left(\frac{\hbar k}{m}\right) = \frac{m^{3/2}V}{2^{1/2}\pi^2\hbar^{3/2}}\omega^{1/2}$$

16.6-3. For excitations obeying the spectral relation $\omega = Ak^n$, $n > 0$, find the density of orbital states function $D'(\omega)$.

16-7   THE DEBYE MODEL OF NONMETALLIC CRYSTALS

At the conclusion of Section 16.2 we reviewed the Einstein model of a crystalline solid, and we observed that the canonical formalism makes more sophisticated models practical. The "Debye model" is moderately more sophisticated and enormously more successful.

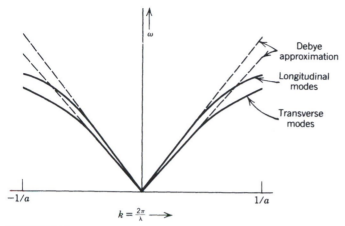

FIGURE 16.1
Dispersion relation for vibrational modes, schematic. The shortest wave length is of the order of the interatomic distance. There are $\tilde{N}$ longitudinal modes and $2\tilde{N}$ transverse modes. The Debye approximation replaces the physical dispersion relation with the linear extrapolation of the long wave length region, or $\omega = v_L k$ and $\omega = v_t k$ for longitudinal and transverse modes respectively.

Again consider $\tilde{N}$ atoms on a lattice, each atom being bound to its neighbors by harmonic forces ("springs"). The vibrational modes consist of $\tilde{N}$ longitudinal and $2\tilde{N}$ transverse normal modes, each of which has a sinusoidal or "wavelike" structure. The shortest wave lengths are of the order of twice the interatomic distance. The very long wave length longitudinal modes are not sensitive to the crystal structure and they are identical to sound waves in a continuous medium. The dispersion curves of ω versus $k(= 2\pi/\lambda)$ are accordingly linear in the long wave length limit, as shown in Fig. 16.1. For shorter wave lengths the dispersion curves "flatten out," with a specific structure that reflects the details of the crystal structure. P. Debye[1], following the lead of Einstein, bypassed the mechanical complications and attempted only to capture the general features in a simple, tractable approximation. The Debye model assumes that the modes all lie on linear "dispersion curves" (Fig. 16.1), as they would in a continuous medium. The slope of the longitudinal dispersion curve is v_L, the velocity of sound in the medium. The slope of the transverse dispersion curve is v_t.

The thermodynamic implications of the model are obtained by calculating the partition sum. The energy is additive over the modes, so that the partition sum factorizes. For each mode the possible energies are $n\hbar\omega(\lambda)$ with $n = 1, 2, 3, \ldots$, where $\omega(\lambda) = 2\pi\nu(\lambda)$ is given by the dotted linear

[1]P. Debye, *Ann. Phys.* **39**, 789 (1912).

curves in Fig. 16.1. As in the Einstein model (equations 16.22 and 16.23)

$$z(\lambda) = \frac{1}{1 - e^{-\beta \hbar \omega(\lambda)}} \tag{16.48}$$

and

$$Z = \prod_{\text{modes}} z(\lambda) = \prod_{\text{modes}} \left[1 - e^{-\beta \hbar \omega(\lambda)} \right]^{-1} \tag{16.49}$$

where $\prod_{\text{modes}}$ denotes a product over all $3\tilde{N}$ modes. The Helmholtz potential is

$$F = k_B T \sum_{\text{modes}} \ln \left(1 - e^{-\beta \hbar \omega(\lambda)} \right) \tag{16.50}$$

It is left to the reader to show that the molar heat capacity is

$$c_v = \beta^2 \hbar^2 k_B \sum_{\text{modes}} \frac{\omega^2 e^{\beta \hbar \omega}}{\left(e^{\beta \hbar \omega} - 1 \right)^2} \tag{16.51}$$

The summation over the modes is best carried out by replacing the sum by an integral

$$c_v = \frac{\hbar^2}{k_B T^2} \int_0^{\omega_{\text{max}}} \frac{\omega^2 e^{\beta \hbar \omega}}{\left(e^{\beta \hbar \omega} - 1 \right)^2} D'(\omega) \, d\omega \tag{16.52}$$

where $D'(\omega) \, d\omega$ is the number of modes in the interval $d\omega$. To evaluate $D'(\omega)$ we turn to equation 16.47. For the longitudinal modes the functional relation $k(\omega)$ is (Fig. 16.1)

$$k = \omega / v_L \tag{16.53}$$

and similarly for the two polarizations of transverse modes. It follows, from equation 16.47, that

$$D'(\omega) = \frac{v}{2\pi^2} \left(\frac{1}{v_L^3} + \frac{2}{v_t^3} \right) \omega^2 \tag{16.54}$$

The maximum frequency[2] ω_{max} is determined by the condition that the

---

[2] In the literature ω_{max} is often specified in terms of the "Debye temperature," defined by $\hbar \omega_{\text{max}} / k_B$ and conventionally designated by θ_D.

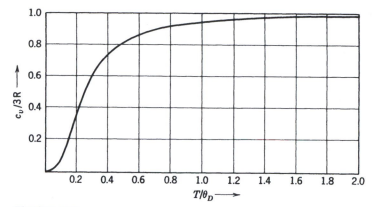

FIGURE 16.2
Vibrational heat capacity of a crystal according to the Debye approximation.

total number of modes be $3N_A$

$$\int_0^{\omega_{max}} D'(\omega)\, d\omega = 3N_A \tag{16.55}$$

from which it follows that

$$\omega_{max}^3 = \frac{18N_A\pi^2}{v}\left(\frac{1}{v_L^3} + \frac{2}{v_t^3}\right)^{-1} \tag{16.56}$$

Inserting $D'(\omega)$ in the integral 16.52 and changing the integration variable from ω to $u(=\beta\hbar\omega)$

$$c_v = \frac{9N_A k_B}{u_m^3} \int_0^{u_m} \frac{u^4 e^u}{(e^u - 1)^2}\, du \tag{16.57}$$

The molar heat capacity, computed from this equation, is shown schematically in Fig. 16.2.

At high temperature ($k_B T \gg \hbar\omega_{max}$) the behavior of c_v is best explored by examining equation 16.51. In this limit $u^2 e^u/(e^u - 1)^2 \rightarrow 1$. Hence each mode contributes k_B to the molar heat capacity (a result of much more general validity, as we shall see subsequently). The molar heat capacity in the high temperature limit is $3N_A k_B$, or $3R$.

At low temperature, where $\beta\hbar\omega_m \equiv u_m \gg 1$, the upper limit in the integral in equation 16.57 can be replaced by infinity; the integral is then simply a constant, and the temperature dependence of c_v arises from the u_m^3 in the denominator. Hence $c_v \sim T^3$ in the low temperature region, a result in excellent agreement with observed heat capacities of nonmetallic crystals. The detailed shape of the heat capacity curve in the intermediate region is less accurate of course. The qualitative shape is similar to that of the Einstein model, Fig. 15.2, except that the sharp exponential rise at low temperature is replaced by the more gentle T^3 dependence.

PROBLEMS

16.7-1. Calculate the energy of a crystal in the Debye approximation. Show that the expression for U leads, in turn, to equation 16.57 for the molar heat capacity.

16.7-2. Calculate the entropy of a crystal in the Debye approximation, and show that your expression for S leads to equation 16.57 for the molar heat capacity.

16.7-3. The frequency $\omega(\lambda)$ of the vibrational mode of wave length λ is altered if the crystal is mechanically compressed. To describe this effect Gruneisen introduced the "Gruneisen parameter"

$$\gamma \equiv - \frac{V}{\omega(\lambda)} \frac{d\omega(\lambda)}{dV}$$

Taking γ as a constant (independent of $\lambda, V, T, \dots$) calculate the mechanical equation of state $P(T, V, N)$ for a Debye–Gruneisen crystal.

 Show that for a Debye–Gruneisen crystal

$$v\alpha = \gamma\kappa_T c_v$$

16-8 ELECTROMAGNETIC RADIATION

 The derivation of the fundamental equation (3.57) of electromagnetic radiation is also remarkably simple in the canonical formalism. Assume the radiation to be contained within a closed vessel, which we may think of as a cubical cavity with perfectly conducting walls. Then the energy resides in the resonant electromagnetic modes of the cavity. As in the Einstein and Debye models, the possible energies of a mode of frequency ω are $n\hbar\omega$, with $n = 0, 1, 2, \dots$. Equations 16.48 and 16.49 are again valid, and

$$F = k_B T \sum_{\text{modes}} \ln\left(1 - e^{-\beta\hbar\omega(\lambda)}\right) \tag{16.58}$$

The sum can be calculated by replacing the sum by an integral (the modes are densely distributed in energy)

$$F = k_B T \int_0^\infty \ln\left(1 - e^{-\beta\hbar\omega}\right) D'(\omega)\, d\omega \tag{16.59}$$

The sole new feature here is that there is no maximum frequency (such as that in the Debye model). Whereas the shortest wavelength (and therefore the largest frequency) of vibrational modes in a solid is determined by the interatomic distance, there is no minimum wavelength of electromagnetic waves. The dispersion relation is again linear, as in the Debye model, and as there are two polarization modes

$$D'(\omega) = \frac{V}{\pi^2 c^3} \omega^2 \tag{16.60}$$

where c is the velocity of light (2.998×10^8 m/s). Then the fundamental equation is

$$F = \frac{V k_B T}{\pi^2 c^3} \int_0^\infty \omega^2 \ln\left(1 - e^{-\beta\hbar\omega}\right) d\omega \tag{16.61}$$

To calculate the energy we use the convenient identity (recall equation 16.13)

$$U = F + TS = F - T\frac{\partial F}{\partial T} = \frac{\partial(\beta F)}{\partial \beta} \tag{16.62}$$

from which

$$U = \frac{V\hbar}{\pi^2 c^3} \int_0^\infty \frac{e^{-\beta\hbar\omega}}{1 - e^{-\beta\hbar\omega}} \omega^3 \, d\omega \tag{16.63}$$

The integral $\int_0^\infty x^3 (e^x - 1)^{-1} dx$ is $3!\zeta(4) = \pi^4/15$, where ζ is the Riemann zeta function[3], whence

$$U = \frac{\pi^2 k_B^4}{15\hbar^3 c^3} VT^4 \tag{16.64}$$

This is the "Stefan–Boltzmann Law," as introduced in equation 3.52. By a simple statistical mechanical calculation we have evaluated the constant b of equation 3.52 in terms of fundamental constants.

PROBLEMS

16.8-1. Show that including the "zero-point energies" of the electromagnetic modes (i.e., $E_n = (n + 1/2)\hbar\omega$) leads to an infinite energy density U/V! This infinite energy density is presumably constant and unchangeable and hence physically unobservable.

---

[3] *cf.* M. Abromowitz and I. A. Stegun, *Handbook of Mathematical Functions*, National Bureau of Standards Applied Mathematics Series, No. 55, 1964. [See equation 23.2.7.]

16.8-2. Show that the energy per unit volume of electromagnetic radiation in the frequency range $d\omega$ is given by the "Planck Radiation Law"

$$\frac{U_\omega}{V}\,d\omega = \frac{\hbar\omega^3}{\pi^2 c^3}\,\left(e^{\beta\hbar\omega} - 1\right)^{-1} d\omega$$

and that at high temperature ($k_B T \gg \hbar\omega$) this reduces to the "Rayleigh–Jeans Law"

$$\frac{U_\omega}{V}\,d\omega \simeq \frac{\omega^2}{\pi^2 c^3}k_B T\,d\omega$$

16.8-3. Evaluating the number of photons per unit volume in the frequency range $d\omega$, as

$$(N_\omega/V)\,d\omega = (U_\omega/V)\,d\omega/\hbar\omega$$

where U_ω is given in problem 16.8-2, calculate the total number of photons per unit volume. Show that the average energy per photon (U/N) is approximately $2.2k_B T$. Note that the integral encountered can be written in terms of the Riemann zeta function, as in the preceding footnote.

16.8-4. Since radiation within a cavity propagates isotropically with velocity c, the flux of energy impinging on unit area of the wall (or passing in one direction through an imaginary unit surface within the cavity) is given by the "Stefan–Boltzmann Law":

$$\text{Energy flux per unit area} = \frac{1}{4}c(U/V) = \frac{1}{4}cbT^4 \equiv \sigma_B T^4$$

The factor of $c/4$ arises as $\frac{1}{2}(c/2)$; the factor of $\frac{1}{2}$ selecting only the radiation crossing the imaginary area from "right" to "left" (or vice versa), and the factor of $c/2$ representing the average component of the velocity normal to the area element. The constant σ_B ($= cb/4$) is known as the "Stefan–Boltzmann constant." As an exercise in elementary kinetic theory, derive the Stefan–Boltzmann law (explicitly demonstrating the averages described).

16-9 THE CLASSICAL DENSITY OF STATES

The basic algorithm for the calculation of a fundamental equation in the canonical formalism requires only that we know the energy of each of the discrete states of the system. Or, if the energy eigenvalues are reasonably densely distributed, it is sufficient to know the density of orbital states. In either case discreteness (and therefore countability) of the states is assumed. This fact raises two questions. First, how can we apply statistical mechanics to classical systems? Second, how did Willard Gibbs invent statistical mechanics in the nineteenth century, long before the birth of quantum mechanics and the concept of discrete states?

As a clue we return to the central equation of the formalism—the equation for the partition sum, which, for a wavelike mode, is (equation 16.47)

$$z = e^{-\beta \bar{F}} = \int e^{-\beta \epsilon} D'(\omega)\, d\omega = \int e^{-\beta \epsilon} \frac{V}{2\pi^2} k^2(\omega)\, dk(\omega) \quad (16.65)$$

We seek to write this equation in a form compatible with classical mechanics, for which purpose we identify $\hbar \mathbf{k}$ with the (generalized) momentum

$$\hbar \mathbf{k} = \mathbf{p} \qquad (16.66)$$

whence

$$z = \frac{1}{2\pi^2 \hbar^3} \int e^{-\beta \epsilon} V p^2\, dp \qquad (16.67)$$

To treat the coordinates and momenta on an equal footing the volume can be written as an integral over the spatial coordinates. Furthermore, the role of the energy E in classical mechanics is played by the Hamiltonian function $\mathcal{H}(x, y, z, p_x, p_y, p_z)$. And finally we shift from $4\pi p^2\, dp$ to $dp_x\, dp_y\, dp_z$ as the "volume element in the momentum subspace," whence the partition function becomes

$$z = \frac{1}{h^3} \int e^{-\beta \mathcal{H}} dx\, dy\, dz\, dp_x\, dp_y\, dp_z \qquad (16.68)$$

Except for the appearance of the classically inexplicable prefactor $(1/h^3)$, this representation of the partition sum (per mode) is fully classical. It was in this form that statistical mechanics was devised by Josiah Willard Gibbs in a series of papers in the *Journal of the Connecticut Academy* between 1875 and 1878. Gibbs' postulate of equation 16.68 (with the introduction of the quantity h, for which there was no a priori classical justification) must stand as one of the most inspired insights in the history of physics. To Gibbs, the numerical value of h was simply to be determined by comparison with empirical thermophysical data.

The expression 16.68 is written as if for a single particle, with three position coordinates and three momentum coordinates. This is purely symbolic. The x, y, and z can be any "generalized coordinates" $(q_1, q_2, \ldots)$, and the momenta p_x, p_y, and p_z are then the "conjugate momenta." The number of coordinates and momenta is dictated by the structure of the system, and more generally we can write

$$Z = \int e^{-\beta \mathcal{H}} \prod_j \left(\frac{dq_j}{h^{1/2}} \frac{dp_j}{h^{1/2}} \right) \qquad (16.69)$$

This is the basic equation of the statistical mechanics of classical systems.

Finally we take note of a simple heuristic interpretation of the "classical density of orbital states" function. *In the classical phase space* (coordinate–momentum space) *each hypercube of "linear dimension" $h^{\frac{1}{2}}$ corresponds to one quantum mechanical state*. It is as if the orbital states are "squeezed as closely together" in phase space as is permitted by the Heisenberg uncertainty principle $\Delta q_j \Delta p_j \geq h$.

Whatever the interpretation, and quite independently of the plausibility arguments of this section, classical statistical mechanics is defined by equation 6.68 or 6.69.

16-10   THE CLASSICAL IDEAL GAS

The monatomic classical ideal gas provides a direct and simple application of the classical density of states and of the classical algorithm (16.69) for the calculation of the partition function.

The model of the gas is a collection of $\tilde{N}$ $(= NN_A)$ point mass "atoms" in a container of volume V, maintained at a temperature T by diathermal contact with a thermal reservoir. The energy of the gas is the sum of the energies of the individual atoms. Interactions between molecules are disbarred (unless such interactions make no contribution to the energy—as, for instance, the instantaneous collisions of hard mass points).

The energy is the sum of one-particle "kinetic energies," and the partition sum factors. We undertake to calculate z_{transl}, the one-particle translational partition sum, and from the classical formulation (16.69) we find directly that

$$z_{\text{transl}} = \frac{1}{h^3} \iiint dx\, dy\, dz \int_{-\infty}^{\infty} \int_{-\infty}^{\infty} \int_{-\infty}^{\infty} dp_x\, dp_y\, dp_z\, e^{-\beta(p_x^2 + p_y^2 + p_z^2)/2m}$$

$$= \frac{V}{h^3} [2\pi m k_B T]^{3/2} \tag{16.70}$$

It is of interest to note that we could have obtained this result by treating the particle *quantum mechanically*, by summing over its discrete states, and by approximating the summation by an integral. This exercise is left to the reader (Problem 16.10-4).

Having now calculated z we might expect to evaluate Z as $z^{\tilde{N}}$, and thereby to calculate the Helmholtz potential F. If we do so we find a Helmholtz potential that is not extensive! We could have anticipated this impending catastrophe, for the one-particle partition function z is extensive (equation 16.70) whereas we expect it to be intensive ($F = -\tilde{N}k_B T \ln z$). The problem lies not in an error of calculation, but in a fundamental principle. *To identify Z as $z^{\tilde{N}}$ is to assume the particles to be* *distinguishable*, as if each bears an identifying label or number (like a set of billiard balls). Quantum mechanics, unlike classical mechanics, gives a profound meaning to the concept of indistinguishability. Indistinguishability does *not* imply merely that the particles are "identical"—it requires that the identical particles behave under interchange in ways that have no classical analogue. Identical particles must obey either Fermi–Dirac or Bose–Einstein "permutational parity"; concepts with statistical mechanical consequences which we shall study in greater detail in Chapter 17. Now, however, we seek only a classical solution. We do so by recognizing that $z^{\tilde{N}}$ is the partition sum of a set of *distinguishable* particles. We therefore attempt to correct this partition sum by division by $\tilde{N}!$. The rationale is that all $\tilde{N}!$ permutations of the "labels" among the $\tilde{N}$ distinguishable particles should be counted as a single state for indistinguishable particles. Thus we finally arrive at the partition sum for a classical monatomic ideal gas

$$Z = (1/\tilde{N}!)z_{\text{transl}}^{\tilde{N}} \tag{16.71}$$

with z_{transl} as calculated in equation 16.70.

The Helmholtz potential is

$$F = -k_B T \ln Z = -\tilde{N}k_B T \ln\left[\frac{V}{\tilde{N}}\left(\frac{2\pi mk_B T}{h^2}\right)^{3/2}\right] - \tilde{N}k_B T \tag{16.72}$$

where we have utilized the Stirling approximation ($\ln \tilde{N}! \simeq \tilde{N} \ln \tilde{N} - \tilde{N}$) which holds for large $\tilde{N}$.

To compare this equation with the fundamental equation introduced in Chapter 3 we make a Legendre transform to entropy representation, finding

$$S = \tilde{N}k_B\left[\frac{5}{2} - \frac{3}{2}\ln(3\pi\hbar^2/m)\right] + \tilde{N}k_B \ln\left(U^{3/2}V/\tilde{N}^{5/2}\right) \tag{16.73}$$

This is precisely the form of the monatomic ideal gas equation with which we have become familiar. The constant s_0, undetermined in the thermodynamic context, has now been evaluated in terms of fundamental constants.

Reflection on the problem of counting states reveals that division by $\tilde{N}!$ is a rather crude classical attempt to account for indistinguishability. The error can be appreciated by considering a model system of two identical particles, each of which can exist in either of two orbital states (Fig. 16.3). Classically we find four states for the distinguishable particles, and we then divide by 2! to "correct" for indistinguishability. If the particles are fermions only one particle is permitted in a single one-particle state, so that there is only *one* permissible state of the system. For bosons, in

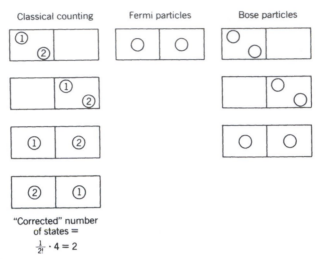

FIGURE 16.3
States of a two-particle system according to classical, Fermi and Bose counting.

contrast, any number of particles are permitted in a one-particle state; consequently there are *three* permissible states of the system (Fig. 16.3). "Corrected classical counting" is incorrect for either type of real particle!

At sufficiently high temperature the particles of a gas are distributed over many orbital states, from very low to very high energies. The probability of two particles being in the same orbital state becomes very small at high temperature. The error of classical counting then becomes insignificant, as that error is associated with the occurrence of more than one particle in a one-particle state. *All gases approach ideal gas behavior at sufficiently high temperature.*

Consider now a mixture of two monatomic ideal gases. The partition sum is factorizable and, as in equation 16.71

$$Z = Z_1 Z_2 = \frac{1}{\tilde{N}_1!} z_1^{\tilde{N}_1} \frac{1}{\tilde{N}_2!} z_2^{\tilde{N}_2} \tag{16.74}$$

The Helmholtz potential is the sum of the Helmholtz potentials for the two gases. The volume appearing in the Helmholtz potential of *each* gas is the common volume occupied by *both*. The temperature is, of course, the common temperature. The fundamental equation so obtained is equivalent to that introduced in Section 3.4 (equation 3.40), but again we have evaluated the constants that were arbitrary in the thermodynamic context.

PROBLEMS

16.10-1. Show that the calculation of $Z = z^{\tilde{N}}$, with z given by equation 16.70, is correct for an ensemble of individual atoms each in a (different) volume V. Show that the fundamental equation obtained from $Z = z^{\tilde{N}}$ is properly extensive when so interpreted.

16.10-2. Show that the fundamental equation of a "multicomponent simple ideal gas," which follows from equation 16.74, is identical to that of equation 3.40.

16.10-3. The factors $(1/\tilde{N}_1!)(1/\tilde{N}_2!)$ in equation 16.74 give an additive contribution to the Helmholtz potential that does not depend in any way on the forms of z_1 and z_2. Show that these factors lead to a "mixing" term in the entropy (*not* in the Helmholtz potential!) of the form

$$\tilde{S}_{mixing} = (-x_1 \ln x_1 - x_2 \ln x_2)k_B$$

This mixing term appears in fluids as well as in ideal gases. It accounts for the fact that the mixing of two fluids is an irreversible process (recall Example 2 of Section 4.5).

16.10-4. Consider a particle of mass m in a cubic container of volume V. Show that the separation of successive energy levels is given approximately by $\Delta E \simeq \pi^2 \hbar^2 / 2mV^{2/3}$, and roughly evaluate ΔE for helium atoms in a container of volume one m^3. Show that, for any temperature higher than $\simeq 10^{-8}$ K, the quantum mechanical partition sum can be approximated well by an integral. Show that this "approximation" leads to equation 16.70.

16.10-5. A single particle is contained in a vessel of volume $2V$ which is divided into two equal sub-volumes by a partition with a small hole in it. The particle carries an electric charge, and the hole in the partition is the site of a localized electric field; the net effect is that the particle has a potential energy of zero on one side of the partition, and of ε_e on the other side. What is the probability that the particle will be found in the zero-potential half of the vessel, if the system is maintained in equilibrium at temperature T? How would this result be affected by internal modes of the particles? How would the result be affected if the dispersion relation of the particles were such that the energy was proportional to the momentum, rather than to its square? If the container were to contain one mole of an ideal gas (non-interacting particles despite the electric charge on each!) what would be the pressure in each sub-volume?

16-11   HIGH TEMPERATURE PROPERTIES— THE EQUIPARTITION THEOREM

The evaluation of z_{transl} in equation 16.70, in which z_{transl} was found to be proportional to $T^{\frac{3}{2}}$, is but a special case of a general theorem of wide applicability. Consider some normal mode of a system—the mode may be translational, vibrational, rotational, or perhaps of some other more abstract nature. Let a generalized coordinate associated with the mode be q and let the associated (or "conjugate") momentum be p. Suppose the energy (Hamiltonian) to be of the form

$$E = Aq^2 + Bp^2 \qquad (16.75)$$

Then the classical prescription for calculating the partition function will contain a factor of the form

$$z \sim \int\int \frac{dq\, dp}{h} e^{-\beta(Aq^2 + Bp^2)} \qquad (16.76)$$

or, as in equation 16.70, if $A \neq 0$ and $B \neq 0$

$$z \sim \left(\frac{\pi k_B T}{hA}\right)^{1/2}\left(\frac{\pi k_B T}{hB}\right)^{1/2} \qquad (16.77)$$

If either A or B is equal to zero the corresponding integral is a (bounded) constant determined by the limits on the associated integral. The integration over x in equation 16.70 is an example of such a case, and the corresponding integral is $V^{\frac{1}{3}}$.

The significant result in 16.77 is that, *at sufficiently high temperature* (so that the classical density of states is applicable) *every quadratic term in the energy contributes a factor of $T^{\frac{1}{2}}$ to the partition function*.

Equivalently, at sufficiently high temperature every quadratic term in the energy contributes a term $\left(\frac{1}{2}\tilde{N} \ln T\right)$ to $-\beta F$ or a term $\left(-\frac{1}{2}\tilde{N}k_B T \ln T\right)$ to the Helmholtz potential F, or a term $\frac{1}{2}\tilde{N}k_B T(1 + \ln T)$ to the entropy.

Or finally, the result in its most immediately significant form is: *At sufficiently high temperature every quadratic term in the energy contributes a term $\frac{1}{2}\tilde{N}k_B$ to the heat capacity*. This is the "equipartition theorem" of classical statistical mechanics.

A gas of point mass particles has three quadratic terms in the energy: $(p_x^2 + p_y^2 + p_z^2)/2m$. The heat capacity at constant volume of such a gas, at high temperature, is $\frac{3}{2}\tilde{N}k_B$, or $\frac{3}{2}R$ per mole.

Application of the equipartition theorem to a gas of polyatomic molecules is best illustrated by several examples. Consider first a heteronuclear diatomic molecule. It has three translational modes; each such mode has a quadratic kinetic energy but no potential energy; these three modes contribute $\frac{3}{2}k_B$ to the high temperature molar heat capacity. In addition the molecule has one vibrational mode; this mode has both kinetic and potential energy (both quadratic) and the mode therefore contributes $\frac{2}{2}k_B$. Finally the molecule has two rotational modes (i.e., it requires two angles to specify its orientation). These rotational modes have quadratic kinetic energy but no potential energy terms; they contribute $\frac{2}{2}k_B$. Thus the heat capacity per molecule is $\frac{7}{2}k_B$ at high temperature (or $\frac{7}{2}R$ per mole).

In general the total number of modes must be three times the number of atoms in the molecule. This is true because the mode amplitudes are a substitute set of coordinates that can replace the set of cartesian coordinates of each atom in the molecule. The number of the latter clearly is triple the number of atoms.

Consider a heteronuclear triatomic molecule. There are nine modes. Of these, three are translational modes; each contributes $\frac{1}{2}k_B$ to the heat capacity. There are three rotational modes, corresponding to the three angles required to orient a general object in space. Each rotational mode has only a kinetic energy term, and each contributes $\frac{1}{2}k_B$ to the heat capacity. By subtraction there remain three vibrational modes, each with kinetic and potential energy, and each contributing $\frac{2}{2}k_B$. Thus the high temperature heat capacity is $6k_B$ per molecule.

If the triatomic molecule is linear there is one less rotational mode and therefore one additional vibrational mode. The high temperature heat capacity is increased to $\frac{13}{2}k_B$. Note that the *shape* of the molecule can be discerned by measurement of the heat capacity of the gas!

In all of the preceding discussion we have neglected contributions that may arise from the internal structure of the atoms. These contributions generally have much higher energy and they contribute only at enormously high temperature.

If the molecules are *homonuclear* (indistinguishable atoms), rather than heteronuclear, additional quantum mechanical symmetry requirements again complicate the counting of states. Nevertheless, the analogous form of the equipartition theorem emerges at high temperature. The classical partition function simply contains a factor of $\left(\frac{1}{2}\right)^N$ to account for the indistinguishability of the two atoms within each of the $\tilde{N}$ molecules, and it contains a factor of $1/\tilde{N}!$ to account for the indistinguishability of the $\tilde{N}$ molecules.

17

ENTROPY AND DISORDER: GENERALIZED CANONICAL FORMULATIONS

17-1 ENTROPY AS A MEASURE OF DISORDER

In the two preceding chapters we have considered two types of physical situations. In one the system of interest is isolated; in the other the system is in diathermal contact with a thermal reservoir. Two very different expressions for the entropy in terms of the state probabilities $\{f_j\}$ result.

If the system is isolated it spends equal time in each of the permissible states (the number of which is Ω):

$$f_j = \frac{1}{\Omega} \tag{17.1}$$

and the entropy is

$$S = k_B \ln \Omega \tag{17.2}$$

If the system is in diathermal contact with a thermal reservoir, the fraction of time that it spends in the state j is

$$f_j = \frac{e^{-\beta E_j}}{Z}, \qquad Z = \sum_j e^{-\beta E_j} \tag{17.3}$$

and the entropy is $(U/T - F/T)$ which we write in the form

$$S = k_B \beta \sum_j f_j E_j + k_B \ln Z \tag{17.4}$$

We now pause to inquire as to whether these results reveal some underlying significance of the entropy. Are they to be taken purely formally as particular computational results, or can we infer from them some intuitively revealing insights to the significance of the entropy concept?

In fact the conceptual framework of "information theory," erected by Claude Shannon[1] in the late 1940s, provides a basis for interpretation of the entropy in terms of Shannon's measure of *disorder*.

The concept of "order" (or its negation, "disorder") is qualitatively familiar. A neatly built brick wall is evidently more ordered than a heap of bricks. Or a "hand" of four playing cards is considered to be more ordered if it consists of four aces than if it contains, for instance, neither pairs nor a straight. A succession of groups of letters from the alphabet is recognized as more ordered if each group concords with a word listed in the dictionary rather than resembling the creation of a monkey playing with a typewriter.

Unfortunately the "heap of bricks" may be the prized creation of a modern artist, who would be outraged by the displacement of a single brick! Or the hand of cards may be a winning hand in some unfamiliar game. The apparently disordered text may be a perfectly ordered, but coded, message. The order that we seek to quantify must be an order with respect to some prescribed criteria; the standards of architecture, the rules of poker, or the corpus of officially recognized English words. Disorder within one set of criteria may be order within another set.

In statistical mechanics we are interested in the disorder in the distribution of the system over the permissible microstates.

Again we attempt to clarify the problem with an analogy. Let us suppose that a child is told to settle down in any room of his choice, and to wait in that room until his parents' return (this is the rule defining order!). But of course the child does not stay in a single room—he wanders restlessly throughout the house spending a fraction of time f_j in the jth room.

The problem solved by Shannon is the definition of a quantitative measure of the disorder associated with a given distribution $\{f_j\}$.

Several requirements of the measure of disorder reflect our qualitative concepts:

(*a*) The measure of disorder should be defined entirely in terms of the set of numbers $\{f_j\}$.

(*b*) If any one of the f_j is unity (and all the rest consequently are zero) the system is completely ordered. The quantitative measure of disorder should then be zero.

(*c*) The maximum disorder corresponds to each f_j being equal to $1/\Omega$ —that is, to the child showing no preference for any of the rooms in the house, among which he wanders totally randomly.

[1] C. E. Shannon and W. Weaver, *The Mathematical Theory of Communications* (Univ. of Illinois Press, Urbana, 1949).

(**d**)   The maximum disorder should be an increasing function of Ω (being greater for a child wandering randomly through a large house rather than through a small house).

(**e**)   The disorder should compound additively over "partial disorders." That is, let $f^{(1)}$ be the fraction of time the child spends on the first floor, and let Disorder$^{(1)}$ be the disorder of his distribution over the first floor rooms. Similarly for $f^{(2)}$ and Disorder$^{(2)}$. Then the total disorder should be

$$\text{Disorder} = f^{(1)} \times \text{Disorder}^{(1)} + f^{(2)} \times \text{Disorder}^{(2)} \quad (17.5)$$

These qualitatively reasonable attributes uniquely determine the measure of disorder[2]. Specifically

$$\text{Disorder} = -k \sum_j f_j \ln f_j \quad (17.6)$$

where k is an arbitrary positive constant.

We can easily verify that the disorder vanishes, as required, if one of the f_j is unity and all others are zero. Also the maximum value of the disorder (when each $f_j = 1/\Omega$) is $k \ln \Omega$ (see **Problem 17.1-1**), and this does increase monotonically with Ω as required in (d) above.

The maximum value of the disorder, $k \ln \Omega$, is precisely the result (equation 17.1) previously found for the entropy of a closed system. Complete concurrence requires only that we choose the constant k to be Boltzmann's constant k_B. *For a closed system the entropy corresponds to Shannon's quantitative measure of the maximum possible disorder in the distribution of the system over its permissible microstates.*

We then turn our attention to systems in diathermal contact with a thermal reservoir, for which $f_j = \exp(-\beta E_j)/Z$ (equation 17.3). Inserting this value of the f_j into the definition of the disorder (equation 17.6), we find the disorder to be

$$\text{Disorder} = k_B \beta \sum_j f_j E_j + k_B \ln Z \quad (17.7)$$

Again the disorder of the distribution is precisely equal to the entropy (recall equation 17.4).

This agreement between entropy and disorder is preserved for all other boundary conditions—that is for systems in contact with pressure reservoirs, with particle reservoirs, and so forth.

Thus we recognize that the physical interpretation of the entropy is that *the entropy is the quantitative measure of the disorder in the relevant distribution of the system over its permissible microstates.*

[2] For a proof see A. I. Khinchin, *Mathematical Foundations of Information Theory* (Dover Publications, New York, 1957).

It should not be surprising that this result emerges. Our basic assumption in statistical mechanics was that the random perturbations of the environment assure equal fractional occupation of all microstates of a closed system—that is, *maximum disorder*. In thermodynamics the entropy enters as a quantity that is maximum in equilibrium. Identification of the entropy as the disorder simply brings these two viewpoints into concurrence for closed systems.

PROBLEMS

17.1-1. Consider the quantity $x \ln x$ in the limit $x \to 0$. Show by L'Hôpital's rule that $x \ln x$ vanishes in this limit. How is this related to the assertion after equation 17.6, that the disorder vanishes when one of the f_j is equal to unity?

17.1-2. Prove that the disorder, defined in equation 17.6, is nonnegative for all physical distributions.

17.1-3. Prove that the quantity $-k\Sigma_j f_j \ln f_j$ is maximum if all the f_j are equal by applying the mathematical inequality valid for any continuous convex function $\Phi(x)$

$$\Phi\left(\frac{1}{\Omega} \sum_{k=1}^{\Omega} a_k\right) \leq \frac{1}{\Omega} \sum_{k=1}^{\Omega} \Phi(a_k)$$

Give a graphical interpretation of the inequality.

17-2 DISTRIBUTIONS OF MAXIMAL DISORDER

The interpretation of the entropy as the quantitative measure of disorder suggests an alternate perspective in which to view the canonical distribution. This alternative viewpoint is both simple and heuristically appealing, and it establishes an approach that will be useful in discussions of other distributions.

We temporarily put aside the perspective of Legendre transformations and even of temperature, returning to the most primitive level, at which a thermodynamic system is described by its extensive parameters $U, V, N_1, \ldots, N_r$. We then consider a system within walls restrictive with respect to $V, N_1, \ldots, N_r$, but nonrestrictive with respect to the energy U. The values of $V, N_1, \ldots, N_r$ restrict the possible microstates of the system, but it is evident that states of any energy consistent with $V, N_1, \ldots, N_r$ are permitted. Nevertheless a thermodynamic measurement of the energy yields a value U. This observed value is the average energy, weighted by the (as yet unknown) probability factors f_j

$$U = \sum_j f_j E_j \tag{17.8}$$

As a "matter of curiosity" let us explore the following question: *What distribution* $\{f_j\}$ *maximizes the disorder subject only to the requirement that it yields the observed value of* U (equation 17.8)?

The disorder is

$$\text{Disorder} = -k_B \sum_j f_j \ln f_j \qquad (17.9)$$

and if this is to be maximum

$$\delta(\text{Disorder}) = -k_B \sum_j (\ln f_j + 1) \delta f_j = 0 \qquad (17.10)$$

Now if the f_j were independent variables we could equate each term in the sum separately to zero. But the factors f_j are not independent. They are subject to the auxiliary condition (17.8) and to the normalization condition

$$\sum_j f_j = 1 \qquad (17.11)$$

The mathematical technique for coping with these auxiliary conditions is the method of Lagrange multipliers[3]. The prescription is to calculate the differentials of each of the auxiliary conditions

$$\sum_j \delta f_j = 0 \qquad (17.12)$$

$$\sum_j E_j \delta f_j = 0 \qquad (17.13)$$

to multiply each by a "variational parameter" (λ_1 and λ_2), and to add these to equation 17.10

$$-k_B \sum_j (\ln f_j + 1 + \lambda_1 + \lambda_2 E_j) f_j = 0 \qquad (17.14)$$

The method of Lagrange multipliers guarantees that each term in equation 17.14 then can be put individually and independently equal to zero, providing that the variational parameters are finally chosen so as to satisfy the two auxiliary conditions 17.8 and 17.11.

Thus, for *each j*

$$\ln f_j + 1 + \lambda_1 + \lambda_2 E_j = 0 \qquad (17.15)$$

---

[3] *cf.* G. Arfken, *Mathematical Methods for Physicists* (Academic Press, New York, 1960) or any similar reference on mathematical methods for scientists.

or

$$f_j = e^{-(1 + \lambda_1 + \lambda_2 E_j)} \tag{17.16}$$

We now must determine λ_1 and λ_2 so as to satisfy the auxiliary conditions. That is, from 17.11

$$e^{-(1 + \lambda_1)} \sum_j e^{-\lambda_2 E_j} = 1 \tag{17.17}$$

and from 17.8

$$e^{-(1 + \lambda_1)} \sum_j E_j e^{-\lambda_2 E_j} = U \tag{17.18}$$

These are identical in form with the equations of the canonical distribution! The quantity λ_2 is merely a different notation for

$$\lambda_2 \equiv \beta = \frac{1}{k_B T} \tag{17.19}$$

and then, from 17.18 and 16.12

$$e^{-(1 + \lambda_1)} = \frac{1}{\sum_j e^{-\beta E_j}} = \frac{1}{Z} \tag{17.20}$$

That is, *except for a change in notation, we have rediscovered the canonical distribution.*

The canonical distribution is the distribution over the states of fixed $V, N_1, \ldots, N_r$ that maximizes the disorder, subject to the condition that the average energy has its observed value. This conditional maximum of the disorder is the entropy of the canonical distribution.

Before we turn to the generalization of these results it may be well to note that we refer to the f_j as "probabilities." The concept of probability has two distinct interpretations in common usage. "Objective probability" refers to a *frequency*, or a *fractional occurrence*; the assertion that "the probability of newborn infants being male is slightly less than one half" is a statement about census data. "Subjective probability" is a measure of *expectation based on less than optimum information*. The (subjective) probability of a *particular yet unborn* child being male, *as assessed by a physician*, depends upon that physician's knowledge of the parents' family histories, upon accumulating data on maternal hormone levels, upon the increasing clarity of ultrasound images, and finally upon an educated, but still subjective, guess.

The "disorder," a function of the probabilities, has two corresponding interpretations. The very term *disorder* reflects an objective interpretation, based upon objective fractional occurrences. The same quantity, based on the subjective interpretation of the f_j's, is a measure of the uncertainty of a prediction that may be based upon the f_j's. If one f_j is unity the uncertainty is zero and a perfect prediction is possible. If all the f_j are equal the uncertainty is maximum and no reliable prediction can be made.

There is a school of thermodynamicists[4] who view thermodynamics as a subjective science of prediction. If the energy is known, it constrains our guess of any other property of the system. If *only* the energy is known the most valid guess as to other properties is based on a set of probabilities that maximize the residual uncertainty. *In this interpretation the maximization of the entropy is a strategy of optimal prediction.*

To repeat, we view the probabilities f_j as objective fractional occurrences. The entropy is a measure of the objective disorder of the distribution of the system among its microstates. That disorder arises by virtue of random interactions with the surroundings or by other random processes (which may be dominant).

PROBLEMS

17.2-1. Show that the maximum value of the disorder, as calculated in this section, does agree with the entropy of the canonical distribution (equation 17.4).

17.2-2. Given the identification of the disorder as the entropy, and of f_j as given in equation 17.16, prove that $\lambda_2 = 1/(k_B T)$ (equation 17.19).

17-3   THE GRAND CANONICAL FORMALISM

Generalization of the canonical formalism is straightforward, merely substituting other extensive parameters in place of the energy. We illustrate by focusing on a particularly powerful and widely used formalism, known as the "grand canonical" formalism.

Consider a system of fixed volume in contact with both energy and particle reservoirs. The system might be a layer of molecules adsorbed on a surface bathed by a gas. Or it may be the contents of a narrow necked but open bottle lying on the sea floor.

Considering the system plus the reservoir as a closed system, for which every state is equally probable, we conclude as in equation 16.1, that the fractional occupation of a state of the system of given energy E_j and mole

---

[4]*cf.* M. Tribus, *Thermostatistics and Thermodynamics* (D. Van Nostrand and Co., New York, 1961). E. T. Jaynes, *Papers on Probability, Statistics, and Statistical Physics*, Edited by R. D. Rosenkrantz, (D. Reidel, Dordrecht and Boston, 1983).

number N_j is

$$f_j = \frac{\Omega^{\text{res}}(E_{\text{total}} - E_j, N_{\text{total}} - N_j)}{\Omega^{\text{total}}(E_{\text{total}}, N_{\text{total}})} \tag{17.21}$$

But again, expressing Ω in terms of the entropy

$$f_j = \exp\left[\left(\frac{1}{k_B}\right)S^{\text{res}}(E_{\text{total}} - E_j, N_{\text{total}} - N_j) - \left(\frac{1}{k_B}\right)S^{\text{tot}}(E_{\text{total}}, N_{\text{total}})\right] \tag{17.22}$$

Expanding as in equations 16.3 to 16.5

$$f_j = e^{\beta\Psi}e^{-\beta(E_j - \mu N_j)} \tag{17.23}$$

where Ψ is the "grand canonical potential"

$$\Psi = U - TS - \mu N = U[T, \mu] \tag{17.24}$$

The factor $e^{\beta\Psi}$ plays the role of a normalizing factor

$$e^{\beta\Psi} = \mathcal{Z}^{-1} \tag{17.25}$$

where $\mathcal{Z}$, the "grand canonical partition sum," is

$$\mathcal{Z} = \sum_j e^{-\beta(E_j - \mu N_j)} \tag{17.26}$$

The algorithm for calculating a fundamental equation consists of evaluating the grand canonical partition sum $\mathcal{Z}$ as a function of T and μ (and implicitly as a function also of V). Then $\beta\Psi$ is simply the logarithm of $\mathcal{Z}$. This functional relationship can be viewed in two ways, summarized in the mnemonic squares of Fig. 17.1.

The conventional view is that $\Psi(T, V, \mu)$ is the Legendre transform of U, or $\Psi(T, V, \mu) = U[T, \mu]$. The thermodynamics of this Legendre transformation is exhibited in the first mnemonic square of Fig. 17.1. It is evident that this square is isomorphic with the familiar square, merely replacing the extensive parameter V by N and reversing the corresponding arrow.

The more fundamental, and far more convenient view, is based on Massieu functions, or transforms of the entropy (Section 5.4). The second and third squares exhibit this transform; the third square merely alters the scale of temperature from T to k_BT, or from $1/T$ to β. The logarithm of the grand canonical partition sum $\mathcal{Z}$ is the Massieu transform $\beta\Psi$.

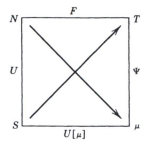

$$\Psi = U[T,\mu] = U - TS - \mu N$$

$$\frac{\partial \Psi}{\partial T} = -S \qquad \frac{\partial \Psi}{\partial \mu} = -N$$

$$-\frac{1}{T}\Psi = S\left[\frac{1}{T}, \frac{\mu}{T}\right] = S - \frac{1}{T}U + \frac{\mu}{T}N$$

$$\frac{\partial(\Psi/T)}{\partial(1/T)} = U \qquad \frac{\partial(\Psi/T)}{\partial(\mu/T)} = -N$$

$$-\beta\Psi = S[\beta, \beta\mu] = S - \beta U + \beta\mu N$$

$$\frac{\partial(\beta\Psi)}{\partial\beta} = U \qquad \frac{\partial(\beta\Psi)}{\partial(\beta\mu)} = -N$$

FIGURE 17.1
Mnemonic squares of the grand canonical potential.

A particularly useful identity which follows from these relationships is

$$U = \frac{\partial(\beta\Psi)}{\partial\beta} = -\left(\frac{\partial \ln \mathcal{Z}}{\partial\beta}\right)_{\beta\mu} \tag{17.27}$$

This relationship also follows directly from the probability interpretation of the f_j (see Problem 17.3-1). In carrying out the indicated differentiation (after having calculated $\mathcal{Z}$ or $\beta\Psi$) we must pair a factor β with every factor μ, and we then maintain all such $\beta\mu$ products constant as we differentiate with respect to the remaining β's.

Before illustrating the application of the grand canonical formalism it is interesting to corroborate that it, too, can be obtained as a distribution of maximal disorder. We maximize the disorder (entropy)

$$S = -k_B \sum_j f_j \ln f_j \qquad (17.28)$$

subject to the auxiliary conditions that

$$\sum_j f_j = 1 \qquad (17.29)$$

$$\sum_j f_j E_j = E \qquad (17.30)$$

and

$$\sum_j f_j N_j = N \qquad (17.31)$$

Then

$$\delta S = -k_B \sum_j (\ln f_j + 1)\, \delta f_j = 0 \qquad (17.32)$$

Taking differentials of equations 17.29 to 17.31, multiplying by Lagrange multipliers λ_1, λ_2, and λ_3, and adding

$$\sum_j (\ln f_j + 1 + \lambda_1 + \lambda_2 E_j + \lambda_3 N_j) = 0 \qquad (17.33)$$

Each term then may be equated separately to zero (as in equation 17.15), and

$$f_j = e^{-(1 + \lambda_1 + \lambda_2 E_j + \lambda_3 N_j)} \qquad (17.34)$$

The Lagrange multipliers must now be evaluated by equations 17.29 to 17.31. Doing so identifies them in terms of $\beta\ (=\lambda_2)$, $\beta\mu\ (= -\lambda_3)$, and $\beta\Psi\ (= -1 - \lambda_1)$, again establishing equation 17.23.

It should be noted that the mole number N_j can be replaced by the particle number $\tilde{N}_j$ (where $\tilde{N}_j = N_j \times$ Avagadro's number). In that case μ, the Gibbs potential per mole, is replaced by the Gibbs potential per particle. Although a rational notation for the latter quantity would be $\tilde{\mu}$, *we shall henceforth write μ for either the Gibbs potential per mole or the Gibbs potential per particle, permitting the distinction to be established by the context.*

Example: Molecular Adsorption on a Surface

Consider a gas in contact with a solid surface. The molecules of the gas can adsorb on specific sites on the surface, the sites being determined by the molecular structure of the surface. We assume, for simplicity, that the sites are sparsely enough distributed over the surface that they do not directly interact. There are $\tilde{N}$ such sites, and each can adsorb zero, one, or two molecules. Each site has an energy that we take as zero if the site is empty, as ε_1 if the site is singly occupied, and as ε_2 if the site is doubly occupied. The energies ε_1 and ε_2 may be either positive or negative; positive adsorption energies favor empty sites, and negative adsorption energies favor adsorption. The surface is bathed by a gas of temperature T and pressure P, and of sufficiently large mole number that it acts as a reservoir with respect to energy and particle number. We seek the "fractional coverage" of the surface, or the ratio of the number of adsorbed molecules to the number of adsorption sites.

The solution of this problem by the grand canonical potential permits us to focus our attention entirely on the surface sites. These sites can be populated by both energy and particles, which play completely analogous roles in the formalism.

The gaseous phase which bathes the surface establishes the values of T and μ, being both a thermal and a particle reservoir. The given data may be (and generally is) unsymmetric, specifying T and P of the gas rather than T and μ. In such a case μ, the Gibbs potential per particle of the gas, must first be evaluated from the fundamental equation of the gas, if known, or from integration of the Gibbs–Duhem relation if the equations of state are known. We assume that this preliminary thermodynamic calculation has been carried out and that T and μ of the gas are specified. Thenceforth the analysis is completely symmetric between energy and particles.

Because the surface sites do not interact, the grand partition sum factors

$$\mathcal{Z} = \mathfrak{z}^{\tilde{N}}$$

The grand partition sum for a single site contains just three terms, corresponding to the empty, the single occupied, and the doubly occupied states

$$\mathfrak{z} = 1 + e^{-\beta(\varepsilon_1 - \mu)} + e^{-\beta(\varepsilon_2 - 2\mu)}$$

Each of the three terms in $\mathfrak{z}$, divided by $\mathfrak{z}$, is the probability of the corresponding state. Thus the mean number of molecules adsorbed per site is

$$\bar{n} = \frac{e^{-\beta(\varepsilon_1 - \mu)} + 2e^{-\beta(\varepsilon_2 - 2\mu)}}{\mathfrak{z}}$$

and the mean energy per site is

$$\bar{\varepsilon} = \frac{\varepsilon_1 e^{-\beta(\varepsilon_1 - \mu)} + \varepsilon_2 e^{-\beta(\varepsilon_2 - 2\mu)}}{\mathfrak{z}}$$

An alternative route to these latter two results, and to the general thermodynamics of the system, is via calculation of the grand canonical potential, $\Psi =$

$-k_B T \log \mathscr{Z}$ (equation 17.25).

$$\Psi = -\tilde{N}k_B T \log \left(1 + e^{-\beta(\varepsilon_1 - \mu)} + e^{-\beta(\varepsilon_2 - 2\mu)}\right)$$

The number $\hat{N}$ of adsorbed atoms on the $\tilde{N}$ sites is obtained thermodynamically by differentiation of Ψ

$$\hat{N} = -\frac{\partial \Psi}{\partial \mu}$$

and, of course, such a differentiation is equal to $\tilde{N}\bar{n}$ with $\bar{n}$ as previously found. Similarly the energy of the surface system is found by equation 17.27, and this gives a result identical to $\tilde{N}\bar{\varepsilon}$.

The reader is strongly urged to do Problem 17.3-4.

PROBLEMS

17.3-1. Calculate $(\partial \log \mathscr{Z}/\partial \beta)_{\beta\mu}$ directly from equation 17.26 and show that the result is consistent with equation 17.27.

17.3-2. A system is contained in a cylinder with diathermal impermeable walls, fitted with a freely moveable piston. The external temperature and pressure are constant. Derive an appropriate canonical formalism for this system. Identify the logarithm of the corresponding partition sum.

17.3-3. For the surface adsorption model of the preceding Example, investigate the mean number of molecules adsorbed per site $(\bar{n})$ in the limit $T \to 0$, for all combinations of signs and relative magnitudes of $(\varepsilon_1 + \mu_0)$ and $(\varepsilon_2 + 2\mu_0)$, where μ_0 is the value of the μ of the gas at $T = 0$. Explain these results heuristically.

17.3-4. Suppose the adsorption model to be augmented by assuming that two adsorbed molecules on the same site interact in a vibrational mode of frequency ω. Thus the energy of an empty site is zero, the energy of a singly occupied site is ε_1, and the energy of a doubly occupied site can take any of the values $\varepsilon_2 + n'\hbar\omega$, with $n' = 0, 1, 2, \ldots$ Calculate

a) The grand canonical partition sum
b) The grand canonical potential
c) The mean occupation number, as computed directly from (a)
d) The mean occupation number, as computed directly from (b)
e) The probability that the system is in the state with $n = 2$ and $n' = 3$

Answer: Denoting $\varepsilon_j - \mu$ by ε_j',

$$(b) \qquad \Psi = \tilde{N}k_B T \ln\left(1 + e^{-\beta\varepsilon_1'} + \frac{e^{-\beta\varepsilon_2'}}{1 - e^{-\beta\hbar\omega}}\right)$$

$$(c, d) \qquad \bar{n} = \frac{(1 - e^{-\beta\hbar\omega})e^{-\beta\varepsilon_1'} + 2e^{-\beta\varepsilon_2'}}{(1 - e^{-\beta\hbar\omega})(1 + e^{-\beta\varepsilon_1'}) + e^{-\beta\varepsilon_2'}}$$

$$(e) \qquad f_{2,3} = e^{-\beta(\varepsilon_2' + 2\hbar\omega)}/\mathscr{Z}$$

17.3-5. Calculate the fundamental equation of the polymer model of Section 15.4 in a formalism canonical with respect to length and energy. Note that the "weight" in Fig. 15.4 plays the role of a "tension reservoir." Also recall Problem 17.3-2, the results of which may be helpful if the volume there is replaced by the length as an extensive parameter (as if the two transverse dimensions of the system are formally taken as constant).

17.3-6. A system contains $\tilde{N}$ sites and $\tilde{N}$ electrons. At a given site there is only one accessible orbital state, but that orbital state can be occupied by zero, one, or two electrons (of opposite spin). The site energy is zero if the site is either empty or singly occupied, and it is ε if the site is doubly occupied. In addition there is an externally applied magnetic field which acts only on the spin coordinates.

a) Calculate the chemical potential μ as a function of the temperature and the magnetic field.

b) Calculate the heat capacity of the system.

c) Calculate the initial magnetic susceptibility of the system (i.e., the magnetic susceptibility in small magnetic field).

17.3-7. Carbon monoxide molecules (CO) can be adsorbed at specific sites on a solid surface. The oxygen atom of an adsorbed molecule is immobilized on the adsorption site; the axis of the adsorbed molecule thereby is fixed perpendicular to the surface so that the rotational degree of freedom of the adsorbed molecule is suppressed. In addition the vibrational frequency of the molecule is altered, the effective mass changing from the "reduced mass" $m_C m_O / (m_C + m_O)$ to m_C. Only one molecule can be adsorbed at a given site. The binding energy of an adsorbed molecule is E_b. The surface is bathed by CO gas at temperature T and pressure P. Calculate the fraction (f) of occupied adsorption sites if the system is in equilibrium. Assume the temperature to be of the order of one or two hundred Kelvin, and assume the pressure to be sufficiently low that the CO vapor can be regarded as an ideal diatomic gas.

Hint: Recall the magnitudes of characteristic rotational and vibrational frequencies, as expressed in equivalent temperatures, in Section 16.3.

18

QUANTUM FLUIDS

18-1 QUANTUM PARTICLES: A "FERMION PRE-GAS MODEL"

At this point we might be tempted to test the grand canonical formalism on the ideal gas, not to obtain new results of course, but to compare the analytic convenience and power of the various formalisms. Remarkably, the grand canonical formalism proves to be extremely *un*congenial to the classical ideal gas model! The catastrophe of nonextensivity that plagued the calculation in the canonical formalism becomes even more awkward in the grand canonical formalism[1].

As so often happens in physics, the formalism points the way to reality. The awkwardness of the formalism is a signal that the *model* is unphysical —that there are no classical particles in nature! There are only fermions and bosons, two types of quantum mechanical particles. For these the grand canonical formalism becomes extremely simple!

Fermions are the quantum analogues of the material particles of classical physics. Electrons, protons, neutrons, and a panoply of more esoteric particles are fermions. The nineteenth century "law of impenetrability of matter" is replaced by an antisymmetry condition on the quantum mechanical wave function[2]. This condition implies (as the only consequence of which we shall have need) that *only a single fermion can occupy a given orbital state*.

Bosons are the quantum analogues of the "waves" of classical physics. Photons, the quanta of light, are typical bosons. Just as waves can be freely superposed classically, *so an arbitrary number of bosons can occupy a single orbital state*. Furthermore, there exist bosons with zero rest mass—such bosons, like classical waves, can be freely created or annihi-

---

[1] The root of the difficulty lies in the fact that the grand canonical formalism focusses not on the particles, but on the *orbital states*. There is then no natural way to count the states "as if the particles had labels" (later to be corrected by division by $\bar{N}!$).

[2] The wave function must be antisymmetric under interchange of two fermions, thereby interposing a node between the fermions and preventing two fermions (of the same spin state) from occupying the same spatial position.

lated. The radiation of electromagnetic waves by a hot body is described in quantum terminology as the creation and emission of photons.

The fundamental particles in nature possess intrinsic angular momentum, or "spin." The (immutable) magnitude of this intrinsic angular momentum is necessarily a multiple of $\hbar/2$; those particles with odd multiples of $\hbar/2$ are fermions, and those with even multiples of $\hbar/2$ are bosons.

The orientation of the intrinsic angular momentum is also quantized. For fermions of "spin $\frac{1}{2}$" (angular momentum $= \hbar/2$) the angular momentum can have either of two orientations (along any arbitrarily designated axis). These two orientations are designated by *up* and *down*, or by the two values $m_s = \frac{1}{2}$ and $m_s = -\frac{1}{2}$ of the "magnetic quantum number" m_s.

Finally, an orbital state of a quantum particle is labeled by the quantum numbers of its spatial wave function *and* by the magnetic quantum number m_s of its spin orientation. For a particle in a cubic container the three spatial quantum numbers are the three components of the wave vector $\mathbf{k}$ (recall equation 16.37), so that an orbital state is completely labeled by $\mathbf{k}$ and m_s.

Preparatory to the application of the grand canonical formalism to Fermi and Bose ideal gases, it is instructive to consider a simpler model that exhibits the physics in greater clarity. This model has only three energy levels, so that all summations over states can be exhibited explicitly. Except for this simplification, the analysis stands in strict step by step correspondence with the analysis of quantum gases to be developed in the following sections; hence the name *pre-gas model*.

We consider first the spin-$\frac{1}{2}$ fermion pre-gas model. The model system is such that only three spatial orbits are permitted; particles in these spatial orbits have energies ε_1, ε_2, and ε_3. The model system is in contact with a thermal reservoir and with a reservoir of spin-$\frac{1}{2}$ Fermi particles; the reservoirs impose fixed values of the temperature T and of the molar Gibbs potential μ (which, for fermion systems, is also known as the *Fermi level*).

Each spatial orbit corresponds to two orbital states, one of spin up and one of spin down. There are therefore six orbital states, which can be numbered (n, m_s) with $n = 1, 2, 3$ and $m_s = -\frac{1}{2}, +\frac{1}{2}$.

The grand canonical partition sum factors with respect to the six orbital states

$$\mathbf{Z} = z_{1,-1/2} z_{1,1/2} z_{2,-1/2} z_{2,1/2} z_{3,-1/2} z_{3,1/2} \tag{18.1}$$

and each orbital state partition sum has two terms, corresponding to the state being either empty or occupied. In the absence of a magnetic field,

$$z_{n,m_s} = 1 + e^{-\beta(\varepsilon_n - \mu)} \tag{18.2}$$

Alternatively we can pair the two orbital states with the same n but with $m_s = \pm \frac{1}{2}$

$$z_{n,1/2} z_{n,-1/2} = \left[1 + e^{-\beta(\epsilon_n - \mu)}\right]^2 = 1 + 2e^{-\beta(\epsilon_n - \mu)} + e^{-2\beta(\epsilon_n - \mu)}$$

$$(18.3)$$

This product can be interpreted in terms of the *four* states of given n: the empty state, two singly occupied states, and one doubly occupied state.

The probability that the orbital state (n, m_s) is empty is $1/z_{n,m}$, and the probability that it is occupied is

$$f_{n,m} = \frac{e^{-\beta(\epsilon_n - \mu)}}{z_{n,m}} = \frac{1}{e^{\beta(\epsilon_n - \mu)} + 1} \qquad (18.4)$$

The fundamental equation follows directly from equations 18.1 to 18.3

$$e^{-\beta\Psi} = \mathbf{Z} = \left[1 + e^{-\beta(\epsilon_1 - \mu)}\right]^2 \left[1 + e^{-\beta(\epsilon_2 - \mu)}\right]^2 \left[1 + e^{-\beta(\epsilon_3 - \mu)}\right]^2 (18.5)$$

We can find the mean number of particles in the system by differentiation ($\tilde{N} = -\partial\Psi/\partial\mu$). Alternatively we can sum the probability of occupation $f_{n,m}$ over all six orbital states

$$\tilde{N} = \sum_{n,m} f_{n,m} = \frac{2}{e^{\beta(\epsilon_1 - \mu)} + 1} + \frac{2}{e^{\beta(\epsilon_2 - \mu)} + 1} + \frac{2}{e^{\beta(\epsilon_3 - \mu)} + 1} \quad (18.6)$$

The entropy of the system can be obtained by differentiation of the fundamental equation ($S = -\partial\Psi/\partial T$). Alternatively it can be calculated from the occupation probabilities (Problem 18.1-1).

The energy is found thermodynamically by differentiation: $U = (\partial\beta\Psi/\partial\beta)_{\beta\mu}$ (equation 17.27). Alternatively, from the probability interpretation of $f_{n,m}$

$$U = \sum_{n,m} \epsilon_{n,m} f_{n,m} = \frac{2\epsilon_1}{e^{\beta(\epsilon_1 - \mu)} + 1} + \frac{2\epsilon_2}{e^{\beta(\epsilon_2 - \mu)} + 1} + \frac{2\epsilon_3}{e^{\beta(\epsilon_3 - \mu)} + 1}$$

$$(18.7)$$

If the system of interest is actually in contact with T and μ reservoirs, these results are in convenient form. But it may happen that the physical system that we wish to describe is enclosed in nonpermeable walls that impose constancy of the particle number $\tilde{N}$ rather than of μ. Nevertheless *the fundamental equation is an attribute of the thermodynamic system, independent of boundary conditions*, so that the preceding formalism re-

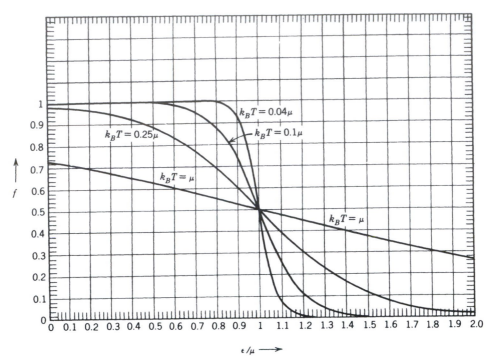

FIGURE 18.1
The probability of occupation, by a fermion, of an orbital state of energy ε at temperature T.

mains valid. However, the Fermi level μ is not a known quantity. Instead the value of μ adjusts to a change in temperature in such a way as to maintain $\tilde{N}$ constant—a response governed by equation 18.6.

Unfortunately equation 18.6 does not lend itself easily to explicit solution for μ as a function of T and $\tilde{N}$. However the solution can be obtained numerically or by series expansions in certain temperature regions, as we shall soon see. It is instructive first to reconsider the preceding analysis in more pictorial terms.

The occupation probability f of an orbital state of energy ε (as given by equation 18.4) is shown in Fig. 18.1. This occupation probability is more general than the present model, of course. It applies to any orbital state of a fermion. In the limit of zero temperature, any state of energy $\varepsilon < \mu$ is occupied and any state of energy $\varepsilon > \mu$ is empty. As the temperature is raised the states with energies slightly less than μ become partially depopulated, and the states with energies slightly greater than μ become populated. The range of energies within which this population transfer occurs is of the order of $4k_BT$ (see Problems 18.1-4, 18.1-5, 18.1-6).

The probability of occupation of a state with energy equal to μ is always one half, and a plot of $f(\varepsilon, T)$ as a function of ε (such as in Fig. 18.1) is symmetric under inversion through the point $\varepsilon = \mu$, $f = \frac{1}{2}$ (see Problem 18.1-5).

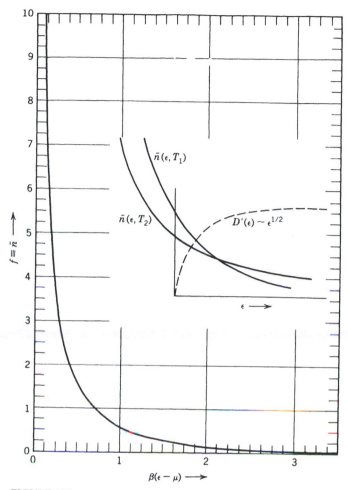

FIGURE 18.2

The Bose mean occupation number $\bar{n}$ of an orbital state of energy ε, at given T and μ. The insert is schematic, for $T_2 < T_1$ and $\mu_2 < \mu_1$.

With these pictorial insights we can explore the dependence of μ on T for the fermion pre-gas model. For definiteness suppose the system to contain four fermions. Furthermore, suppose that two of the energy levels coincide, with $\varepsilon_1 = \varepsilon_2$, and with $\varepsilon_3 > \varepsilon_2$. At $T = 0$ the four fermions fill the four orbital states of energy ε_1 ($= \varepsilon_2$), and the two states of energy ε_3 are empty. The Fermi level must lie somewhere between ε_2 and ε_3, but the precise value of μ must be found by considering the limiting value as $T \to 0$. For very low T

$$ f \equiv \frac{1}{e^{\beta(\varepsilon-\mu)} + 1} \simeq \begin{cases} e^{-\beta(\varepsilon-\mu)} & \text{for } \varepsilon > \mu \text{ and } T \simeq 0 \\ 1 - e^{\beta(\varepsilon-\mu)} & \text{for } \varepsilon < \mu \text{ and } T \simeq 0 \end{cases} \quad (18.8) $$

Thus, if $\varepsilon_1 = \varepsilon_2 < \varepsilon_3$, and $\tilde{N} = 4$, equation 18.6 becomes, for $T \simeq 0$

$$4 = 4\left(1 - e^{\beta(\varepsilon_1 - \mu)}\right) + 2e^{-\beta(\varepsilon_3 - \mu)} \tag{18.9}$$

or

$$\mu = \frac{\varepsilon_1 + \varepsilon_3}{2} + \frac{1}{2}k_B T \ln 2 + \cdots \tag{18.10}$$

In this case μ is midway between ε_1 and ε_3 at $T = 0$, and μ increases linearly as T increases.

It is instructive to compare this result with another special case, in which $\varepsilon_1 < \varepsilon_2 = \varepsilon_3$. If we were to have four fermions in the system the Fermi level (μ) would coincide with ε_2 at $T = 0$. More interesting is the case in which there are only two fermions. Then at $T = 0$ the Fermi level lies between ε_1 and ε_2 ($= \varepsilon_3$). We proceed as previously. Equation 18.9 is replaced, for $T \simeq 0$, by

$$2 = 2\left(1 - e^{\beta(\varepsilon_1 - \mu)}\right) + 4e^{-\beta(\varepsilon_3 - \mu)} \tag{18.11}$$

and

$$\mu = \frac{\varepsilon_1 + \varepsilon_3}{2} - \frac{1}{2}k_B T \ln 2 + \cdots \tag{18.12}$$

In each of the cases the Fermi level moves away from the doubly degenerate energy level. The reader should visualize this effect in the pictorial terms of Fig. 18.1, recognizing the centrality of the inversion symmetry of f relative to the point at $\varepsilon = \mu$.

From these several special cases it now should be clear that the general principles that govern the temperature dependence of μ (for a system of constant $\tilde{N}$) are:

(**a**) The occupation probability departs from zero or unity over a region of $\Delta\varepsilon \simeq \pm 2k_B T$ around μ.

(**b**) As T increases, the Fermi level μ is "repelled" by high densities of states within this region.

PROBLEMS

18.1-1. Obtain the mean number of particles in the fermion pre-gas model by differentiating Ψ, as given in equation 18.5. Show that the result agrees with $\tilde{N}$ as given in equation 18.6.

18.1-2. The entropy of a system is given by $S = -k_B \Sigma_j f_j \ln f_j$, where f_j is the probability of a microstate of the system. Each microstate of the fermion pre-gas model is described by specifying the occupation of *all six* orbital states.

a) Show that there are $2^6 = 64$ possible microstates of the model system, and that there are therefore 64 terms in the expression for the entropy.

b) Show that this expression reduces to

$$S = -k_B \sum_{n,m} f_{nm} \ln f_{nm}$$

and that this equation contains only six terms. What special properties of the model effect this drastic reduction?

18.1-3. Apply equation 17.27 for U to the fundamental equation of the fermion pre-gas model, and show that this gives the same result for U as in equation 18.7.

18.1-4. Show that $df/d\varepsilon = -\beta/4$ at $\varepsilon = \mu$. With this result show that f falls to $f = 0.25$ at *approximately* $\varepsilon \simeq \mu + k_B T$ and that f rises to $f = 0.75$ at *approximately* $\varepsilon \simeq \mu - k_B T$ (check this result by Fig. 18.1). This rule of thumb gives a qualitative and useful picture of the range of ε over which f changes rapidly.

18.1-5. Show that Fig. 17.2 [of $f(\varepsilon, T)$ as a function of ε] is symmetric under inversion through the point $\varepsilon = \mu$, $f = \frac{1}{2}$. That is, show that $f(\varepsilon, T)$ is subject to the symmetry relation

$$f(\mu + \Delta, T) = 1 - f(\mu - \Delta, T)$$

or

$$f(\varepsilon, T) = 1 - f(2\mu - \varepsilon, T)$$

and explain why this equation expresses the symmetry alluded to.

18.1-6. Suppose $f(\varepsilon, T)$ is to be approximated as a function of ε by three linear regions, as follows. In the vicinity of $\varepsilon \simeq \mu$, $f(\varepsilon, \mu)$ is to be approximated by a straight line going through the point $(\varepsilon = \mu,\ f = \frac{1}{2})$ and having the correct slope at that point. For low ε, $f(\varepsilon, \mu)$ is to be taken as unity. And at high ε, $f(\varepsilon, \mu)$ is to be taken as zero.

What is the slope of the central straight line section? What is the "width," in energy units, of the central straight line section? Compare this result with the "rule of thumb" given in Problem 18.1-4.

18-2 THE IDEAL FERMI FLUID

We turn our attention to the "ideal Fermi fluid," a model system of wide applicability and deep significance. The ideal Fermi fluid is a quantum analogue of the classical ideal gas; it is a system of fermion particles between which there are no (or negligibly small) interaction forces.

Conceptually, the simplest ideal Fermi fluid is a collection of neutrons, and such a fluid is realized in neutron stars and in the nucleus of heavy atoms (as one component of the neutron–proton "two-component fluid").

Composite "particles," such as atoms, behave as fermion particles if they contain an odd number of fermion constituents. Thus helium-three (^{3}He) atoms (containing two protons, one neutron, and two electrons) behave as fermions. Accordingly, a gas of ^{3}He atoms can be treated as an "ideal Fermi fluid." In contrast, ^{4}He atoms, containing an additional neutron, behave as bosons. The spectacular difference between the properties of ^{3}He and ^{4}He fluids at low temperatures, despite the fact that the two types of atoms are chemically indistinguishable, is a striking confirmation of the statistical mechanics of these quantum fluids.

Electrons in a metal are another Fermi fluid of great interest, to which we shall address our attention in Section 18.4.

We first consider the statistical mechanics of a general ideal, Fermi fluid. The analysis will follow the pattern of the fermion pre-gas model of the preceding section. Since the number of orbital states of the fluid is very large, rather than being the mere six orbital states of the pre-gas model, summations will be replaced by integrals. But otherwise the analyses stand in strict step by step correspondence.

To calculate the fundamental relation of an ideal fermion fluid we *choose* to consider it as being in interaction with a thermal and a particle reservoir, of temperature T and electrochemical potential μ. We stress again that the particular system being studied in the laboratory may have different boundary conditions—it may be closed, or it may be in diathermal contact only with a thermal reservoir, and so forth. But thermodynamic fundamental relations do not refer to any particular boundary condition, and we are free to choose any convenient boundary condition that facilitates the calculation. We choose the boundary conditions appropriate to the grand canonical formalism.

The orbital states available to the fermions are specified by the wave vector $\mathbf{k}$ of the wave function (recall equation 16.43) and by the orientation of the spin ("up" or "down" for a spin-$\frac{1}{2}$ fermion). The partition sum factors over the possible orbital states

$$\mathcal{Z} = \prod_{\mathbf{k}, m_s} z_{\mathbf{k}, m_s} \tag{18.13}$$

where m_s can take two values, $m_s = \frac{1}{2}$ implying spin up and $m_s = -\frac{1}{2}$ implying spin down. *Each orbital state can be either empty or singly occupied.* The energy of an empty orbital state is zero, and the energy of an occupied orbital state $\mathbf{k}, m_s$ is

$$\varepsilon_{\mathbf{k}, m_s} = \frac{p^2}{2m} = \frac{\hbar^2 k^2}{2m} \qquad \text{(independent of } m_s) \tag{18.14}$$

so that the partition sum of the orbital state $\mathbf{k}, m_s$ is

$$z_{\mathbf{k}, m_s} = 1 + e^{-\beta((\hbar^2 k^2/2m) - \mu)} \tag{18.15}$$

It is conventional to refer to the product $z_{\mathbf{k}, 1/2} \cdot z_{\mathbf{k}, -1/2}$ as $z_{\mathbf{k}}$, the "partition sum of the mode k"

$$\mathcal{Z} = \prod_{k, m_s} z_{k, m_s} = \prod_{k} z_{k, 1/2} z_{k, -1/2}$$

$$= \prod_{k} \left[1 + 2e^{-\beta((\hbar^2 k^2/2m) - \mu)} + e^{-\beta((2\hbar^2 k^2/2m) - 2\mu)} \right] \quad (18.16)$$

The three terms refer then to the totally empty mode, to the singly occupied mode (with two possible spin orientations), and to the doubly occupied mode (with one spin up and one down).

Each orbital state (k, m_s) is independent, and the probability of occupation is

$$f_{k, m_s} = \frac{e^{-\beta((\hbar^2 k^2/2m) - \mu)}}{z_{k, m_s}} = \frac{1}{e^{\beta((\hbar^2 k^2/2m) - \mu)} + 1} \quad (18.17)$$

This function is shown in Fig. 18.1.

At this point we can proceed by either of two routes. The fundamental algorithm instructs us to calculate the grand canonical potential Ψ ($= -k_B T \ln \mathcal{Z}$), thereby obtaining a fundamental relation. Alternatively, we can calculate all physical quantities of interest directly from equation 18.17. We shall first calculate the fundamental relation and then return to explore the (parallel) information available from knowledge of the "orbital-state distribution function" f_{k, m_s}.

The grand canonical potential is

$$\Psi = -k_B T \sum_{k} \ln z_k = -k_B T \sum_{k} \ln \left[1 + e^{-\beta((\hbar^2 k^2/2m) - \mu)} \right]^2 \quad (18.18)$$

The density of orbital states (of a *single* spin orientation) is $D(\varepsilon) d\varepsilon$, which has been calculated in Equation 16.47.

$$D(\varepsilon) d\varepsilon = \frac{V}{2\pi^2} k^2 \frac{dk}{d\varepsilon} d\varepsilon = \frac{V}{4\pi^2} \left(\frac{2m}{\hbar^2} \right)^{3/2} \varepsilon^{1/2} d\varepsilon \quad (18.19)$$

Inserting a factor of 2 to account for the two possible spin orientations, Ψ can then be written as

$$\Psi = -2k_B T \int_0^\infty \ln \left(1 + e^{-\beta(\varepsilon - \mu)} \right) D(\varepsilon) d\varepsilon$$

$$= -k_B T \frac{V}{2\pi^2} \left(\frac{2m}{\hbar^2} \right)^{3/2} \int_0^\infty \varepsilon^{1/2} \ln \left(1 + e^{-\beta(\varepsilon - \mu)} \right) d\varepsilon \quad (18.20)$$

Unfortunately the integral cannot be evaluated in closed form. Quantities of direct physical interest, obtained by differentiation of Ψ, must also be expressed in terms of integrals. Such quantities can be calculated to any desired accuracy by numerical quadrature or by various approximation schemes. In principle the statistical mechanical phase of the problem is completed with equation 18.20.

It is of interest to calculate the number of particles $\tilde{N}$ in the gas. By differentiation of Ψ

$$\tilde{N} = -\frac{\partial \Psi}{\partial \mu} = 2 \int_0^\infty \frac{1}{e^{\beta(\varepsilon - \mu)} + 1} D(\varepsilon) \, d\varepsilon$$

$$= \frac{V}{2\pi^2} \left(\frac{2m}{\hbar^2} \right)^{3/2} \int_0^\infty \frac{\varepsilon^{1/2}}{e^{\beta(\varepsilon - \mu)} + 1} \, d\varepsilon \qquad (18.21)$$

The first form of this equation reveals most clearly that it is identical to a summation of occupation probabilities over all states. Similarly the energy obtained by differentiation is identical to a summation of εf over all states

$$U = \left(\frac{\partial \beta \Psi}{\partial \beta} \right)_{\beta \mu} = 2 \int_0^\infty \frac{\varepsilon}{e^{\beta(\varepsilon - \mu)} + 1} D(\varepsilon) \, d\varepsilon$$

$$= \frac{V}{2\pi^2} \left(\frac{2m}{\hbar^2} \right)^{3/2} \int_0^\infty \frac{\varepsilon^{3/2}}{e^{\beta(\varepsilon - \mu)} + 1} \, d\varepsilon \qquad (18.22)$$

A flow-chart for the statistical mechanics of quantum fluids is shown in Table 18.1. Bose fluids are included, although we shall consider them explicitly only in later sections. The analysis differs only in several changes in sign, as will emerge in Section 18.5.

Before exploring these general results in specific detail it is wise to corroborate that for high temperature they do reduce to the classical ideal gas, and to explore the criterion that separates the classical from the quantum mechanical regime.

PROBLEMS

18.2-1. Prove equations c, g, h, i, and j of Table 18.1 (for fermions only).

18-3   THE CLASSICAL LIMIT AND THE QUANTUM CRITERION

The hallmark of the quantum regime is that a fermion particle is not free to occupy any arbitrarily chosen orbital state, for some states may already be filled. However at low density or high temperature the probability of occupation of each orbital state is small, thereby minimizing the

TABLE 18.1

Statistical Mechanics of Quantum Fluids. The upper sign refers to fermions and the lower to bosons.

(a) The partition sum factors. The number of spin orientations is $g_0 = 2S + 1$ ($g_0 = 1$ for bosons of spin zero; $g_0 = 2$ for fermions of spin $\frac{1}{2}$, etc.)

(b) z_k is the partition sum of a single orbital state (of definite k and m_s).

(c) $f_{k, m}$ is the mean occupation number (or "occupation probability") of the orbital state k, m_s.

(d, e, and f) $D(\varepsilon)$ is the density of orbital states of a single spin orientation.

(g) $\Psi(T, \mu)$ is a fundamental relation.

(h, i, and j) $P = P(U, V)$ is an equation of state, common to both fermion and bosons.

$$\mathcal{Z} = \prod_{k, m_s} z_{k, m_s} = \prod_k z_k^{g_0} \tag{a}$$

$$z_k = \left[1 \pm e^{-\beta(\varepsilon_k - \mu)}\right]^{\pm 1} \tag{b}$$

$$f_{k, m_s} = \frac{1}{e^{\beta(\varepsilon_k - \mu)} \pm 1} \tag{c}$$

$$-\beta\Psi = \ln \mathcal{Z} = g_0 \sum_k \ln z_k = \pm g_0 \sum_k \ln\left[1 \pm e^{-\beta(\varepsilon_k - \mu)}\right] \tag{d}$$

$$= \pm g_0 \int_0^\infty \ln\left[1 \pm e^{-\beta(\varepsilon - \mu)}\right] D(\varepsilon)\, d\varepsilon \tag{e}$$

$$D(\varepsilon) = \frac{V}{(2\pi)^2}\left(\frac{2m}{\hbar^2}\right)^{3/2} \varepsilon^{1/2} \tag{f}$$

Integrating by parts

$$\Psi = -\frac{2}{3}\frac{g_0 V}{(2\pi)^2}\left(\frac{2m}{\hbar^2}\right)^{3/2}\int_0^\infty \frac{\varepsilon^{3/2}}{e^{\beta(\varepsilon - \mu)} \pm 1}\, d\varepsilon \quad \text{(Fundamental Equation)} \tag{g}$$

Note

$$\Psi = -\frac{2}{3}\int_0^\infty \varepsilon f(\varepsilon)\, g_0 D(\varepsilon)\, d\varepsilon = -\frac{2}{3} U \tag{h}$$

Also

$$\Psi = -PV \quad \text{(for simple systems)} \tag{i}$$

$$\therefore \quad P = \frac{2}{3}\frac{U}{V} \quad \text{(equation of state)} \tag{j}$$

effect of the fermion prohibition against multiple occupancy. All gases become classical at low density or high temperature, in which conditions relatively few particles are distributed over many states.

The probability of occupancy of a state of energy ε is $[e^{\beta(\varepsilon - \mu)} + 1]^{-1}$, and this is small (for all ε) if $e^{-\beta\mu}$ is large, or if the *fugacity* $e^{\beta\mu}$ is small:

$$e^{\beta\mu} \ll 1 \quad \text{(classical regime)} \tag{18.23}$$

In this classical regime the occupation probability reduces to

$$f_{k, m_s} \simeq e^{\beta\mu} e^{-\beta\varepsilon} \tag{18.24}$$

In terms of Fig. 18.1, the classical region corresponds to the recession of the Fermi level μ to such large negative values that all physical orbitals lie on the "tail" of the $f(\varepsilon, T)$ curve.

We first corroborate that the occupation probability of equation 18.24 does reproduce classical results, and we then explore the physical condition that leads to a small fugacity.

The number of particles $\tilde{N}$ is expressed by equation 18.21 which, for small fugacity, becomes

$$\tilde{N} \simeq \frac{g_0 V}{(2\pi)^2} \left(\frac{2m}{\hbar^2}\right)^{3/2} e^{\beta\mu} \int_0^\infty e^{-\beta\varepsilon} \varepsilon^{1/2} d\varepsilon = \frac{g_0 V}{\lambda_T^3} e^{\beta\mu} \tag{18.25}$$

where λ_T (a quantity to be given a physical interpretation momentarily) is defined by

$$\lambda_T = \frac{h}{\sqrt{2\pi m k_B T}} \tag{18.26}$$

and where $g_0 = 2S + 1$ is the number of permissible spin orientations (equal to two for the spin $\frac{1}{2}$ case). Similarly the energy, as expressed in equation 17.62, becomes

$$U = \frac{2V}{(2\pi)^2} \left(\frac{2m}{\hbar^2}\right)^{3/2} e^{\beta\mu} \int_0^\infty e^{-\beta\varepsilon} \varepsilon^{3/2} d\varepsilon = \frac{3}{2} k_B T \frac{g_0 V}{\lambda_T^3} e^{\beta\mu} \tag{18.27}$$

Dividing

$$U = \tfrac{3}{2} \tilde{N} k_B T \tag{18.28}$$

This is the well-known equation of state of the classical ideal gas. In addition the individual equations 18.25 and 18.27 can be corroborated as valid for the classical ideal gas.

With the reassurance that the Fermi gas does behave appropriately in the classical limit, we may inquire as to the criterion that divides the quantum and classical regimes. It follows from our discussion that this division occurs when the fugacity is of the order of unity

$$e^{\beta\mu} \simeq 1 \quad \text{(classical–quantum boundary)} \tag{18.29}$$

or, from equation 18.25

$$\lambda_T^3 \bigg/ \left(\frac{g_0 V}{\tilde{N}} \right) \simeq 1 \quad \text{(classical–quantum boundary)} \quad (18.30)$$

This "quantum criterion" acquires a revealing pictorial interpretation when we explore the significance of λ_T. In fact λ_T is the quantum mechanical wave length of a particle with kinetic energy $k_B T$ (see Problem 18.3-2), whence λ_T is known as the "thermal wave length." From equation 18.25 we see that *in the classical limit the fugacity is the ratio of the "thermal volume"* λ_T^3 *to the volume per particle (of a single spin orientation)* $V/(\tilde{N}/g_0)$. *The system is in the quantum regime if the thermal volume is larger than the actual volume per particle (of a single spin orientation) either by virtue of large $\tilde{N}$ or by virtue of low T (and consequently of large λ_T).*

PROBLEMS

18.3-1. Calculate the definite integrals appearing in equations 18.25 and 18.26 by letting $\varepsilon = x^2$ and noting that each of the resulting integrals is the derivative (with respect to β) of a simpler integral.

18.3-2. Validate the interpretation of λ_T as the "thermal wavelength" by identifying the wavelength with the momentum p by the quantum mechanical definition $p = h/\lambda$, and by comparing the energy $p^2/2m$ to $k_B T$.

18-4 THE STRONG QUANTUM REGIME: ELECTRONS IN A METAL

The electrons in a metal would appear, at first thought, to be a very poor example of an ideal Fermi fluid, for the charges on the electrons ostensibly imply strong interparticle forces. However the background positive charges of the fixed ions tend to neutralize the negative charges of the electrons, at least on the average. And the very long range of the Coulomb force ensures that the average effect is the dominant effect, for the potential at any point is the resultant of contributions from enormously many electrons and positive ions—some nearby and many further removed in space. All of this can be made quantitative, and the accuracy of the approximation can be estimated and controlled by the methodology of solid state physics. We proceed by simply accepting the model of electrons in a metal as an ideal fermion gas, on the basis of the slender plausibility of these remarks.

An estimate of the Fermi level (to be made shortly) will reveal that for all reasonable temperatures $\mu \gg k_B T$. Thus electrons in a metal are an example of an ideal Fermi gas in the strong quantum regime. The analysis of this section is simply an examination of the Fermi gas in this strong quantum regime, with the allusion to electrons in a metal only to provide a physical context for the more general discussion.

Consider first the state of the electrons at zero temperature, and denote the value of the Fermi level at $T = 0$ as μ_0 (the "Fermi energy"). The occupation probability f is unity for $\varepsilon < \mu_0$ and is zero for $\varepsilon > \mu_0$, so that (from equation 18.21)

$$\tilde{N} = \frac{\sqrt{2}\, m^{3/2} V}{\pi^2 \hbar^3} \int_0^{\mu_0} \varepsilon^{1/2}\, d\varepsilon = \frac{(2m)^{3/2} V}{3\pi^2 \hbar^3}\, \mu_0^{3/2} \tag{18.31}$$

or

$$\mu_0 = \frac{\hbar^2}{2m}\left(3\pi^2 \frac{\tilde{N}}{V}\right)^{2/3} \tag{18.32}$$

The number of conduction electrons per unit volume in metals is of the order of 10^{22} to 10^{23} electrons/cm^3 (corresponding to one or two electrons per ion and an interionic distance of ≈ 5 Å). Consequently for electrons in metals the Fermi energy μ_0 (or the "Fermi temperature" μ_0/k_B) is of the magnitude

$$\frac{\mu_0}{k_B} \approx 10^4 \text{ K to } 10^5 \text{ K} \tag{18.33}$$

For other previously cited Fermi fluids the Fermi temperature may be even higher—of the order of 10^9 K for the electrons in white dwarf stars or 10^{12} K for the nucleons in heavy atomic nuclei and in neutron stars.

The enormously high Fermi temperature implies that the energy of the electron gas is correspondingly high. The energy at zero temperature is

$$U(T = 0) = 2 \int_0^{\mu_0} \varepsilon D(\varepsilon)\, d\varepsilon = \frac{3}{5}\tilde{N}\mu_0 \tag{18.34}$$

Thus the energy per particle is $\frac{3}{5}\mu_0$, or approximately 10^4 K in equivalent temperature units.

As the temperature rises, the Fermi level decreases (being "repelled" by the higher density of states at high energy, as we observed in the "fermion pre-gas model" of Section 18.1). Furthermore some electrons are "promoted" from orbitals below μ to orbitals above μ, increasing the energy of the system. To explore these effects quantitatively it is convenient to invoke a general result for integrals of the form $\int \phi(\varepsilon) f(\varepsilon, T) \, d\varepsilon$, where $\phi(\varepsilon)$ is an arbitrary function and $f(\varepsilon, T)$ is the Fermi occupation probability. This integral can be expanded in a power series in the temperature by invoking the step-function shape of $f(\varepsilon, T)$ at low temperatures (Problem 18.4-2), giving

$$\int_0^\infty \phi(\varepsilon) f(\varepsilon, T) \, d\varepsilon = \int_0^\mu \phi(\varepsilon) \, d\varepsilon + \frac{\pi^2}{6} (k_B T)^2 \phi'(\mu)$$

$$+ \frac{7\pi^4}{360} (k_B T)^4 \phi'''(\mu) + \cdots \qquad (18.35)$$

where ϕ' and ϕ''' are the first and third derivatives of ϕ with respect to ε, evaluated at $\varepsilon = \mu$. It should be noted that μ is the temperature dependent Fermi level (not the zero-temperature Fermi energy μ_0).

We first find the dependence of the Fermi energy on the temperature. The Fermi energy is determined by equation 18.21

$$\tilde{N} = 2 \int_0^\infty f(\varepsilon, T) D(\varepsilon) \, d\varepsilon = \frac{V}{2\pi^2} \left(\frac{2m}{\hbar^2} \right)^{3/2} \int_0^\infty \varepsilon^{1/2} f(\varepsilon, T) \, d\varepsilon$$

$$(18.36)$$

Then taking $\phi(\varepsilon) = \varepsilon^{1/2}$ in equation 18.35

$$\tilde{N} = \frac{V}{3\pi^2} \left(\frac{2m}{\hbar^2} \right)^{3/2} \mu^{3/2} \left[1 + \frac{\pi^2}{8} \left(\frac{k_B T}{\mu} \right)^2 + \cdots \right] \qquad (18.37)$$

At zero temperature we recover equation (18.32) for μ_0. To carry the solution to second order in T it is sufficient to replace μ by μ_0 in the second-order term, whence

$$\mu(T) = \mu_0 \left[1 - \frac{\pi^2}{12} \left(\frac{k_B T}{\mu_0} \right)^2 + \cdots \right] \qquad (18.38)$$

This result corroborates our expectation that the Fermi level decreases with increasing temperature. But for a typical value of μ_0/k_B (on the order of 10^4 K) the Fermi level at room temperature is decreased by only around 0.1% from its zero-temperature value!

The energy is given in an identical fashion, merely replacing $\varepsilon^{1/2}$ by $\varepsilon^{3/2}$, giving

$$U = \frac{V}{5\pi^2} \left(\frac{2m}{\hbar^2} \right)^{3/2} \mu^{5/2} \left[1 + \frac{5}{8} \pi^2 \left(\frac{k_B T}{\mu} \right)^2 + \cdots \right] \qquad (18.39)$$

Comparison with equation 18.32 corroborates that at $T = 0$ we recover the relationship $U = \frac{3}{5}\tilde{N}\mu_0$ (equation 18.34). This suggests dividing equation 18.39 by equation 18.37, giving

$$U = \frac{3}{5}\tilde{N}\mu\left[1 + \frac{1}{2}\pi^2\left(\frac{k_BT}{\mu}\right)^2 + \cdots\right] \tag{18.40}$$

Replacing $\mu(T)$ by equation 18.38 we finally find

$$U = \frac{3}{5}\tilde{N}\mu_0\left[1 + \frac{5}{12}\pi^2\left(\frac{k_BT}{\mu_0}\right)^2 + \cdots\right] \tag{18.41}$$

and the heat capacity is

$$C = \frac{3}{2}\tilde{N}k_B\left(\frac{\pi^2}{3}\frac{k_BT}{\mu_0}\right) + O(T^3) \tag{18.42}$$

The prefactor $\frac{3}{2}\tilde{N}k_B$ is the classical result, and the factor in parentheses is the "quantum correction factor" due to the quantum properties of the fermions. The quantum correction factor is of the order of $\frac{1}{10}$ at room temperature (for $\mu_0/k_B \simeq 10^4$ K). This drastic reduction of the heat capacity from its classically expected value is in excellent agreement with experiment for essentially all metals.

In order to compare the observed heat capacity of metals with theory it must be recalled (Section 16.6) that the lattice vibrations also contribute a term proportional to T^3, in addition to the linear and cubic terms contributed by the electrons

$$C = AT + BT^3 + \cdots \tag{18.43}$$

The coefficient A is equal to the coefficient in equation 18.42 whereas B arises both from the cubic terms in equation 18.42 and (predominately) from the coefficient in the Debye theory. It is conventional to plot experimental data in the form C/T versus T^2, so that the coefficient A is obtained as the $T = 0$ intercept and the coefficient B is the slope of the straight line. In fact such plots of experimental data do give excellent straight lines, with values of A and B in excellent agreement with equation 18.42 and the Debye theory (16.51).

The heat capacity (18.42) can be understood semiquantitatively and intuitively. As the temperature rises from $T = 0$, electrons are "promoted" from energies just below μ_0 to energies just above μ_0. This population transfer occurs primarily within a range of energies of the order of $2k_BT$ (recall Fig. 18.1 and Problem 18.1-7). The number of electrons so promoted is then of the order of $D(\mu_0)2k_BT$, and each increases its energy by roughly k_BT. Thus the increase in energy is of the order of

$$U - U_0 \simeq 2D(\mu_0)(k_BT)^2 \tag{18.44}$$

But $D(\mu_0) = 3\tilde{N}/2\mu_0$, so that

$$U - U_0 \simeq \frac{3\tilde{N}(k_BT)^2}{\mu_0} \tag{18.45}$$

and

$$C \simeq \frac{3}{2}\tilde{N}k_B\left(2\frac{k_BT}{\mu_0}\right) \tag{18.46}$$

This rough estimate is quite close to the quantitative result calculated in equation 18.42, which merely substitutes $\pi^2/3$ for the factor 2 in the parentheses of equation 18.46.

PROBLEMS

18.4-1. Show that equation 18.32 can be interpreted as $\mu_0 = \hbar^2 k_F^2/2m$ where k_F is the radius of the sphere in **k**-space such that one octant contains $2\tilde{N}$ particles (recall Section 16.6). Why $2\tilde{N}$ rather than $\tilde{N}$ particles?

18.4-2. Derive equation 18.35 by the following sequence of operations:
a) Denoting the integral in equation 18.35 by I, first integrate by parts and let $\Phi \equiv \int_0^\varepsilon \phi(\varepsilon')\,d\varepsilon'$. Then expanding $\Phi(\varepsilon)$ in a power series in $(\varepsilon - \mu)$ to third order, show that

$$I = -\sum_{m=0}^{\infty} \frac{1}{m!} \frac{d^m\Phi(\mu)}{d\mu^m} I_m$$

with

$$I_m = \int_0^\infty (\varepsilon - \mu)^m \frac{df}{d\varepsilon}\,d\varepsilon = -\beta^{-m}\int_{-\beta\mu}^\infty \frac{e^x}{(e^x+1)^2}x^m\,dx$$

b) Show that only an exponentially small error is made by taking the lower limit of integration as $-\infty$, and that then all terms with m odd vanish.
c) Evaluate the first two nonvanishing terms and show that these agree with equation 18.35.

18-5 THE IDEAL BOSE FLUID

The formalism for the ideal Bose fluid bears a strikingly close similarity to that for the ideal Fermi fluid. As was anticipated in Table 18.1, and as we shall validate here, the formalisms differ only in several changes in sign. But the consequences are dramatically different. Whereas fermions at low temperatures tend to "saturate" orbital states up to some specific Fermi energy, bosons all tend to "condense" into the single lowest orbital state. This condensation happens precipitously, at (and below) a sharply defined "condensation temperature." The resultant phase transition leads to superfluidity in ^{4}He (a phenomenon *not* seen in ^{3}He, which is a fermion fluid) and it leads to superconductivity in lead and in various other metals.

We consider an ideal Bose fluid, composed of particles of integral spin. The number of spin orientations is then $g_0 = 2S + 1$, where S is the magnitude of the spin.

The possible orbital states of the bosons in the fluid are labeled by $\mathbf{k}$ and m_s, precisely as in the fermion case, and again the grand canonical partition sum factors with respect to the orbital states (as in line a of Table 18.1).

The partition sum of a single orbital state is independent of m_s, and is, for *each* value of m_s

$$z_{\mathbf{k}} = z_{\mathbf{k}, m_s} = 1 + e^{-\beta(\varepsilon_k - \mu)} + e^{-\beta(2\varepsilon_k - 2\mu)} + e^{-\beta(3\varepsilon_k - 3\mu)} + \cdots$$

$$= \frac{1}{1 - e^{-\beta(\varepsilon_k - \mu)}} \tag{18.47}$$

This validates line (b) of Table 18.1.

The average number of bosons in the orbital state $\mathbf{k}, m_s$ is

$$\bar{n}_{\mathbf{k}, m_s} = \left[e^{-\beta(\varepsilon_k - \mu)} + 2e^{-\beta(2\varepsilon_k - 2\mu)} + 3e^{-\beta(3\varepsilon_k - 3\mu)} + \cdots \right] / z_{\mathbf{k}, m_s}$$

$$= k_B T \frac{\partial}{\partial \mu} \ln z_{\mathbf{k}, m_s} \tag{18.48}$$

which is just the analogue of the relation $\beta N = \partial/\partial \mu \ln \mathcal{Z}$, but is now applied to a single orbital state. Carrying out the differentiation we find

$$\bar{n}_{\mathbf{k}, m_s} \equiv f_{\mathbf{k}, m_s} = \frac{1}{e^{\beta(\varepsilon_k - \mu)} - 1} \tag{18.49}$$

and this is the result listed in line c of Table 18.1. It is important to note that, in contrast to the fermion case, $f_{\mathbf{k}, m_s}$ is not necessarily less than (or equal to) unity. The quantity $f_{\mathbf{k}, m_s}$ is frequently referred to as an "occupation probability," but it is more properly identified as a "mean occupation number" $\bar{n}_{\mathbf{k}, m_s}$.

A moment's reflection on the form of $\bar{n}_{\mathbf{k}, m_s}$ reveals that for a gas of material Bose particles the molar Gibbs function must be negative. For if μ were positive the orbital state with $\varepsilon_{\mathbf{k}}$ equal to μ would have an infinite occupation number! We thus conclude that for a gas with a bounded number of particles (and with a choice of energy scale in which the lowest energy orbital has zero energy) *the molar Gibbs potential μ is always negative*.

The form of $\bar{n}$ as a function of $\beta(\varepsilon - \mu)$ is shown in Fig. 18.2. The occupation number falls from an infinite value at $\varepsilon = \mu$ to unity at $\varepsilon = \mu + 0.693 k_B T$. In the insert of Fig. 18.2, the orbital occupation number is shown schematically as a function of ε for two different temperatures $(T_2 > T_1)$ and for two choices of μ.

If the system of interest is in contact with a particle reservoir, so that μ is constant, then the curve of $\bar{n}(\varepsilon, T_2)$ in the insert should be shifted to the right. The number of particles in such a system increases with temperature. If the system of interest is maintained at constant particle number, the integral of $\bar{n}(\varepsilon, T) D(\varepsilon)$ is conserved. As is evident from the figure, the molar Gibbs potential μ then must decrease with increasing temperature (just at it does in the Fermi gas).

The grand canonical potential Ψ is the logarithm of $\mathcal{Z}$ which, in turn, is the product of the $z_{\mathbf{k}, m_s}$ given in equation 18.47. Thus, as in Table 18.1 (lines d to g),

$$\beta \Psi = g_0 \int_0^\infty \ln \left[1 - e^{-\beta(\varepsilon - \mu)} \right] D(\varepsilon) \, d\varepsilon \tag{18.50}$$

or, integrating by parts

$$\Psi = -\frac{2}{3} \frac{g_0 V}{(2\pi)^2} \left(\frac{2m}{\hbar^2} \right)^{3/2} \int_0^\infty \frac{\varepsilon^{3/2}}{e^{\beta(\varepsilon - \mu)} - 1} \, d\varepsilon \tag{18.51}$$

and again the mechanical equation of state is $P = 2U/3V$ (lines i and j of Table 18.1).

For a system of particles maintained at constant μ by a particle reservoir the thermodynamics follows in a straightforward fashion. But for a system at constant $\tilde{N}$ the apparently innocuous formalism conceals some startling and dramatic consequences, with no analogues in either fermion or classical systems. As a preliminary to such considerations it is useful to turn our attention to systems in which the particle number is physically nonconserved.

18-6 NONCONSERVED IDEAL BOSON FLUIDS: ELECTROMAGNETIC RADIATION REVISITED

As we observed in Section 18.1, bosons are the quantum analogues of the "waves" of classical physics. A residue of this classical significance is that, unlike fermions, bosons need not be conserved. In some cases, as in a fluid of ^{4}He atoms, the boson particles are conserved; in other cases, as in a "photon gas" (recall Section 3.6), the bosons are not conserved. There exist processes, for instance, in which two photons interact through a nonlinear coupling to produce three photons. How then are we to adapt the formalism of the ideal Bose fluid to this possibility of nonconservation?

We recall the reasoning in Sections 17.2 and 17.3, leading to the grand canonical formalism. We there maximized the disorder subject to auxiliary constraints on the energy (equation 17.30) and on the number of particles (equation 17.31). These constraints introduced Lagrange parameters λ_2 and λ_3 (equation 17.33), which were then physically identified as $\lambda_2 = \beta$ and as $\lambda_3 = \beta\mu$. *Treatment of nonconserved particles simply requires that we omit the constraint equation on particle number.* Omission of the parameter λ_3 is equivalent to taking $\lambda_3 = 0$, or to taking $\mu = 0$. We thus arrive at the conclusion that *the molar Gibbs potential of a nonconserved Bose gas is zero.*

For $\mu = 0$ the grand canonical formalism becomes identical to the canonical formalism. Hence the grand canonical analysis of the photon gas simply reiterates the canonical treatment of electromagnetic radiation as developed in Section 16.7. The reader should trace this parallelism through in step by step detail. referring to Table 18.1 and Section 16.7 (see also Problem 18.6-2).

It is instructive to reflect on the different viewpoints taken in Section 16.7 and in this section. In the previous analysis our focus was on the *normal modes of the electromagnetic field*, and this led us to the canonical formalism. In this section our focus shifted to the *quanta of the field*, or the *photons*, for which the grand canonical formalism is the more natural. But the nonconservation of the particles requires μ to vanish and thereby achieves exact equivalence between the two formalisms. Only the language changes!

The number of photons of energy ε is $(e^{\beta\varepsilon} - 1)^{-1}$, where the permitted energies are given by

$$\varepsilon = \hbar\omega = \hbar c \frac{2\pi}{\lambda} = \frac{hc}{\lambda} \tag{18.52}$$

Here c is the velocity of light and λ is the quantum mechanical wavelength of the photon (or the wavelength of the normal mode, in the mode language of Section 16.7). The population of bosons of infinitely long wavelength is unbounded[3]. The energy of these long wavelength photons vanishes, so that no divergence of the energy is associated with the formal divergence of the boson number.

To recapitulate, electromagnetic radiation can be conceptualized either in terms of the normal modes or in terms of the quanta of excitation of these modes. The former view leads to a canonical formalism. The latter leads to the concept of a nonconserved Bose gas, to the conclusion that the molar Gibbs potential of the gas is zero, and to an unbounded population of (unobservable) zero energy bosons in the lowest orbital state.

All of this might appear to be highly contrived and formally baroque were it not to have a direct analogy in *conserved* boson systems, giving rise to such startling physical effects as superfluidity in ^{4}He and superconductivity in metals, to which we now turn.

PROBLEMS

18.6-1. Calculate the number of photons in the lowest orbital state in a cubic vessel of volume 1 m^3 at a temperature of 300 K. What is the total energy of these photons? What is the number of photons in a single orbital state with a wavelength of 5000 Å, and what is the total energy of these photons?

18.6-2.
(*a*) In applying the grand canonical formalism to the photon gas can we use the density of orbital states function $D(\varepsilon)$ as in equation (f) of Table 18.1? Explain.
(*b*) Denoting the velocity of light by c, show that writing $c = $ (wavelength/period) implies $\omega = ck$. From this relation and from Section 16.5 find the density of orbital states $D(\varepsilon)$.
(*c*) Show that the grand canonical analysis of the photon gas corresponds precisely with the theory given in Section 16.8.

18-7 BOSE CONDENSATION

Having the interlude of Section 18.6 to provide perspective, we focus on a system of conserved particles enclosed in impermeable walls. Then, as we saw in Fig. 18.2 and the related discussion, the molar Gibbs potential μ must increase as the temperature decreases (just as in the fermion case).

Assuming the bosons to be material particles of which the kinetic energy is $\varepsilon = p^2/2m$, the density of orbital states is proportional to $\varepsilon^{1/2}$

---

[3] Of course such infinite-wavelength photons can be accommodated only in a infinitely large container, but the number of photons can be increased beyond any preassigned bound in a finite container of sufficiently large size.

(equation f of Table 18.1) and the number of particles is

$$\tilde{N}_e = \frac{g_0 V}{(2\pi)^2} \left(\frac{2m}{\hbar^2} \right)^{3/2} \int_0^\infty \frac{\varepsilon^{1/2}}{\xi^{-1} e^{\beta \varepsilon} - 1} \, d\varepsilon \qquad (18.53)$$

where ξ is the fugacity

$$\xi \equiv e^{\beta \mu} \qquad (18.54)$$

and where the subscript e is affixed to $\tilde{N}_e$ for reasons that will become understandable only later; for the moment $\tilde{N}_e$ is simply another notation for $\tilde{N}$. The molar Gibbs potential is always negative (for conserved particles) so that the fugacity lies between zero and unity.

$$0 < \xi < 1 \qquad (18.55)$$

This observation encourages us to expand the integral in equation 18.53 in powers of the fugacity, giving

$$\tilde{N}_e = \left[\frac{g_0 V}{(2\pi)^2} \left(\frac{2m}{\hbar^2} \right)^{3/2} \right] \frac{\sqrt{\pi}}{2} (k_B T)^{3/2} F_{3/2}(\xi) = \frac{g_0 V}{\lambda_T^3} F_{3/2}(\xi) \quad (18.56)$$

where λ_T is the "thermal wavelength" (equation 18.26) and

$$F_{3/2}(\xi) = \sum_{r=1}^\infty \frac{\xi^r}{r^{3/2}} = \xi + \frac{\xi^2}{2\sqrt{2}} + \frac{\xi^3}{3\sqrt{3}} + \cdots \qquad (18.57)$$

At high temperature the fugacity is small and $F_{3/2}(\xi)$ can be replaced by ξ (its leading term), in which case equation 18.56 reduces to its classical form 18.25.
    Similarly

$$U = \left[\frac{g_0 V}{(2\pi)^2} \left(\frac{2m}{\hbar^2} \right)^{3/2} \right] \frac{3\sqrt{\pi}}{4} (k_B T)^{5/2} F_{5/2}(\xi) = \frac{3}{2} k_B T \frac{g_0 V}{\lambda_T^3} F_{5/2}(\xi)$$

$$(18.58)$$

where

$$F_{5/2}(\xi) \equiv \sum_{r=1}^\infty \frac{\xi^r}{r^{5/2}} = \xi + \frac{\xi^2}{4\sqrt{2}} + \frac{\xi^3}{9\sqrt{3}} + \cdots \qquad (18.59)$$

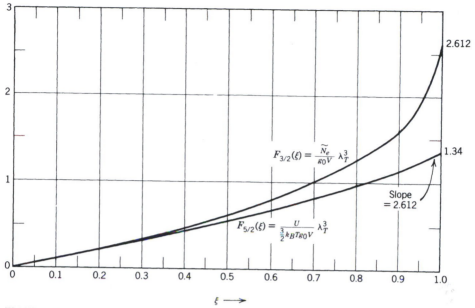

FIGURE 18.3

The functions $F_{3/2}(\xi)$ and $F_{5/2}(\xi)$ that characterize the particle number and the energy (equations 18.57–18.60) of a gas of conserved bosons.

Again the equation for U reduces to its classical form 18.27 if $F_{5/2}(\xi)$ is replaced by ξ, the leading term in the series.

Dividing 18.58 by 18.56

$$U = \frac{3}{2}\tilde{N}_e k_B T \frac{F_{5/2}(\xi)}{F_{3/2}(\xi)} \qquad (18.60)$$

so that the ratio $F_{5/2}(\xi)/F_{3/2}(\xi)$ measures the deviation from the classical equation of state.

For both $F_{3/2}(\xi)$ and $F_{5/2}(\xi)$ all the coefficients in their defining series are positive, so that both functions are monotonically increasing functions of ξ, as shown in Fig. 18.3. Each function has a slope of unity at $\xi = 0$. At $\xi = 1$ the functions $F_{3/2}$ and $F_{5/2}$ have the value 2.612 and 1.34, respectively.

The two functions satisfy the relation

$$\frac{d}{d\xi}F_{5/2}(\xi) = \frac{1}{\xi}F_{3/2}(\xi) \qquad (18.61)$$

from which it follows that the slope of $F_{5/2}(\xi)$ at $\xi = 1$ is equal to $F_{3/2}(1)$, or 2.612. The slope of $F_{3/2}(\xi)$ at $\xi = 1$ is infinite (Problem 18.7-2).

The formal procedure in analyzing a given gas is now explicit. Let us suppose that $\tilde{N}_e$, V, and T are known. Then $F_{3/2}(\xi) = \tilde{N}_e\lambda_T^3/g_0V$ is known, and the fugacity ξ can be determined directly from Fig. 18.3. Given the fugacity all thermodynamic functions are determined in the grand canonical formalism. The energy, for example, can be evaluated by Fig. 18.3 and equations 18.58 or 18.60.

All of the previous discussion seems to be reasonable and straightforward until one suddenly recognizes that given values of $\tilde{N}_e$, V, and T may result in the quantity $\tilde{N}_e\lambda_T^3/g_0V$ being greater than 2.612. Then Fig. 18.3 permits no solution for the fugacity ξ! The analysis fails in this "extreme quantum limit"!

A moment's reflection reveals the source of the problem. As $\tilde{N}_e\lambda_T^3/g_0V$ ($= F_{3/2}(\xi)$) approaches 2.612 the fugacity approaches unity, or the molar Gibbs potential μ approaches zero. But we have noted earlier that at $\mu = 0$ the occupation number $\bar{n}$ of the orbital state of zero energy diverges. This pathological behavior of the ground-state orbital was lost in the transition from a sum over orbital states to an integral (weighted by the density of orbital states that vanishes at $\mu = 0$). This formalism is acceptable for $g_0V/\tilde{N}_e\lambda_T^3 < 2.612$, but if this quantity is greater than 2.612 we must treat the replacement of a sum over states by an integral with greater care and delicacy.

We postpone briefly the corrections to the analysis that are required if $g_0V/\tilde{N}_e\lambda_T^3 \geq 2.612$, to first evaluate the temperature at which the failure of the "integral analysis" (as opposed to the "summation analysis") occurs. Setting $g_0V/\tilde{N}_e\lambda_T^3 = 2.612$ we find

$$k_BT_c = \frac{2\pi\hbar^2}{m}\left(\frac{1}{2.612}\frac{\tilde{N}}{g_0V}\right)^{2/3} \tag{18.62}$$

where T_c is called the Bose condensation temperature. *For temperature greater than* T_c *the "integral analysis" is valid. At and below* T_c *a "Bose condensation" occurs, associated with an anomalous population of the orbital ground state.*

If the atomic mass m and the observed number density $\tilde{N}_e/g_0V$ of liquified ⁴He are inserted in equation 18.62 one finds a condensation temperature reasonably close ($\simeq 3$ K) to the temperature (2.17 K) at which superfluidity and other nonclassical effects occur. This agreement is reasonable in light of the gross approximation involved in treating ⁴He liquid as an ideal noninteracting gas.

To explore the population of the orbital ground state, and of other low-lying excited orbital states, we recall that the total number of particles is

$$\tilde{N}_e = \sum_{k,m_s} \bar{n}(\varepsilon_k) = g_0\sum_k \left[e^{\beta(\varepsilon_k - \mu)} - 1\right]^{-1} \tag{18.63}$$

and the allowed values of ε_k are

$$\varepsilon_{n_x, n_y, n_z} = \frac{p^2}{2m} = \frac{h^2}{2m}\left(\frac{1}{\lambda_x^2} + \frac{1}{\lambda_y^2} + \frac{1}{\lambda_z^2}\right) = \frac{h^2}{8mV^{2/3}}\left(n_x^2 + n_y^2 + n_z^2\right)$$

$$(18.64)$$

where we have again invoked the quantum mechanical relationship between momentum and wavelength ($p = h/\lambda$), assumed a cubic "box" of length $V^{1/3}$, and required that an integral number of half wavelengths "fit" along each axis ($\frac{1}{2}n_x\lambda_x = V^{1/3}$, etc.). The energies of the discrete quantum mechanical states are precisely those from which we inferred the density of orbital states function in Section 16.5. The ground state energy is that in which $n_x = n_y = n_z = 1$ (and we normally choose the energy scale relative to this state). The first excited state has two of the n's equal to unity and one equal to two—this state is three-fold degenerate. The difference in energy is $\varepsilon_{211} - \varepsilon_{111} = 6h^2/mV^{2/3}$. For a container of volume 1 liter ($V = 10^{-3}\text{m}^3$), and with m taken as the atomic mass of ^{4}He ($\approx 6.6 \times 10^{-27}$ Kg), the energy of the first excited state (relative to the ground state energy is)

$$\varepsilon_{211} - \varepsilon_{111} = 6h^2/mV^{2/3} \approx 2.5 \times 10^{-37} \text{ J}$$

or

$$\left(\varepsilon_{211} - \varepsilon_{111}\right)/k_B \approx 2 \times 10^{-14} \text{ K} \qquad (18.65)$$

Thus the discrete states are indeed *very* closely spaced in energy—far closer than k_BT at any reasonable temperature. We might well have felt confident in replacing the sum by an integral!

But let us examine more closely the population of each state as the chemical potential approaches ε_{111} from below. In particular we inquire as to the value of μ for which the population of the orbital ground state alone is comparable to the entire number of particles in the gas. Let n_0 be the number of particles in the ground state orbital, so that $[\exp \beta(\varepsilon_{111} - \mu) - 1]^{-1} = n_0$. Then if $n_0 \gg 1$ it follows that $\beta(\varepsilon_{111} - \mu) \ll 1$ and we can expand the exponential to first order, so that $n_0 \sim k_BT/(\varepsilon_{111} - \mu)$. Thus the population of the orbital ground state becomes comparable to the entire number of particles in the system (say $n_0 \simeq 10^{22}$) if $\beta(\varepsilon_{111} - \mu) \sim 10^{-22}$.

What, then, is the population of the first excited orbital state? The energy difference $(\varepsilon_{111} - \mu)/k_B$ is $\simeq 10^{-21}$ K (for $T \simeq 10$ K) whereas $(\varepsilon_{211} - \varepsilon_{111})/k_B \simeq 10^{-14}$ K (equation 18.65). It follows that $n_{211}/n_0 \simeq 10^{-7}$. The population of higher states continues to fall extremely rapidly.

As the temperature decreases in a Bose gas the molar Gibbs potential increases and approaches the energy of the ground state orbital. The population of the ground state orbital increases, becoming a nonnegligible fraction of the total number of bosons in the gas at the critical temperature T_c. The occupation number of any individual other state is relatively negligible.

As the temperature decreases further μ cannot approach closer to the ground state energy than $\beta(\mu - \varepsilon_{111}) = 1/\tilde{N} \approx 10^{-23}$ (at which value the ground state alone would host all N particles in the gas!). Hence the ground state shields all other states from too close an approach of μ, and each other state individually can host only a relatively small number of particles. Together, of course, the remaining states host all the particles not in the ground state.

With this understanding of the mechanism of the Bose condensation it is a simple matter to correct the analysis. All orbital states other than the ground state are adequately represented by the integral over the density of orbital states function. The ground state energy must be separately and explicitly listed in the sum over states.

The number of particles is, then

$$\tilde{N} = n_0 + \tilde{N}_e \tag{18.66}$$

where n_0 is the number of particles in the ground state orbital

$$n_0 = \left(e^{-\beta\mu} + 1 \right)^{-1} = \frac{\xi}{1 - \xi} \tag{18.67}$$

and where $\tilde{N}_e$ is the number of particles in "excited states" (i.e., in all orbital states other than the ground orbital state). The number of "excited particles" $\tilde{N}_e$ is as given in equation 18.54.

The expression 18.59 for the energy remains correct, since the population of the zero energy orbitals makes no contribution to the energy. Thus *the entire correction to the theory consists of the reinterpretation of $\tilde{N}_e$ as the number of excited particles, and the adjuncture of the two additional equations* 18.67 *and* 18.68.

Equivalently, we can simply add the ground state term to our previous expression for the grand canonical potential (equation 18.51), giving the fundamental relation

$$\Psi = g_0 k_B T \ln\left(1 - \xi\right) - g_0 k_B T \frac{V}{\lambda_T^3} F_{5/2}(\xi) \tag{18.68}$$

where, of course, ξ is the fugacity $e^{\beta\mu}$.

With equations 18.56 to 18.60 and 18.66 to 18.67, we can explore a variety of observable properties of Bose fluids. These properties are summarized in Table 18.2 and illustrated schematically in Fig. 18.4.

TABLE 18.2
Properties of the Ideal Bose Fluid

Fundamental equation

$$\Psi = k_B T \ln(1 - \xi) - k_B T \left(V/\lambda_T^3 \right) F_{5/2}(\xi)$$

Condensation temperature

$$k_B T_c = \frac{2\pi\hbar^2}{m} \left(\frac{1}{2.612} \frac{\tilde{N}}{g_0 V} \right)^{3/2}$$

Condensed and excited bosons

$$\tilde{N} = n_0 + \tilde{N}_e, \quad n_0 = \frac{\xi}{1 - \xi}, \quad \tilde{N}_e = \frac{V}{\lambda^3} F_{3/2}(\xi)$$

$$T > T_c: \quad n_0 \lll \tilde{N}, \quad \tilde{N}_e \simeq \tilde{N} = \frac{V}{\lambda^3} F_{3/2}(\xi)$$

$$T < T_c: \quad n_0/\tilde{N} = 1 - \tilde{N}_e/\tilde{N} = 1 - \left(\frac{T}{T_c} \right)^{3/2}$$

Energy

$$T > T_c: \quad U = \frac{3}{2} \tilde{N} k_B T \frac{F_{5/2}(\xi)}{F_{3/2}(\xi)}$$

$$T < T_c: \quad U = \frac{3}{2} \tilde{N} k_B T \frac{F_{5/2}(1)}{F_{3/2}(1)} \left(\frac{T}{T_c} \right)^{3/2} = 0.76 \tilde{N} k_B T_c \left(\frac{T}{T_c} \right)^{5/2}$$

Heat capacity c_v (per particle)

$$T > T_c: \quad c_v = \frac{3}{2} k_B \left[\frac{5}{2} \frac{F_{5/2}(\xi)}{F_{3/2}(\xi)} - \frac{3}{2} \frac{F'_{5/2}(\xi)}{F'_{3/2}(\xi)} \right]$$

$$T < T_c: \quad c_v = 1.9 k_B \left(\frac{T}{T_c} \right)^{3/2}$$

Entropy

$$T > T_c: \quad S = \frac{5}{2} k_B \frac{V}{\lambda_T^3} F_{5/2}(\xi) - \tilde{N} k_B \ln \xi$$

$$T < T_c: \quad S = \frac{5}{2} k_B \frac{V}{\lambda_T^3} F_{5/2}(1) = 3.35 k_B \frac{V}{\lambda_T^3}$$

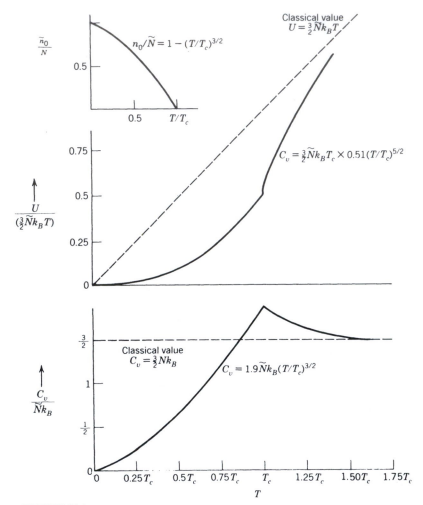

FIGURE 18.4
Properties of an ideal Bose fluid. The energy and heat capacity for $T > T_c$ are schematic.

First, consider the temperature dependence of the number of bosons in the orbital ground state. For $T < T_c$ the maximum number of bosons that can be accommodated in excited states is

$$\tilde{N}_e = \frac{g_0 V}{\lambda_T^3} F_{3/2}(1), \qquad T < T_c \qquad (18.69)$$

and in particular, as $T \to T_c$, $\tilde{N}_e \to \tilde{N}$, so that

$$\tilde{N} = \frac{g_0 V}{\lambda_c^3} F_{3/2}(1) \qquad (18.70)$$

where λ_c is the value of λ_T at $T = T_c$. Dividing

$$\frac{\tilde{N}_e}{\tilde{N}} = \left(\frac{\lambda_c}{\lambda_T}\right)^3 = \left(\frac{T}{T_c}\right)^{3/2} \tag{18.71}$$

The number of particles in the ground state is then

$$\frac{n_0}{\tilde{N}} = 1 - \frac{\tilde{N}_e}{\tilde{N}} = 1 - \left(\frac{T}{T_c}\right)^{3/2} \tag{18.72}$$

This dependence is sketched in Fig. 18.4.

The energy of the system is also of great interest as its derivative is the heat capacity, an easily observable quantity. For $T > T_c$ the energy is given by equation 18.60. For $T < T_c$ equation 18.58 can be written in the form

$$U = \frac{3}{2}k_BT\frac{g_0V}{\lambda_T^3}F_{5/2}(1) = \frac{3}{2}k_BT\frac{\tilde{N}_e}{F_{3/2}(1)}F_{5/2}(1)$$

$$= \frac{3}{2}\tilde{N}k_BT\frac{F_{5/2}(1)}{F_{3/2}(1)}\frac{\tilde{N}_e}{\tilde{N}} = \frac{3}{2}\tilde{N}k_BT(0.51)\left(\frac{T}{T_c}\right)^{3/2}$$

$$= 0.76\tilde{N}k_BT_c\left(\frac{T}{T_c}\right)^{5/2}, \qquad T < T_c \tag{18.73}$$

For $T > T_c$ the energy is given by equation 18.60, or $U = \frac{3}{2}\tilde{N}k_BT[F_{5/2}(\xi)/F^{3/2}(\xi)]$, so that the energy is always less than its classical value. The fugacity is determined as a function of T by Fig. 18.2.

Calculation of the molar heat capacity for $T < T_c$ follows directly by differentiation of equation 18.73

$$c_v = 1.9\tilde{N}k_B\left(\frac{T}{T_c}\right)^{3/2}, \qquad T < T_c \tag{18.74}$$

It is of particular interest that $c_v = 1.9Nk_B$ at $T = T_c$, a value well above the classical value $1.5Nk_B$ which is approached in the classical regime at high temperature.

Calculation of the heat capacity at $T > T_c$ requires differentiation of equation 18.60 at constant $\tilde{N}$, and elimination of $(d\xi/dT)_{\tilde{N}}$ by equation 18.56. The results are indicated schematically in Fig. 18.4 and given in Table 18.2.

The unique cusp in the heat capacity at $T = T_c$ is a signature of the Bose condensation. A strikingly similar discontinuity is observed in ^{4}He fluids; its detailed shape appears to be in agreement with the renormalization group predictions for the universality class of a two-dimensional order parameter (recall the penultimate paragraph of Chapter 12).

Finally we note that the Bose condensation in ^{4}He is accompanied by striking physical properties of the fluid. Below T_c the fluid flows freely through the finest capillary tubes. It runs up and over the side of breakers. It is, as its name denotes, "superfluid." The explanation of these properties lies outside the scope of statistical mechanics. It is sufficient to say that it is the "condensed phase," or the *ground state component*, that alone flows so freely through narrow tubes. This component cannot easily dissipate energy through friction, as it is already in the ground state. More significantly, the condensed phase has a *quantum coherence* with no classical analogue; the bosons that share a single state are correlated in a fashion totally different from the excited particles (which are randomly distributed over enormously many states).

A similar Bose condensation occurs in the electron fluid in certain metals. By an interaction involving phonons, pairs of electrons bind together in correlated motion. These electron pairs then act as bosons. The Bose condensation of the pairs leads to superconductivity, the analogue of the superfluidity of ^{4}He.

PROBLEMS

18.7-1. Show that equations 18.56 and 18.58, for $\tilde{N}_e$ and U, respectively, approach their proper classical limits in the classical regime.

18.7-2. Show that $F_{3/2}(1)$, $F_{5/2}(1)$, and $F'_{5/2}(1)$ are all finite, whereas $F'_{3/2}(1)$ is infinite. Here $F'_{3/2}(1)$ denotes the derivative of $F_{3/2}(x)$, evaluated at $x = 1$.
Hint: Use the integral test of convergence of infinite series, whereby $\sum_{n=1}^{\infty} f(n)$ converges or diverges with $\int_1^{\infty} f(x)\,dx$ (if $0 < f_{n+1} < f_n$ for all n).

18.7-3. Show that the explicit inclusion of the orbital ground state contributes $g_0 k_B T \ln(1 - \xi)$ to the grand canonical partition sum, thereby validating equation 18.68.

19

FLUCTUATIONS

19-1 THE PROBABILITY DISTRIBUTION OF FLUCTUATIONS

A thermodynamic system undergoes continual random transitions among its microstates. If the system is composed of a subsystem in diathermal contact with a thermal reservoir, the subsystem and the reservoir together undergo incessant and rapid transitions among their joint microstates. These transitions lead sometimes to states of high subsystem energy and sometimes to states of low subsystem energy, as the constant total energy is shared in different proportions between the subsystem and the reservoir. The subsystem energy thereby fluctuates around its equilibrium value. Similarly there are fluctuations of the volume of a system in contact with a pressure reservoir.

The "subsystem" may, in fact, be a small portion of a larger system, the remainder of the system then constituting the "reservoir." In that case the fluctuations are *local fluctuations* within a nominally homogeneous system.

Both the volume and the energy simultaneously fluctuate in a system that is in open contact with pressure and thermal reservoirs. If the microstates of small volume tend to have relatively large (or small) energy the fluctuations of volume and energy will be negatively (or positively) correlated.

Gross macroscopic observations of an open system generally reveal only the thermodynamic values of the extensive parameters. Only near the critical point do the fluctuations become so large that they become evident to simple macroscopic observations, as by the "critical opalescence" alluded to in Section 10.1. Farther from the critical point the fluctuations can be observed, with increasing difficulty, using increasingly sophisticated instruments of high temporal and spatial resolving power. Furthermore, as we shall see shortly, theory reveals interesting relationships between the fluctuations and thermodynamic quantities such as the heat capacities. These relationships are exploited by materials scientists to

423

provide a convenient method of calculation of the heat capacities and of similar properties.

The statistical mechanical form of the probability distribution for a fluctuating extensive parameter is now familiar. If the subsystem is in diathermal contact with a thermal reservoir the probability that the system occupies *a particular microstate* of energy $\hat{E}$ is $e^{\beta F - \beta \hat{E}}$. If the subsystem is in contact with both a pressure and a thermal reservoir the probability that the system occupies *a particular microstate* of energy $\hat{E}$ and volume $\hat{V}$ is $\exp[\beta G - \beta(\hat{E} + P\hat{V})]$. And, more generally, *for a system in contact with reservoirs corresponding to the extensive parameters* $X_0, X_1, \ldots, X_s$, *the probability that the system occupies a particular microstate with parameters* $\hat{X}_0, \hat{X}_1, \ldots, \hat{X}_s$ *is*

$$f_{\hat{X}_0, \hat{X}_1, \ldots, \hat{X}_s} = \exp\left\{ -k_B^{-1} S[F_0, \ldots, F_s] - k_B^{-1}(F_0 \hat{X}_0 + \cdots + F_s \hat{X}_s) \right\}$$

$$(19.1)$$

Here $S[F_0, \ldots, F_s]$ is the Massieu function (the Legendre transform of the entropy) and $F_0, \ldots, F_s$ are the entropic intensive parameters (with values equal to those of the reservoirs).

19-2 MOMENTS OF THE ENERGY FLUCTUATIONS

Let us suppose (temporarily) that the energy $\hat{E}$ is the only fluctuating variable, all other extensive parameters being constrained by restrictive walls. The deviations $(\hat{E} - U)$ of $\hat{E}$ from its average value $\langle \hat{E} \rangle \equiv U$ is itself a fluctuating variable, of average value zero. The *mean square deviation* $\langle (\hat{E} - U)^2 \rangle$, or the "second central moment," is a measure of the width of the energy fluctuations. A full description of the energy fluctuations requires knowledge of all the central moments $\langle (\hat{E} - U)^n \rangle$, with $n = 2, 3, 4, \ldots$

The second central moment of the fluctuations follows directly from the form of the canonical probability distribution, for

$$\langle (\hat{E} - U)^2 \rangle = \sum_j (E_j - U)^2 e^{\beta(F - E_j)} \qquad (19.2)$$

But we recall that

$$\frac{\partial}{\partial \beta}(\beta F) = U \qquad (19.3)$$

so that equation 19.2 can be written as

$$\langle (\hat{E} - U)^2 \rangle = - \sum_j (E_j - U) \frac{\partial}{\partial \beta} e^{\beta(F - E_j)} \tag{19.4}$$

$$= - \frac{\partial}{\partial \beta} \sum_j (E_j - U) e^{\beta(F - E_j)} - \frac{\partial U}{\partial \beta} \tag{19.5}$$

The first central moment vanishes, and the derivative $\partial U / \partial \beta$ is related to the heat capacity, whence

$$\langle (\hat{E} - U)^2 \rangle = - \frac{\partial U}{\partial \beta} = k_B T^2 N c_v \tag{19.6}$$

There are several attributes of this result that should be noted. Most important is the fact that the mean square energy fluctuations are proportional to the size of the system. Therefore, the *relative root mean square dispersion* $\langle (\hat{E} - U)^2 \rangle^{1/2} / U$, which measures the amplitude of the fluctuations relative to the mean energy[1], is proportional to $N^{-1/2}$. *For large systems ($N \rightarrow \infty$) the fluctuation amplitudes become negligible relative to the mean values, and thermodynamics becomes precise.*

For systems in which a large amount of heat is required to produce an appreciable change in temperature (c_v large) the fluctuations in energy are correspondingly large. Furthermore the energy fluctuations in all systems become very small at low temperatures (where $c_v \rightarrow 0$). Finally we recall that both the heat capacity and the fluctuation amplitudes diverge at the critical point, consistent with equation 19.6.

Calculation of higher moments of the energy fluctuations recapitulates equations 19.4 to 19.6.

$$\langle (\hat{E} - U)^{n+1} \rangle = - \sum_j (E_j - U)^n \frac{\partial}{\partial \beta} e^{\beta(F - E_j)}$$

$$= - \frac{\partial}{\partial \beta} \sum_j (E_j - U)^n e^{\beta(F - E_j)} + \sum_j e^{\beta(F - E_j)} \frac{\partial}{\partial \beta} (E_j - U)^n$$

$$= - \frac{\partial}{\partial \beta} \langle (\hat{E} - U)^n \rangle + n \sum_j e^{\beta(F - E_j)} (E_j - U)^{n-1} \frac{\partial U}{\partial \beta}$$

$$= - \frac{\partial}{\partial \beta} \langle (\hat{E} - U)^n \rangle + n \langle (E - U)^{n-1} \rangle \frac{\partial U}{\partial \beta} \tag{19.7}$$

[1] For definiteness, the energy U is here taken as zero in the $T = 0$ state of the system.

The higher-order moments of the energy fluctuations can be generated from the lower-order moments by the recursion relation 19.7. In particular the third moment is

$$\langle (\hat{E} - U)^3 \rangle = N k_B^2 T^2 \left(2 T c_v + T^2 \frac{\partial c_v}{\partial T} \right) \tag{19.8}$$

PROBLEMS

19.2-1. A molecule has a vibrational mode of natural frequency ω. The molecule is embedded in a macroscopic system of temperature T. Calculate the second central moment $\langle \hat{E}^2 \rangle - \langle E \rangle^2$ of the energy of the vibrational mode, as a function of ω, T, and fundamental constants.

19.2-2. Calculate the third central moment for the molecule in the preceding problem.

19.2-3. Calculate the mean square deviation of the energy contained within a fixed volume V' in a radiation field (recall Section 3.6). Assume the volume V' to be small compared to the volume V of the radiation, and assume the radiation to be in equilibrium at temperature T. Note that the product $(N c_v)$ in equation 19.6 is the total heat capacity of the sample considered.

19-3   GENERAL MOMENTS AND CORRELATION MOMENTS[2]

In the general case we are interested not only in the fluctuation moment of variables other than the energy, but also in combined moments that measure the *correlation* of two or more fluctuating variables. We consider first the general second moments of the form

$$\langle \Delta \hat{X}_j \Delta \hat{X}_k \rangle \equiv \sum_{\text{states}} (\hat{X}_j - X_j)(\hat{X}_k - X_k) f_{\hat{X}_0, \hat{X}_1, \ldots, \hat{X}_s} \tag{19.9}$$

where $f_{\hat{X}_0, \hat{X}_1, \ldots, \hat{X}_s}$ is given by equation 19.1. This second moment measures the correlation of the fluctuations of the two variables $\hat{X}_j$ and $\hat{X}_k$ in a system in contact with reservoirs of constant $F_0, F_1, \ldots, F_s$.

To carry out the summation over the microstates we first observe that, because of the form of f (equation 19.1)

$$\frac{\partial f}{\partial F_k} = \frac{1}{k_B} \left(-\hat{X}_k - \frac{\partial}{\partial F_k} S[F_0, \ldots, F_s] \right) f = -\frac{1}{k_B} (\hat{X}_k - X_k) f \tag{19.10}$$

[2] On the Formalism of Thermodynamic Fluctuation Theory, R. F. Greene and H. B. Callen, *Phys. Rev.* **85**, 16 (1956).

so that

$$\langle \Delta \hat{X}_j \Delta \hat{X}_k \rangle = -k_B \sum (\hat{X}_j - X_j) \frac{\partial f}{\partial F_k} \qquad (19.11)$$

$$= -k_B \frac{\partial}{\partial F_k} \langle \hat{X}_j - X_j \rangle + k_B \left\langle \frac{\partial}{\partial F_k} (\hat{X}_j - X_j) \right\rangle \quad (19.12)$$

$$= -k_B \frac{\partial}{\partial F_k} \langle \Delta \hat{X}_j \rangle - k_B \frac{\partial X_j}{\partial F_k} \qquad (19.13)$$

The first term vanishes because $\langle \Delta \hat{X}_j \rangle$ vanishes independently of the value of F_k, so that

$$\langle \Delta \hat{X}_j \Delta \hat{X}_k \rangle = -k_B \left(\frac{\partial X_j}{\partial F_k} \right)_{F_0 \ldots F_{k-1}, F_{k+1} \ldots F_s, X_{s+1} \ldots X_t} \qquad (19.14)$$

This equation is the most significant general result of the theory of thermodynamic fluctuations.

Particular note should be taken of the variables to be held constant in the derivative in equation 19.14. These are precisely the variables held constant in the physical system; the intensive parameters $F_0, \ldots, F_s$ of the reservoirs (except for F_k) and the extensive parameters $X_{s+1}, \ldots, X_t$ which are constrained by the walls. It should also be noted that the right-hand side of equation 19.14 is symmetric in j and k by virtue of a Maxwell relation.

If X_j and X_k in equation 19.14 are each taken as the energy we recover equation 19.6 for the fluctuations of energy in a system in contact with a thermal reservoir. But consider the same system in contact simultaneously with thermal and pressure reservoirs, so that both the energy and the volume can fluctuate. Then

$$\langle (\Delta \hat{E})^2 \rangle = -k_B \left(\frac{\partial U}{\partial (1/T)} \right)_{P/T, N_1, N_2, \ldots}$$

$$= k_B T^2 N c_P - k_B T^2 P V \alpha + k_B T P^2 V \kappa_T \qquad (19.15)$$

$$\langle \Delta \hat{E} \Delta \hat{V} \rangle = -k_B \left(\frac{\partial V}{\partial (1/T)} \right)_{P/T, N_1, \ldots} = k_B T^2 V \alpha - k_B T P V \kappa_T$$

$$(19.16)$$

and

$$\langle (\Delta \hat{V})^2 \rangle = -k_B \left(\frac{\partial V}{\partial (P/T)} \right)_{1/T, N_1, \dots}$$

$$= -k_B T \left(\frac{\partial V}{\partial P} \right)_{T, N_1, \dots} = k_B T V \kappa_T \qquad (19.17)$$

The energy fluctuations are indeed quite different from those given in equation 19.6. Furthermore the energy and the volume fluctuations are correlated, as expected.

Finally we can obtain a recursion relation relating higher order correlation moments to lower order moments, fully analogous to equation 19.7. Consider the moment $\langle \phi \Delta \hat{X}_k \rangle$, where ϕ is a product of the form $\Delta \hat{X}_i \Delta \hat{X}_j, \dots$. Then equations 19.9 to 19.12 can be repeated, with ϕ replacing $\Delta \hat{X}_j$, so that

$$\langle \phi \Delta \hat{X}_k \rangle = -k_B \left\langle \phi \frac{\partial f}{\partial F_k} \right\rangle$$

$$= -k_B \frac{\partial}{\partial F_k} \langle \phi \rangle + k_B \left\langle \frac{\partial \phi}{\partial F_k} \right\rangle \qquad (19.18)$$

which permits generation of successively higher correlation moments.

As an example of this procedure take ϕ as $\Delta X_i \Delta X_j$, to obtain the third moment

$$\langle \Delta \hat{X}_i \Delta \hat{X}_j \Delta \hat{X}_k \rangle = -k_B \frac{\partial}{\partial F_k} \langle \Delta \hat{X}_i \Delta \hat{X}_j \rangle - k_B \langle \Delta \hat{X}_i \rangle \frac{\partial X_j}{\partial F_k} - k_B \langle \Delta \hat{X}_j \rangle \frac{\partial X_i}{\partial F_k}$$

$$(19.19)$$

but $\langle \Delta \hat{X}_i \rangle = \langle \Delta \hat{X}_j \rangle = 0$, so that

$$\langle \Delta \hat{X}_i \Delta \hat{X}_j \Delta \hat{X}_k \rangle = -k_B \frac{\partial}{\partial F_k} \langle \Delta \hat{X}_i \Delta \hat{X}_j \rangle$$

$$= k_B^2 \frac{\partial^2 X_i}{\partial F_j \partial F_k} \qquad (19.20)$$

Again, the variables to be held constant in the differentiation reflect the boundary conditions of the fluctuating system.

Finally it should be noted that the fluctuations we have calculated are *thermodynamic* fluctuations. There are additional quantum mechanical fluctuations that can be nonzero even for a system in a single quantum state. For normal macroscopic systems (excluding "quantum systems" such as superconductors or superfluids) the thermodynamic fluctuations totally dominate the quantum mechanical fluctuations.

PROBLEMS

19.3-1. An ideal gas is in contact with a thermal and a pressure reservoir. Calculate the correlation moment $\langle \Delta \hat{E} \Delta \hat{V} \rangle$ of its energy and volume fluctuations.

19.3-2. Repeat Problem 19.3-1 for a van der Waals gas (recall Problem 3.8-3).

19.3-3. A conceptual subsystem of N moles in a single-component simple ideal gas system undergoes energy and volume fluctuations. The total system is at a temperature of 0°C and a pressure of 1 atm. What must be the size of N for the root mean square deviation in energy to be 1% of the average energy of the subsystem?

19.3-4. What is the order of magnitude of the mean square deviation of the volume of a typical metal sample of average volume equal to 1 cm³? The sample is at room temperature and pressure.

19.3-5. Consider a small volume V within a two-component simple system. Let $x_1 = N_1/(N_1 + N_2)$, in which N_1 and N_2 are the mole numbers within V. Show that

$$N^2(\Delta \hat{x}_1)^2 = x_2^2(\Delta \hat{N}_1)^2 - 2x_1x_2(\Delta \hat{N}_1 \Delta \hat{N}_2) + x_1^2(\Delta \hat{N}_2)^2$$

and compute the mean square deviation of concentration $\langle(\Delta \hat{x}_1)^2\rangle$.

19.3-6. Consider a small quantity of matter consisting of a fixed number N moles in a large fluid system. Let ρ_N be the average density of these N moles: the mass divided by the volume. Show that equation 19.17 implies that the density fluctuations are

$$\frac{\langle(\Delta \hat{\rho}_N)^2\rangle}{\rho_N^2} = +\frac{k_B T \kappa_T}{V}$$

in which V is the average volume of the N moles.

19.3-7. Show that the density fluctuations of an ideal gas are given by

$$\frac{\langle(\Delta \hat{\rho}_N)^2\rangle}{\rho_N^2} = \tilde{N}^{-1}$$

That is, the relative mean square density deviation is the reciprocal of the number of molecules in the subsystem.

19.3-8. Show that the relative root mean square deviation in density of 10^{-6} g of air at room temperature and pressure is negligible. Consider air as an ideal gas. Show that the relative rms deviation in density of 10^{-18} g of air at room temperature and pressure is approximately 1%. Show that the average volume of the samples is approximately 1mm³ in the first case and smaller than the cube of the wavelength of visible light in the second case.

19.3-9. The dielectric constant ε of a fluid varies with the density by the relation

$$\frac{\varepsilon - 1}{\varepsilon + 2} = A\rho$$

in which A is a constant. Show that the fluctuations in dielectric constant of a small quantity of N moles of matter in a large system are

$$\langle (\Delta \hat{\varepsilon})^2 \rangle = \frac{k_B T \kappa_T}{9V} (\varepsilon - 1)^2 (\varepsilon + 2)^2$$

in which V is the average volume of N moles.

19.3-10. If light of intensity I_0 is incident on a region of volume V, which has a difference $\delta \hat{\varepsilon}$ of dielectric constant from its average surroundings, the intensity of light I_θ scattered at an angle θ and at a distance r is

$$I_\theta = \frac{\pi^2 V^2 (\Delta \hat{\varepsilon})^2}{2\lambda_0^4} I_0 \frac{1 + \cos^2 \theta}{r^2}$$

in which λ_0 is the vacuum wavelength of the incident light. This is called Rayleigh scattering.

In a fluid each small volume V scatters incoherently, and the total scattered intensity is the same as the scattered intensities from each region.

From problem 19.3-9 we have

$$V^2 \langle (\Delta \hat{\varepsilon})^2 \rangle = \tfrac{1}{2} k_B T \kappa_T (\varepsilon - 1)^2 (\varepsilon + 2)^2 V$$

and summing this quantity over the total fluid we find

$$\sum V^2 \langle (\Delta \hat{\varepsilon})^2 \rangle = \tfrac{1}{2} k_B T \kappa_T (\varepsilon - 1)^2 (\varepsilon + 2)^2 V_{total}$$

where V_{total} is the total volume of the fluid. Consequently, the total scattered intensity at an angle θ and at a distance r from the scattering system is

$$I_\theta = \frac{\pi^2}{18} \frac{k_B T \kappa_T}{\lambda_0^4} (\varepsilon - 1)^2 (\varepsilon + 2)^2 I_0 V_{total} \frac{1 + \cos^2 \theta}{r^2}$$

By integrating over the surface of a sphere, show that the total scattered intensity is

$$I_{scattered} = \frac{8\pi^3}{27\lambda_0^4} k_B T \kappa_T (\varepsilon - 1)^2 (\varepsilon + 2)^2 I_0 V_{total}$$

Discuss the relevance of this result to critical opalescence (Section 10.1).

It is interesting to note that, because of the λ_0^{-4} dependence of the scattering, blue light is much more strongly scattered than red. The sun appears red when it is low on the horizon because the blue light is selectively scattered, leaving the direct rays from the sun deficient in blue. On the other hand, the diffuse light of the daytime sky, composed of the indirectly scattered sunlight, is predominantly blue. The color of the sky accordingly is everyday evidence of the existence of thermodynamic fluctuations.

19.3-11. The classical theory of fluctuations, due to Einstein, proceeds from equation 19.1 which, in general form, is

$$f_{\hat{X}_0, \hat{X}_1, \ldots} = e^{-S(F_0, \ldots)} e^{-\frac{1}{k_B}(\hat{S} - \sum_j F_j \hat{X}_j)}$$

Expanding $\hat{S}$ around its equilibrium value S, in powers of the deviations $\Delta X_j = \hat{X}_j - X_j$, and keeping terms only to second order

$$f_{\hat{X}_0, \hat{X}_1, \ldots} = A \exp \frac{1}{2k_B} \left[\sum_0^s \sum_0^s S_{jk} \Delta X_j \Delta X_k \right]$$

where $S_{jk} = \partial^2 S / \partial X_j \partial X_k$ and there A is a normalizing constant. This is a multidimensional Gaussian probability distribution. By direct integration calculate the second moments and show that they are correctly given. (The third and higher moments are not correct!)

20

VARIATIONAL PROPERTIES, PERTURBATION EXPANSIONS, AND MEAN FIELD THEORY

20-1 THE BOGOLIUBOV VARIATIONAL THEOREM

To calculate the fundamental equation for a particular system we must first evaluate the permissible energy levels of the system and then, given those energies, we must sum the partition sum. Neither of these steps is simple, except for a few "textbook models." In such models, several of which we have studied in preceding chapters, the energy eigenvalues follow a simple sequence and the partition sum is an infinite series that can be summed analytically. But for most systems both the enumeration of the energy eigenvalues and the summation of the partition sum pose immense computational burdens. Approximation techniques are required to make the calculations practical. In addition these approximation techniques provide important heuristic insights to complex systems.

The strategy followed in the approximation techniques to be described is first to identify a soluble model that is somewhat similar to the model of interest, and then to apply a method of controlled corrections to calculate the effect of the difference in the two models. Such an approach is a statistical "perturbation method."

Because perturbation methods rest upon the existence of a library of soluble models, there is great stress in the statistical mechanical literature on the invention of new soluble models. Few of these have direct physical relevance, as they generally are devised to exploit some ingenious mathematical trick of solution rather than to mirror real systems (thereby giving rise to the rather abstract flavor of some statistical mechanical literature).

The first step in the approximation strategy is to identify a practical criterion for the choice of a soluble model with which to approximate a given system. That criterion is most powerfully formulated in terms of the Bogoliubov variational theorem.

Consider a system with a Hamiltonian $\mathcal{H}$, and a soluble model system with a Hamiltonian $\mathcal{H}_0$. Let the difference be $\mathcal{H}_1$, so that $\mathcal{H} = \mathcal{H}_0 + \mathcal{H}_1$. It is then convenient to define

$$\mathcal{H}(\lambda) = \mathcal{H}_0 + \lambda \mathcal{H}_1 \tag{20.1}$$

where λ is a parameter inserted for analytic convenience. By permitting λ to vary from zero to unity we can smoothly bridge the transition from the soluble model system ($\mathcal{H}_0$) to the system of interest; $\mathcal{H}(1) = \mathcal{H}_0 + \mathcal{H}_1$.

The Helmholtz potential corresponding to $\mathcal{H}(\lambda)$ is $F(\lambda)$, where

$$-\beta F(\lambda) = \ln \sum_j e^{-\beta E_j(\lambda)} \equiv \ln \operatorname{tr} e^{-\beta \mathcal{H}(\lambda)} \tag{20.2}$$

Here the symbol $\operatorname{tr} e^{-\beta \mathcal{H}(\lambda)}$ (to be read as the "trace" of $e^{-\beta \mathcal{H}(\lambda)}$) is *defined* by the second equality; *the trace of any quantity is the sum of its quantum eigenvalues.* We use the notation "tr" simply as a convenience.

We now study the dependence of the Helmholtz potential on λ. The first derivative is[1]

$$\frac{dF(\lambda)}{d\lambda} = \frac{\operatorname{tr} \mathcal{H}_1 e^{-\beta(\mathcal{H}_0 + \lambda \mathcal{H}_1)}}{\operatorname{tr} e^{-\beta(\mathcal{H}_0 + \lambda \mathcal{H}_1)}} = \langle \mathcal{H}_1 \rangle \tag{20.3}$$

and the second derivative is

$$\frac{d^2 F}{d\lambda^2} = -\beta \left[\frac{\operatorname{Tr} \mathcal{H}_1^2 e^{-\beta(\mathcal{H}_0 + \lambda \mathcal{H}_1)}}{\operatorname{tr} e^{-\beta(\mathcal{H}_0 + \lambda \mathcal{H}_1)}} - \left(\frac{\operatorname{tr} \mathcal{H}_1 e^{-\beta(\mathcal{H}_0 + \lambda \mathcal{H}_1)}}{\operatorname{tr} e^{-\beta(\mathcal{H}_0 + \lambda \mathcal{H}_1)}} \right)^2 \right] \tag{20.4}$$

$$= -\beta \left[\langle \mathcal{H}_1^2 \rangle - \langle \mathcal{H}_1 \rangle^2 \right] \tag{20.5}$$

$$= -\beta \langle \mathcal{H}_1 - \langle \mathcal{H}_1 \rangle \rangle^2 \tag{20.6}$$

where the averages are taken with respect to the canonical weighting factor $e^{-\beta \mathcal{H}(\lambda)}$. The operational meaning of these weighted averages will be clarified by a specific example to follow.

An immediate and fateful consequence of equation 20.6 is that $d^2 F/d\lambda^2$ is negative (or zero) for all λ

$$\frac{d^2 F}{d\lambda^2} \le 0 \qquad \text{(for all λ)} \tag{20.7}$$

---

[1] In the quantum mechanical context the operators $\mathcal{H}_0$ and $\mathcal{H}_1$ are here assumed to commute. The result is independent of this assumption. For the noncommutative case, and for an elegant general discussion see R. Feynman, *Statistical Mechanics—A Set of Lectures* (W. A. Benjamin, Inc., Reading, Massachusetts, 1972).

Consequently a plot of $F(\lambda)$ as a function of λ is everywhere concave. It follows that $F(\lambda)$ lies below the straight line tangent to $F(\lambda)$ at $\lambda = 0$;

$$F(\lambda) \le F(0) + \lambda (dF/d\lambda)_{\lambda=0} \tag{20.8}$$

and specifically, taking $\lambda = 1$

$$F \le F_0 + \langle \mathcal{H}_1 \rangle_0 \tag{20.9}$$

The quantity $\langle \mathcal{H}_1 \rangle_0$ is as defined in equation 20.3, but with $\lambda = 0$; it is the average value of $\mathcal{H}_1$ in the soluble model system. Equation 20.9 is the Bogoliubov inequality. It states that *the Helmholtz potential of a system with Hamiltonian $\mathcal{H} = \mathcal{H}_0 + \mathcal{H}_1$ is less than or equal to the "unperturbed" Helmholtz potential* (corresponding to $\mathcal{H}_0$) *plus the average value of the "perturbation" $\mathcal{H}_1$ as calculated in the unperturbed (or soluble model) system.*

Because the quantity on the right of equation 20.9 is an upper bound to the Helmholtz potential of the ("perturbed") system, it clearly is desirable that this bound be as small as possible. Consequently *any adjustable parameters in the unperturbed system are best chosen so as to minimize the quantity $F_0 + \langle \mathcal{H}_1 \rangle_0$.*

This is the criterion for the choice of the "best" soluble model system. Then F_0 is the Helmholtz potential of the optimum model system, and $\langle \mathcal{H}_1 \rangle_0$ is the leading correction to this Helmholtz potential.

The meaning and the application of this theorem are best illustrated by a specific example, to which we shall turn momentarily. However we first recast the Bogoliubov inequality in an alternative form that provides an important insight. If we write F_0, the Helmholtz potential of the unperturbed system, explicitly as

$$F_0 = \langle \mathcal{H}_0 \rangle_0 - TS_0 \tag{20.10}$$

then equation 20.9 becomes

$$F \le \langle \mathcal{H}_0 \rangle_0 + \langle \mathcal{H}_1 \rangle_0 - TS_0 \tag{20.11}$$

or

$$F \le \langle \mathcal{H} \rangle_0 - TS_0 \tag{20.12}$$

That is, *the Helmholtz potential of a system with Hamiltonian $\mathcal{H} = \mathcal{H}_0 + \mathcal{H}_1$ is less than or equal to the full energy $\mathcal{H}$ averaged over the state probabilities of the unperturbed system, minus the product of T and the entropy of the unperturbed system.*

Example 1

A particle of mass m is constrained to move in one dimension in a quartic potential of the form $V(x) = D(x/a)^4$, where $D > 0$ and where a is a measure of the linear extension of the potential. The system of interest is composed of $\tilde{N}$ such particles in thermal contact with a reservoir of temperature T. An extensive parameter of the system is defined by $X \equiv \tilde{N}a$, and the associated intensive parameter is denoted by P. Calculate the equations of state $U = U(T, X, \tilde{N})$ and $P = P(T, X, \tilde{N})$, and the heat capacity $c_p(T, X, \tilde{N})$.

To solve this problem by the standard algorithm would require first a quantum mechanical calculation of the allowed energies of a particle in a quartic potential, and then summation of the partition sum. Neither of these calculations is analytically tractable. We avoid these difficulties by seeking an approximate solution. In particular we inquire as to the best quadratic potential (i.e., the best simple harmonic oscillator model) with which to approximate the system, and we then assess the leading correction to account for the difference in the two models.

The quadratic potential that, together with the kinetic energy, defines the "unperturbed Hamiltonian" is

$$V_0(x) = \tfrac{1}{2}m\omega_0^2 x^2 \tag{a}$$

where ω_0 is an as-yet-unspecified constant. Then the "perturbing potential," or the difference between the true Hamiltonian and that of the soluble model system, is

$$\mathcal{H}_1 = D\left(\frac{x}{a}\right)^4 - \frac{1}{2}m\omega_0^2 x^2 \tag{b}$$

The Helmholtz potential of the harmonic oscillator model system is (recall equations 16.22 to 16.24)[2]

$$F_0 = -\tilde{N}k_B T \ln z_0 = \tilde{N}\beta^{-1}\ln\left(e^{\beta\hbar\omega_0/2} - e^{-\beta\hbar\omega_0/2}\right) \tag{c}$$

and the Bogoliubov inequality states that

$$F \le \tilde{N}\beta^{-1}\ln\left(e^{\beta\hbar\omega_0/2} - e^{-\beta\hbar\omega_0/2}\right)$$

$$+ \tilde{N}D\left\langle\left(\frac{x}{a}\right)^4\right\rangle_0 - \left(\frac{\tilde{N}}{2}\right)m\omega_0^2\langle x^2\rangle_0 \tag{d}$$

Before we can draw conclusions from this result we must evaluate the second and third terms. It is an elementary result of mechanics (the "virial theorem") that the value of the potential energy ($\tfrac{1}{2}m\omega_0^2 x^2$) in the nth state of a harmonic oscillator is one half the total energy, so that

$$\left(\tfrac{1}{2}m\omega_0^2 x^2\right)_{n\text{th state}} = \tfrac{1}{2}\left(n + \tfrac{1}{2}\right)\hbar\omega_0 \tag{e}$$

---

[2] But note that the zero of energy has been shifted by $\hbar\omega_0/2$, the so-called zero point energy. The allowed energies are $(n + \tfrac{1}{2})\hbar\omega_0$.

and a similar quantum mechanical calculation gives

$$\langle x^4 \rangle_{n\text{th state}} = \left(\frac{3\hbar^2}{2m^2\omega_0^2} \right) \left(n^2 + n + \frac{1}{2} \right) \tag{f}$$

With the values of these quantities in the nth state we must now average over all states n. Averaging equation (e) in the unperturbed system

$$\left\langle \frac{1}{2} m\omega_0^2 x^2 \right\rangle_0 = \frac{1}{2} \left[\langle n \rangle + \frac{1}{2} \right] \hbar\omega_0 = \frac{1}{2} U_0 = \frac{1}{4} \hbar\omega_0 \frac{e^{\beta\hbar\omega_0} + 1}{e^{\beta\hbar\omega_0} - 1} \tag{g}$$

and we also find

$$\frac{D}{a^4} \langle x^4 \rangle_0 = \frac{D}{a^4} \frac{3\hbar^2}{2m^2\omega_0^2} \left[\langle n^2 \rangle + \frac{U}{\hbar\omega_0} \right]$$

$$= \frac{D}{a^4} \frac{3\hbar^2}{2m^2\omega_0^2} \left[\frac{e^{\beta\hbar\omega_0} + 1}{(e^{\beta\hbar\omega_0} - 1)^2} + \frac{e^{\beta\hbar\omega_0} + 1}{2(e^{\beta\hbar\omega_0} - 1)} \right]$$

$$= \frac{3D\hbar^2}{4a^4 m^2\omega_0^2} \left(\frac{e^{\beta\hbar\omega_0} + 1}{e^{\beta\hbar\omega_0} - 1} \right)^2 \tag{h}$$

Inserting these last two results (equations g and h) into the Bogoliubov inequality (equation d)

$$F \le \tilde{N}\beta^{-1} \ln \left(e^{\beta\hbar\omega_0/2} - e^{-\beta\hbar\omega_0/2} \right)$$

$$+ \frac{3\tilde{N}D\hbar^2}{4a^4 m^2\omega_0^2} \left(\frac{e^{\beta\hbar\omega_0} + 1}{e^{\beta\hbar\omega_0} - 1} \right)^2 - \frac{1}{2} \tilde{N}\hbar\omega_0 \frac{e^{\beta\hbar\omega_0} + 1}{e^{\beta\hbar\omega_0} - 1} \tag{i}$$

The first term is the Helmholtz potential of the unperturbed harmonic oscillator system, and the two remaining terms are the leading correction. The inequality states that the sum of all higher-order corrections would be positive, so that the right-hand side of equation (i) is an upper bound to the Helmholtz potential.

The frequency ω_0 of the harmonic oscillator system has not yet been chosen. Clearly the best approximation is obtained by making the upper bound on F as small as possible. Thus *we choose ω_0 so as to minimize the right-hand side of equation i, which then becomes the best available approximation to the Helmholtz potential of the system.* Denote the value of ω_0 that minimizes F by $\tilde{\omega}_0$ (a function of T, X ($= \tilde{N}a$), and $\tilde{N}$). Then ω_0 in equation (i) can be replaced by $\tilde{\omega}_0$ and the "less than or equal" sign ($\le$) can be replaced by an "approximately equal" sign ($\approx$). So interpreted, equation (i) is the (approximate) fundamental equation of the system.

The mechanical equation of state is, then,

$$-\frac{P}{T} = \frac{1}{\tilde{N}}\left(\frac{\partial F}{\partial a}\right)_{T,\tilde{N}} = \frac{1}{\tilde{N}}\left(\frac{\partial F}{\partial a}\right)_{T,\tilde{N},\tilde{\omega}_0} + \frac{1}{\tilde{N}}\left(\frac{\partial F}{\partial \tilde{\omega}_0}\right)_{T,\tilde{N},a}\left(\frac{\partial \tilde{\omega}_0}{\partial a}\right)_{T,\tilde{N}} \qquad (j)$$

At this point the algebra becomes cumbersome, though straightforward in principle. The remaining quantities sought for can be found in similar form. Instead we turn our attention to a simpler version of the same problem.

Example 2

We repeat the preceding Example, but we consider the case in which the coefficient D/a^4 is small (in a sense to be made more quantitative later), permitting the use of classical statistics. Furthermore we now choose a square-well potential as the unperturbed potential

$$V_0(x) = \begin{cases} 0 & \text{if } -\dfrac{L}{2} < x < \dfrac{L}{2} \\[2mm] \infty & \text{if } |x| > \dfrac{L}{2} \end{cases}$$

The optimum value of L is to be determined by the Bogoliubov criterion.

The unperturbed Helmholtz potential is determined by

$$e^{-\beta F_0} = \mathrm{tr}\, e^{-\beta \mathscr{H}_0} = \int\int \frac{dx\,dp_x}{h} e^{-\beta[p_x^2/2m + V_0(x)]}$$

$$= \frac{1}{h}\int_{-L/2}^{L/2} dx \int_{-\infty}^{\infty} dp_x\, e^{-\beta p_x^2/2m}$$

$$= \frac{1}{h}(2\pi m k_B T)^{1/2} L$$

We have here used classical statistics (as in Sections 16.8 and 16.9), tentatively assuming that L and T are each sufficiently large that $k_B T$ is large compared to the energy differences between quantum states.

The quantity $\langle \mathscr{H}_1 \rangle_0$ is, then,

$$\langle \mathscr{H}_1 \rangle_0 = \frac{\mathrm{tr}\left[Da^{-4}x^4 - V_0(x)\right]e^{-\beta \mathscr{H}_0}}{\mathrm{tr}\, e^{-\beta \mathscr{H}_0}}$$

Furthermore $V_0(x) = 0$ for $|x| < L/2$, whereas $e^{-\beta \mathscr{H}_0} = 0$ for $|x| > L/2$, so that the term involving $V_0(x)$ vanishes. Then

$$\langle \mathscr{H}_1 \rangle_0 = \frac{D}{a^4}\langle x^4 \rangle_0 = \frac{D}{a^4 L}\int_{-L/2}^{L/2} x^4\, dx = \frac{D}{80}\left(\frac{L}{a}\right)^4$$

The Bogoliubov inequality now becomes

$$F \le -k_B T \ln\left[\frac{1}{h}(2\pi m k_B T)^{1/2} L\right] + \frac{1}{80} D\left(\frac{L}{a}\right)^4$$

Minimizing with respect to L

$$L/a = [20k_B T/D]^{1/4}$$

This result determines the optimum size of a square-well potential with which to approximate the thermal properties of the system, and it determines the corresponding approximate Helmholtz potential.

Finally we return to the criterion for the use of classical statistics. In Section 16.6 we saw that the energy separation of translational states is of the order of $h^2/2mL^2$, and the criterion of classical statistics is that $k_B T \gg h^2/2mL^2$. In terms of D the analogous criterion is

$$\frac{D}{a^4} \le 20m^2 \frac{(k_B T)^3}{\hbar^4}$$

For larger values of D the procedure would be similar in principle, but the calculation would require summations over the discrete quantum states rather than simple phase-space integration.

Finally we note that if the temperature is high enough to permit the use of classical statistics the original quartic potential problem is itself soluble! Then there is no need to approximate the quartic potential by utilizing a variational theorem. It is left to the reader (Problem 20.1-2) to solve the original quartic potential problem in the classical domain, and to compare that solution with the approximate solution obtained here.

PROBLEMS

20.1-1. Derive equation (h) of Example 1, first showing that for a harmonic oscillator

$$\langle n \rangle = -\frac{1}{z'} \frac{\partial z'}{\partial(\beta\hbar\omega_0)}$$

and

$$\langle n^2 \rangle = \frac{1}{z'} \frac{\partial^2 z'}{\partial(\beta\hbar\omega_0)^2}$$

where

$$z' = e^{\beta\hbar\omega_0/2} z = \sum_{n=0}^{\infty} e^{-\beta\hbar\omega_0 n}$$

20.1-2. Solve the quartic potential problem of Example 2 assuming the temperature to be sufficiently high that classical statistics can be applied. Compare the Helmholtz potential with that calculated in Example 2 by the variational theorem.

20.1-3. Complete Example 2 by writing the Helmholtz potential $F(T, a)$ explicitly. Calculate the "tension" $\mathcal{T}$ conjugate to the "length" a. Calculate the compliance coefficient $a^{-1}(\partial a / \partial \mathcal{T})_T$.

20.1-4. Consider a particle in a quadratic potential $V(x) = Ax^2 / 2a^2$. Despite the fact that this problem is analytically solvable, approximate the problem by a square potential. Assume the temperature to be sufficiently high that classical statistics can be used in solving the square potential. Calculate the "tension" $\mathcal{T}$ and the compliance coefficient $a^{-1}(\partial a / \partial \mathcal{T})_T$.

20-2 MEAN FIELD THEORY

The most important application of statistical perturbation theory is that in which a system of interacting particles is approximated by a system of noninteracting particles. The optimum noninteracting model system is chosen in accordance with the Bogoliubov inequality, which also yields the first-order correction to the noninteracting or "unperturbed" Helmholtz potential. Because very few interacting systems are soluble analytically, and because virtually all physical systems consist of interacting particles, the "mean field theory" described here is the basic tool of practical statistical mechanics.

It is important to note immediately that the term *mean field theory* often is used in a less specific way. Some of the results of the procedure can be obtained by other more ad hoc methods. Landau-type theories (recall Section 11.4) obtain a temperature dependence of the order parameter that is identical to that obtained by statistical mean field theory. Another approximation, known as the "random phase approximation," also predicts the same equation of state. Neither of these provides a full thermodynamic fundamental equation. Nevertheless various such approximations are referred to generically as mean field theories. We use the term in the more restrictive sense.

Certainly the simplest model of interacting systems, and one that has played a key role in the development of the theory of interacting systems, is the "two-state nearest-neighbor Ising model." The model consists of a regular crystalline array of particles, each of which can exist in either of two orbital states, designated as the "up" and "down" states. Thus the states of the particles can be visualized in terms of classical spins, each of which is permitted only to be either up or down; a site variable σ_j takes the value $\sigma_j = +1$ if the spin at site j is up or $\sigma_j = -1$ if the spin at site j is down. The energies of the two states are $-B$ and $+B$ for the up and down states respectively. In addition nearest neighbor spins have an interaction energy $-2J$ if they are both up or both down, or of $+2J$ if one spin is up and one spin is down. Thus the Hamiltonian is

$$\mathscr{H} = -\sum_{i,j} J_{ij}\sigma_i\sigma_j - B\sum_j \sigma_j \tag{20.13}$$

where $J_{ij} = 0$ if i and j are not nearest neighbors, whereas $J_{ij} = J$ if i and j are nearest neighbors. It should be noted that a specific pair of neighbors (say spins #5 and #8) appears twice in the sum ($i = 5$, $j = 8$ and $i = 8$, $j = 5$).

Quite evidently the problem is an insoluble many-body problem, for each spin is coupled indirectly to every other spin in the lattice. An approximation scheme is needed, and we invoke the Bogoliubov inequality. A plausible form of the soluble model system is suggested by focussing on only the jth spin in the Hamiltonian (20.13); the Hamiltonian is then simply linear in σ_j. We correspondingly choose the "unperturbed" model Hamiltonian to be

$$\mathscr{H}_0 = -\sum_j \hat{B}_j \sigma_j - B\sum_j \sigma_j \tag{20.14}$$

where $\hat{B}_j$ is to be chosen according to the Bogoliubov criterion. We anticipate that $\hat{B}_j$ will be independent of j ($\hat{B}_j = \hat{B}$), for all spins are equivalent. Thus

$$\mathscr{H}_0 = -(\hat{B} + B)\sum_j \sigma_j = -B^*\sum_j \sigma_j \tag{20.15}$$

where we define
$$B^* \equiv \hat{B} + B \tag{20.16}$$

Accordingly the "unperturbed" Helmholtz potential is

$$F_0 = -k_B T \ln \operatorname{tr} e^{-\beta\mathscr{H}_0} = -\tilde{N}k_B T \ln\{e^{+\beta B^*} + e^{-\beta B^*}\} \tag{20.17}$$

where $\tilde{N}$ is the number of sites in the lattice. The Bogoliubov inequality assures us that $F \le F_0 + \langle \mathscr{H} - \mathscr{H}_0 \rangle_0$, or

$$F \le F_0 - \sum_{i,j} J_{ij}\langle \sigma_i\sigma_j \rangle_0 + \hat{B}\tilde{N}\langle \sigma \rangle_0 \tag{20.18}$$

and we procede to calculate $\langle \sigma \rangle_0$ and $\langle \sigma_i\sigma_j \rangle_0$. In the unperturbed system the average of products centered on different sites simply factors;

$$\langle \sigma_i\sigma_j \rangle_0 = \langle \sigma_i \rangle_0\langle \sigma_j \rangle_0 = \langle \sigma \rangle_0^2 \tag{20.19}$$

so that
$$F \le F_0 - \tilde{N}Jz_{nn}\langle\sigma\rangle_0^2 + (B^* - B)\tilde{N}\langle\sigma\rangle_0 \qquad (20.20)$$

where z_{nn} is the number of nearest neighbors of a site in the lattice ($z_{nn} = 6$ for a simple cubic lattice, 8 for a body-centered cubic lattice, etc). Furthermore

$$\langle\sigma\rangle_0 = \frac{e^{\beta B^*} - e^{-\beta B^*}}{e^{\beta B^*} + e^{-\beta B^*}} = \tanh(\beta B^*) \qquad (20.21)$$

We must minimize F with respect to $\hat{B}$. But from equations 20.20, 20.17 and 20.21 we observe that $\hat{B}$ appears in F only in the combination $\hat{B} + B \equiv B^*$. Hence we can minimize F with respect to B^*, giving the result that

$$B^* - B = \hat{B} = 2z_{nn}J\langle\sigma\rangle_0 \qquad (20.22)$$

This is a self-consistent condition, as $\langle\sigma\rangle_0$ is expressed in terms of B^* by equation 20.21.

Prior to analyzing this self-consistent solution for $\langle\sigma\rangle_0$, we observe its significance. If we were to seek $\langle\sigma\rangle$ in mean field theory we might proceed by differentiation (with respect to B) of the Helmholtz potential F (as calculated in mean field theory; equation 20.20). The applied field B appears explicitly in eq'n 20.20, *but it is also implicit in* $\langle\sigma\rangle_0$. Fortunately however, $\langle\sigma\rangle_0$ depends on B only in the combination $B + \hat{B} = B^*$, and we have imposed the condition that $\partial F/\partial B^* = 0$. Thus, in differentiation, only the *explicit* dependence of F on B need be considered. With this extremely convenient simplifying observation we immediately corroborate that differentiation of F (equation 20.20) with respect to B does give $\langle\sigma\rangle_0$. *The "spontaneous moment"* $\langle\sigma\rangle$ *in mean field theory is given properly by its zero-order approximation.*

Returning then to equation 20.21 for $\langle\sigma\rangle_0$ (and hence for $\langle\sigma\rangle$) the solution is best obtained graphically, as shown in Fig. 20.1. The abscissa of the graph is βB^*, or from equation 20.22

$$x \equiv \beta B^* = \beta(2z_{nn}J\langle\sigma\rangle + B) \qquad (20.23)$$

so that equation 20.21 can be written as

$$\langle\sigma\rangle = \frac{k_B T}{2z_{nn}J}x - \frac{B}{2z_{nn}J} = \tanh(x) \qquad (20.24)$$

A plot of $\langle\sigma\rangle$ versus x from the first equality is a straight line of slope $k_B T/2z_{nn}J$ and of intercept $-B/2z_{nn}J$. A plot of $\langle\sigma\rangle$ versus x from the second equality is the familiar $\tanh(x)$ curve shown in Fig. 20.1. The intersection of these two curves determines $\langle\sigma\rangle$.

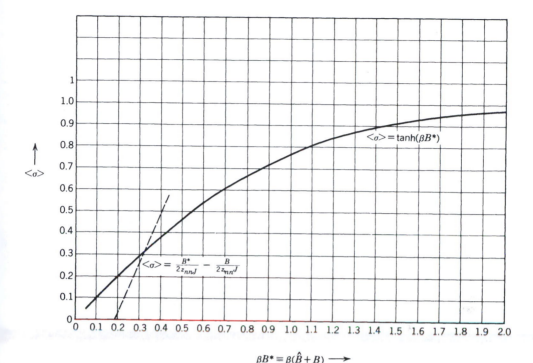

$$\beta B^* \equiv \beta(\hat{B} + B) \longrightarrow$$

FIGURE 20.1

The qualitative behavior of $\langle \sigma(T, B) \rangle$ is evident. For $B = 0$, the straight line passes through the origin, with a slope $k_B T/2 z_{nn} J$. The curve of $\tanh(x)$ has an initial slope of unity. Hence, if $k_B T/2 z_{nn} J > 0$ the straight line and the $\tanh(x)$ curve have only the trivial intersection at $\langle \sigma \rangle = 0$. However, if $k_B T/2 z_{nn} J < 1$ there is an intersection at a positive value of $\langle \sigma \rangle$ and another at a negative value of $\langle \sigma \rangle$, as well as the persistent intersection at $\langle \sigma \rangle = 0$. The existence of three formal solutions for $\langle \sigma \rangle$ is precisely the result we found in the thermodynamic analysis of first-order phase transitions in Chapter 9. A stability analysis there revealed the intermediate value $\langle \sigma \rangle = 0$ to be intrinsically unstable. The positive and negative values of $\langle \sigma \rangle$ are equally stable, and the choice of one or the other is an "accidental" event. We thus conclude that the system exhibits a first-order phase transition at low temperatures, and that the phase transition ceases to exist above the "Curie" temperature T_c given by

$$k_B T_c = 2 z_{nn} J \qquad (20.25)$$

We can also find the "susceptibility" for temperatures above T_c. For small arguments $\tanh y \simeq y$, so that equation 20.24 becomes, for $T > T_c$,

$$\langle \sigma \rangle \simeq \beta(2 z_{nn} J \langle \sigma \rangle + B), \qquad T > T_c \qquad (20.26)$$

or the "susceptibility" is

$$\frac{\langle \sigma \rangle}{B} = \frac{k_B T}{k_B T - 2z_{nn}J} = \frac{T}{T - T_c}, \qquad T > T_c \qquad (20.27)$$

This agrees with the classical value of unity for the critical exponent γ, as previously found in Section 11.4.

To find the temperature dependence of the spontaneous moment $\langle \sigma \rangle$ for temperatures just below T_c we take $B = 0$ in equation 20.21 and 20.22, and we assume $\langle \sigma \rangle$ to be very small. Then the hyperbolic tangent can be expanded in series, whence

$$\langle \sigma \rangle = 2\beta z_{nn}J\langle \sigma \rangle - \tfrac{1}{3}(2\beta z_{nn}J\langle \sigma \rangle)^3 + \cdots$$

or

$$\langle \sigma \rangle = \left(\frac{3}{T_c}\right)^{1/2} \times (T_c - T)^{1/2} + \cdots \qquad (20.28)$$

We thereby corroborate the classical value of $\tfrac{1}{2}$ for the critical exponent α.

It is a considerable theoretical triumph that a first-order phase transition can be obtained by so simple a theory as mean field theory. But it must be stressed that the theory is nevertheless rather primitive. In reality the Ising model does not have a phase transition in one dimension, though it does in both two and three dimensions. Mean field theory, in contrast, predicts a phase transition without any reference to the dimensionality of the crystalline array. And, of course, the subtle details of the critical transitions, as epitomized in the values of the critical exponents, are quite incorrect.

Finally, it is instructive to inquire as to the thermal properties of the system. In particular we seek the mean field value of the entropy $S = -(\partial F/\partial T)_V$. We exploit the stationarity of F with respect to B^* by rewriting equation 20.20, with B^* rewritten as $(k_B T \beta B^*)$

$$F = -\tilde{N}k_B T \ln\left[e^{\beta B^*} + e^{-\beta B^*}\right] - \tilde{N}Jz_{nn}\langle \sigma \rangle^2$$

$$+ \tilde{N}k_B T(\beta B^*)\langle \sigma \rangle - \tilde{N}B\langle \sigma \rangle \qquad (20.29)$$

Then in differentiating F with respect to T we can treat βB^* as a constant

$$S = -\left(\frac{\partial F}{\partial T}\right)_{\beta B^*, \langle \sigma \rangle} = \tilde{N}k_B \ln\left[e^{\beta B^*} + e^{-\beta B^*}\right] - \tilde{N}k_B(\beta B^*)\langle \sigma \rangle$$

$$(20.30)$$

The first term is recognized as $-F_0/T$ (from equation 20.17), and the second term is simply $\langle \mathcal{H}_0 \rangle /T$. Thus

$$S = (\langle \mathcal{H}_0 \rangle_0 - F_0)/T = S_0 \qquad (20.31)$$

The mean field value of the entropy, like the induced moment $\langle \sigma \rangle$, is given correctly in zero order.

The energy U is given by

$$U = F + TS = (F_0 + \langle \mathcal{H} - \mathcal{H}_0 \rangle_0) + TS = \langle \mathcal{H} \rangle_0 \qquad (20.32)$$

The energy is also given correctly in zero order, if interpreted as in 20.32 —but note that this result is quite different from $\langle \mathcal{H}_0 \rangle_0$!

A more general Ising model permits the spin to take the values $-S, -S + 1, -S + 2, \ldots, S - 2, S - 1, S$, where S is an integer or half integer (the "value of the spin"). The theory is identical in form to that of the "two-state Ising model" (which corresponds to $S = \frac{1}{2}$), except that the hyperbolic tangent function appearing in $\langle \sigma \rangle_0$ is replaced by the "Brillouin function":

$$\langle \sigma \rangle_0 = SB_S(\beta B) = \left(S + \frac{1}{2} \right) \coth \left(\frac{2S + 1}{2S} \beta B \right) - \frac{1}{2} \coth \frac{\beta B}{2S}$$

$$(20.33)$$

The analysis follows step-by-step in the pattern of the two-state Ising model considered above — merely replacing equation 20.21 by 20.33. The corroboration of this statement is left to the reader.

In a further generalization, the Heisenberg model of ferromagnetism permits the spins to be quantum mechanical entities, and it associates the external "field" B with an applied magnetic field B_e. Within the mean field theory, however, only the component of a spin along the external field axis is relevant, and the quantum mechanical Heisenberg model reduces directly to the classical Ising model described above. Again the reader is urged to corroborate these conclusions, and he or she is referred to any introductory text on the theory of solids for a more complete discussion of the details of the calculation and of the consequences of the conclusions.

The origin of the name "mean field theory" lies in the heuristic reasoning that led us to a choice of a soluble model Hamiltonian in the Ising (or Heisenberg) problem above. Although each spin interacts with other spins, the mean field approach effectively replaces the bi-linear spin interaction $\sigma_i \sigma_j$ by a linear term $B_j \sigma_j$. The quantity B_j plays the role of an effective magnetic field acting on σ_j, and the optimum choice of B_j is $\langle \sigma \rangle$. Equivalently, the product $\sigma_i \sigma_j$ is "linearized," replacing one factor by its average value. A variety of recipes to accomplish this in a consistent manner exist. However we caution against such recipes, as they generally substitute heuristic appeal for the well-ordered rigor of the Bogoliubov inequality, and they provide no sequence of successive improvements. More immediately, the stationarity of F to variations in B^* greatly simplifies differentiation of F (required to evaluate thermodynamic quantities; recall equation 20.30), and the analogue of this stationarity has no basis in heuristic formulations. But most important, there are applications of the "mean field" formalism (as based on the Bogoliubov inequality) in which products of operators are *not* simply "linearized." For these the very name "mean field" is a misnomer. A simple and instructive case of this type is given in the following Example.

Example

N Ising spins, each capable of taking *three* values ($\sigma = -1, 0, +1$) form a planar triangular array, as shown. Note that there are $2N$ triangles for N spins, and that each spin is shared by six triangles. We assume N to be sufficiently large that edge effects can be ignored.

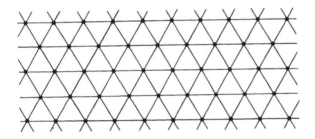

The energy associated with each triangle (a three-body interaction) is

$$-\varepsilon \text{ if two spins are "up"}$$
$$-2\varepsilon \text{ if three spins are "up"}$$
$$0 \text{ otherwise}$$

Calculate (approximately) the number of spins in each spin state if the system is in equilibrium at temperature T.

Solution

The problem differs from the Ising and Heisenberg prototypes in two respects; we are not given an analytic representation of the Hamiltonian (though we could devise one with moderate effort), and a "mean field" type of model Hamiltonian (of the form $B\sum_j \sigma_j$) would not be reasonable. This latter observation follows from the stated condition that the energies of the various possible configurations depend only on the populations of the "up" states, and that there is no distinction in energy between the $\sigma = 0$ and the $\sigma = -1$ states. The soluble model Hamiltonian should certainly preserve this symmetry, which a mean-field type Hamiltonian does not do. Accordingly we take as the soluble model Hamiltonian one in which the energy $-\hat{\varepsilon}$ is associated with each "up" spin in the lattice (the $\sigma = 0$ and -1 states each having zero energy). The energy $\hat{\varepsilon}$ will be the variational parameter of the problem.

The "unperturbed" value of the Helmholtz potential is determined by

$$e^{-\beta F_0} = \left(e^{\beta \hat{\varepsilon}} + 2\right)^N$$

and the probability that a spin is up, to zero order, is

$$f_{0\uparrow} = \frac{e^{\beta \hat{\varepsilon}}}{\left(e^{\beta \hat{\varepsilon}} + 2\right)} = \left(1 + 2e^{-\beta \hat{\varepsilon}}\right)^{-1}$$

whereas

$$f_{0\downarrow} = f_{0\rightarrow} = \frac{(1 - f_{0\uparrow})}{2}$$

Within each triangle the probability of having all three spins up is $f_{0\uparrow}^3$, and the probability of having two spins up is $3f_{0\uparrow}^2(1 - f_{0\uparrow})$. We can now calculate $\langle \mathcal{H} \rangle_0$ and $\langle \mathcal{H}_0 \rangle_0$ directly:

$$\langle \mathcal{H}_0 \rangle_0 = -N\hat{\varepsilon}f_0$$

whereas $\quad \langle \mathcal{H} \rangle_0 = 2N\varepsilon\left\{ -2f_{0\uparrow}^3 - 3f_{0\uparrow}^2(1 - f_{0\uparrow})\right\} = 2N\varepsilon\left\{ f_{0\uparrow}^3 - 3f_{0\uparrow}^2 \right\}$

The variational condition then is

$$F \leq -Nk_BT \ln\left(e^{\beta \hat{\varepsilon}} + 2\right) + 2N\left\{ \varepsilon f_{0\uparrow}^3 - 3\varepsilon f_{0\uparrow}^2 \right\} + N\hat{\varepsilon}f_{0\uparrow}$$

It is convenient to express the argument of the logarithm in terms of f_0;

$$F \leq -Nk_BT \ln\left[\frac{2}{(1 - f_{0\uparrow})}\right] + 2N\left[\varepsilon f_{0\uparrow}^3 - 3f_{0\uparrow}^2 \right] + N\hat{\varepsilon}f_{0\uparrow}$$

The variational parameter $\hat{\varepsilon}$ appears explicitly only in the last term, but it is also implicit in $f_{0\uparrow}$. It is somewhat more convenient to minimize F with respect to $f_{0\uparrow}$ (inverting the functional relationship $f_{0\uparrow}(\hat{\varepsilon})$ to consider $\hat{\varepsilon}$ as a function of $f_{0\uparrow}$)

$$0 = \frac{dF}{df_{0\uparrow}} = \frac{-Nk_BT}{(1 - f_{0\uparrow})} + 6N\varepsilon f_{0\uparrow}^2 - 12N\varepsilon f_{0\uparrow} + N\hat{\varepsilon} + Nf_{0\uparrow}\left(\frac{d\hat{\varepsilon}}{df_{0\uparrow}}\right)$$

The last term is easily evaluated to be $Nk_BT[f_{0\uparrow}^{-1} + (1 - f_{0\uparrow})^{-1}]$, so that the variational condition becomes

$$6\beta\varepsilon f_{0\uparrow}^3 - 12\beta\varepsilon f_{0\uparrow}^2 + f_{0\uparrow} \ln\left[\frac{2f_{0\uparrow}}{(1 - f_{0\uparrow})}\right] + 1 = 0$$

This equation must be solved numerically or graphically. Given the solution for $f_{0\uparrow}$ (as a function of the temperature) the various physical properties of the system can be calculated in a straightforward manner.

PROBLEMS

20.2-1. Formulate the exact solution of the two-particle Ising model with an external "field" (assume that each particle can take only two states; $\sigma = -1$ or $+1$). Find both the "magnetization" and the energy, and show that there is no phase transition in zero external field. Solve the problem by mean field theory, and show that a transition to a spontaneous magnetization in zero external field is predicted to occur at a non-zero temperature T_c. Show that below T_c the spontaneous moment varies as $(T - T_c)^\beta$ and find T_c and the critical exponent β (recall Chapter 11).

20.2-2. Formulate mean field theory for the three state Ising model (in which the variables σ_j in equation 20.13 can take the three values $-1, 0, +1$). Find the "Curie" temperature T_c (as in equation 20.25).

20.2-3. For the Heisenberg ferromagnetic model the Hamiltonian is

$$\mathcal{H} = -\sum_{i,j} J_{ij} \mathbf{S}_i \cdot \mathbf{S}_j - (\mu_B B_e) \sum_j S_{jz}$$

where μ_B is the Bohr magneton and B_e is the magnitude of the external field, which is assumed to be directed along the z axis. The z-components of S_j are quantized, taking the permitted values $S_{jz} = -S, -S + 1, \ldots, S - 1, S$. Show that for $S = \frac{1}{2}$ the mean field theory is identical to the mean field theory for the two-state Ising model if $2S$ is associated with σ and if a suitable change of scale is made in the exchange interaction parameter J_{ij}. Are corresponding changes of scale required for the $S = 1$ case (recall Problem 20.2-1), and if so, what is the transformation?

20.2-4. A metallic surface is covered by a monomolecular layer of $\tilde{N}$ organic molecules in a square array. Each adsorbed molecule can exist in two steric configurations, designated as oblate and prolate. Both configurations have the same energy. However two nearest neighbor molecules mechanically interfere if, and only if, both are oblate. The energy associated with such an oblate–oblate interference is ε (a positive quantity). Calculate a reasonable estimate of the number of molecules in each configuration at temperature T.

20.2-5. Solve the preceding problem if the molecules can exist in three steric configurations, designated as oblate, spherical and prolate. Again all three configurations have the same energy. And again two nearest-neighbor molecules interfere if, and only if, both are oblate; the energy of interaction is ε. Calculate (approximately) the number of molecules in each configuration at temperature T.

Answer:

$\tilde{N}/10$ at $k_B T/\varepsilon \approx 0.266$; $\tilde{N}/5$ at $k_B T/\varepsilon \approx 1.15$

$\tilde{N}/4$ at $k_B T/\varepsilon \approx 2.47$; $3\tilde{N}/10$ at $k_B T/\varepsilon \approx 7.78$

$\tilde{N}/3$ at $k_B T \to \infty$

20.2-6. In the classical Heisenberg model each spin can take any orientation in space (recall that the classical partition function of a single spin in an external field B is $z_{\text{classical}} = \int e^{-\beta BS\cos\theta} \sin\theta \, d\theta \, d\phi$. Show that, in mean field theory,

$$\langle S_z \rangle = S \coth \left[\beta(\hat{B} + B)S \right] - \frac{S}{[\beta(\hat{B} + B)S]}$$

20.2-7. $2\tilde{N}$ two-valued Ising spins are arranged sequentially on a circle, so that the last spin is a neighbor of the first. The Hamiltonian is

$$\mathscr{H} = 2 \sum_{j=1}^{2\tilde{N}} J_j \sigma_j \sigma_{j+1} - B \sum_j \sigma_j$$

where $J_j = J_e$ if j is even and $J_j = J_0$ if j is odd. Assume $J_0 > J_e$.

There are two options for carrying out a mean field theory for this system. The first option is to note that all spins are equivalent. Hence one can choose an unperturbed system of $2\tilde{N}$ single spins, each acted on by an effective field (to be evaluated variationally). The second option is to recognize that we can choose a pair of spins coupled by J_0 (the larger exchange interaction). Each such pair is coupled to two other pairs by the weaker exchange interactions J_e. The unperturbed system consists of $\tilde{N}$ such pairs.

Carry out each of the mean field theories described above. Discuss the relative merits of these two procedures.

20.2-8. Consider a sequence of $2\tilde{N}$ alternating A sites and B sites, the system being arranged in a circle so that the $(2\tilde{N})^{\text{th}}$ site is the nearest neighbor of the first site. Even numbered sites are occupied by two-valued Ising spins, with $\sigma_j = \pm 1$. Odd numbered sites are occupied by three-valued Ising spins, with $\sigma_j = -1, 0, +1$. The Hamiltonian is

$$\mathscr{H} = -2J \sum_j \sigma_j \sigma_{j+1} - B \sum_j \sigma_j$$

a) Formulate a mean field theory by choosing as a soluble model system a collection of independent A sites and a collection of independent B sites, each acted upon by a different mean field.

b) Formulate a mean field theory by choosing as a soluble model system a collection of $\tilde{N}$ independent A–B pairs, with the Hamiltonian of each pair being

$$\mathscr{H}_{\text{pair}} = -2J\sigma_{\text{odd}}\sigma_{\text{even}} + B_{\text{odd}}\sigma_{\text{odd}} + B_{\text{even}}\sigma_{\text{even}}$$

c) Are these two procedures identical? If so, why? If not, which procedure would you judge to be superior, and why?

20-3 MEAN FIELD IN GENERALIZED REPRESENTATION: THE BINARY ALLOY

Mean field theory is slightly more general than it might at first appear from the preceding discussion. The larger context is clarified by a particular example. We consider a binary alloy (recall the discussion of Section 11.3) in which each site of a crystalline array can be occupied by either an A atom or a B atom. The system is in equilibrium with a thermal and particle reservoir, of temperature T and of chemical potentials (i.e., partial molar Gibbs potentials) μ_A and μ_B. The energy of an A atom in the crystal is ε_A, and that of a B atom is ε_B. In addition neighboring A atoms have an interaction energy ε_{AA}, neighboring B atoms have an interaction energy ε_{BB}, and neighboring A–B pairs have an interaction energy ε_{AB}.

We are interested not only in the number of A atoms in the crystal, but in the extent to which the A atoms either segregate separately from the B atoms or intermix regularly in an alternating $ABAB$ pattern. That is, we seek to find the average numbers $\tilde{N}_A$ and $\tilde{N}_B$ of each type of atom, and the average numbers $\tilde{N}_{AA}$, $\tilde{N}_{AB}$, and $\tilde{N}_{BB}$ of each type of nearest neighbor pair. These quantities are to be calculated as a function of T, μ_A and μ_B.

The various numbers $\tilde{N}_A$, $\tilde{N}_{AB}, \ldots$ are not all independent, for

$$\tilde{N}_A + \tilde{N}_B = \tilde{N} \tag{20.34}$$

and by counting the number of "bonds" emanating from A atoms

$$2\tilde{N}_{AA} + \tilde{N}_{AB} = z_{nn}\tilde{N}_A \tag{20.35}$$

Similarly

$$2\tilde{N}_{BB} + \tilde{N}_{AB} = z_{nn}\tilde{N}_B \tag{20.36}$$

where we recall that z_{nn} is the number of nearest neighbors of a single site. Consequently all five numbers are determined by two, which are chosen conveniently to be $\tilde{N}_A$ and $\tilde{N}_{AA}$.

The energy of the crystal clearly is

$$E = \tilde{N}_A\varepsilon_A + \tilde{N}_B\varepsilon_B + \tilde{N}_{AA}\varepsilon_{AA} + \tilde{N}_{AB}\varepsilon_{AB} + \tilde{N}_{BB}\varepsilon_{BB} \tag{20.37}$$

If we associate with each site an Ising spin such that the spin is "up" ($\sigma = +1$) if the site is occupied by an A atom, and the spin is "down" ($\sigma = -1$) if the spin is occupied by a B atom, then

$$\mathcal{H} = C - \sum_i \sum_j J_{ij}\sigma_i\sigma_j - \hat{B}\sum_i \sigma_i \tag{20.38}$$

where

$$J = \tfrac{1}{4}\varepsilon_{AB} - \tfrac{1}{8}\varepsilon_{AA} - \tfrac{1}{8}\varepsilon_{BB} \tag{20.39}$$

$$\hat{B} = \tfrac{1}{2}(\varepsilon_{BB} - \varepsilon_{AA}) + \tfrac{1}{4}z_{nn}(\varepsilon_{AA} + \varepsilon_{BB}) \tag{20.40}$$

$$C = \tfrac{1}{8}\tilde{N}(4\varepsilon_A + 4\varepsilon_B + z_{nn}\varepsilon_{AA} + 2z_{nn}\varepsilon_{AB} + z_{nn}\varepsilon_{BB}) \tag{20.41}$$

These values of J, $\hat{B}$, and C can be obtained in a variety of ways. One simple approach is to compare the values of E (equation 20.37) and of $\mathscr{H}$ (equation 20.38) in the three configurations in which (a) all sites are occupied by A atoms, (b) all sites are occupied by B atoms, and (c) equal numbers of A and B atoms are randomly distributed.

Except for the inconsequential constant C, the Hamiltonian is now that of the Ising model. However, the physical problem is quite different. We must recall that the system is in contact with particle reservoirs of chemical potentials μ_A and μ_B, as well as with a thermal reservoir of temperature T. The problem is best solved in a grand canonical formalism.

The essential procedure in the grand canonical formalism is the calculation of the grand canonical potential $\Psi(T, \mu_A, \mu_B)$ by the algorithm[3]

$$e^{-\beta\Psi} = \operatorname{tr} e^{-\beta(\mathscr{H} - \tilde{\mu}_A \tilde{N}_A - \tilde{\mu}_B \tilde{N}_B)} \tag{20.42}$$

This is isomorphic with the canonical formalism (on which the mean field theory of Section 20.2 was based) if we simply replace the Helmholtz potential F by the grand canonical potential Ψ, and replace the Hamiltonian $\mathscr{H}$ by the "grand canonical Hamiltonian" $\mathscr{H} - \tilde{\mu}_A \tilde{N}_A - \tilde{\mu}_B \tilde{N}_B$.

In the present context we augment the Hamiltonian 20.38 by terms of the form $-\frac{1}{2}[(\tilde{\mu}_A + \tilde{\mu}_B) + (\tilde{\mu}_A - \tilde{\mu}_B)\Sigma_i \sigma_i]$. The grand canonical Hamiltonian is then

$$\mathscr{H}' = C' - \sum_i \sum_j J_{ij}\sigma_i\sigma_j - \hat{B}'\sum_i \sigma_i \tag{20.43}$$

where

$$C' = C - \frac{1}{2}\tilde{N}(\tilde{\mu}_A + \tilde{\mu}_B) \tag{20.44}$$

and

$$\hat{B}' = \hat{B} - \frac{1}{2}(\tilde{\mu}_A - \tilde{\mu}_B) \tag{20.45}$$

The analysis of the Ising model then applies directly to the binary alloy problem (with the Helmholtz potential being reinterpreted as the grand canonical potential). Again mean field theory predicts an order–disorder phase transition. Again that prediction agrees with more rigorous theory in two and three dimensions, whereas a one-dimensional binary crystal should not have an order–disorder phase transition. And again the critical exponents are incorrectly predicted.

More significantly, the general approach of mean field theory is applicable to systems in generalized ensembles, requiring only the reinterpretation of the thermodynamic potential to be calculated, and of the effective "Hamiltonian" on which the calculation is to be based.

---

[3] $\tilde{\mu}_A$ ($\equiv \mu_A/$Avogadro's number) is the chemical potential *per particle*.

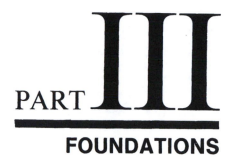

PART III

FOUNDATIONS

21

POSTLUDE: SYMMETRY AND THE CONCEPTUAL FOUNDATIONS OF THERMOSTATISTICS

21-1 STATISTICS

The overall structure of thermostatistics now has been established—of thermodynamics in Part I and of statistical mechanics in Part II. Although these subjects can be elaborated further, the logical basis is essentially complete. It is an appropriate time to reconsider and to reflect on the uncommon form of these atypical subjects.

Unlike mechanics, thermostatistics is not a detailed theory of dynamic response to specified forces. And unlike electromagnetic theory (or the analogous theories of the nuclear "strong" and "weak" forces), thermostatistics is not a theory of the forces themselves. Instead thermostatistics characterizes the equilibrium state of microscopic systems *without reference either to the specific forces or to the laws of mechanical response.* Instead thermostatistics characterizes the equilibrium state as the state that maximizes the *disorder*, a quantity associated with a conceptual framework ("information theory") outside of conventional physical theory. The question arises as to whether the postulatory basis of thermostatistics thereby introduces new principles not contained in mechanics, electromagnetism, and the like or whether it borrows principles in unrecognized form from that standard body of physical theory. In either case, what are the implicit principles upon which thermostatistics rests?

There are, in my view, two essential bases underlying thermostatistical theory. One is rooted in the statistical properties of large complex systems. The second rests in the set of symmetries of the fundamental laws of physics. *The statistical feature veils the incoherent complexity of the atomic dynamics, thereby revealing the coherent effects of the underlying physical symmetries.*

The relevance of the statistical properties of large complex systems is universally accepted and reasonably evident. The essential property is epitomized in the "central limit theorem"[1] which states (roughly) that the probability density of a variable assumes the "Gaussian" form if the variable is itself the resultant of a large number of independent additive subvariables. Although one might naively hope that measurements of thermodynamic fluctuation amplitudes could yield detailed information as to the atomic structure of a system, the central limit theorem precludes such a possibility. It is this insensitivity to specific structural or mechanical detail that underlies the universality and simplicity of thermostatistics.

The central limit theorem is illustrated by the following example.

Example
Consider a system composed of $\tilde{N}$ "elements," each of which can take a value of X in the range $-\frac{1}{2} < X < \frac{1}{2}$. The value of X for each element is a continuous random variable with a probability density that is uniform over the permitted region. The value of X for the system is the sum of the values for each of the elements. Calculate the probability density for the system for the cases $\tilde{N} = 1, 2, 3$. In each case find the standard deviation σ, defined by

$$\sigma^2 = \int f(X) X^2 dX$$

where $f(X)$ is the probability density of X (and where we have given the definition of σ only for the relevant case in which the mean of X is zero). Plot the probability density for $\tilde{N} = 1$, 2, and 3, and in each case plot the Gaussian or "normal" distribution with the same standard deviation.

Note that for even so small a number as $\tilde{N} = 3$ the probability distribution $f(X)$ rapidly approaches the Gaussian form! It should be stressed that in this example the uniform probability density of X is chosen for ease of calculation; *a similar approach to the Gaussian form would be observed for any initial probability density.*

Solution
The probability density for $\tilde{N} = 1$ is $f_1(X) = 1$ for $-\frac{1}{2} < X < \frac{1}{2}$, and zero otherwise. This probability density is plotted in Fig. 21.1*a*. The standard deviation is $\sigma_1 = 1/(2 \cdot \sqrt{3})$. The corresponding Gaussian

$$f_G(X) = (2\pi)^{-1/2} \sigma^{-1} \exp\left(\frac{-X^2}{2\sigma^2}\right)$$

with $\sigma = \sigma_1$ is also plotted in Fig. 21.1*a*, for comparison.

---

[1]*cf.* any standard reference on probability, such as L. G. Parratt, *Probability and Experimental Errors in Science* (Wiley, New York, 1961) or E. Parzen, *Modern Probability Theory and Its Applications* (Wiley, New York, 1960).

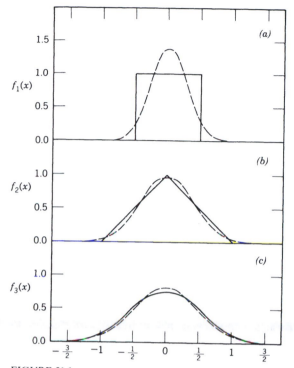

FIGURE 21.1

Convergence of probability density to the Gaussian form. The probability density for systems composed of one, two and three elements, each with the probability density shown in Figure 21.1a. In each case the Gaussian with the same standard deviation is plotted. In accordance with the central limit theorem the probability density becomes Gaussian for large $\tilde{N}$.

To calculate the probability density $f_2(X)$, for $\tilde{N} = 2$, we note (problem 21.1-1) that

$$f_{\tilde{N}+1}(X) = \int_{-\infty}^{\infty} f_{\tilde{N}}(X - X')f_1(X')\,dX'$$

or, with $f_1(X)$ as given

$$f_{\tilde{N}+1}(X) = \int_{-1/2}^{1/2} f_{\tilde{N}}(X - X')\,dX'$$

That is, $f_{\tilde{N}+1}(X)$ is the average value of $f_{\tilde{N}}(X'')$ over a range of length unity centered at X.

This geometric interpretation easily permits calculation of $f_2(X)$ as shown in Fig. 21.1b. From $f_2(X)$, in turn, we find

$$f_3(X) = \begin{cases} \frac{3}{4} - X^2 & \text{if } |X| \leq \frac{1}{2} \\ \frac{9}{8} - \frac{3}{2}X + \frac{1}{2}X^2 & \text{if } \frac{1}{2} \leq |X| \leq \frac{3}{2} \\ 0 & \text{if } |X| \geq \frac{3}{2} \end{cases}$$

The values of σ are calculated to be $\sigma_1 = 1/\sqrt{12}$, $\sigma_2 = 1/\sqrt{6}$ and $\sigma_3 = \frac{1}{2}$. These values agree with a general theorem that for $\tilde{N}$ identical and independent subsystems, $\sigma_{\tilde{N}} = \sqrt{\tilde{N}} \, \sigma_1$. The Gaussian curves of Fig. 21.1 are calculated with these values of the standard deviations. For even so small a value of $\tilde{N}$ as 3 the probability distribution is very close to Gaussian, losing almost all trace of the initial shape of the single-element probability distribution.

PROBLEMS

21.1-1. The probability of throwing a "seven" on two dice can be viewed as the sum of a) the probability of throwing a "one" on the first die multiplied by the probability of throwing a "six" on the second, plus b) the probability of throwing a "two" on the first die multiplied by the probability of throwing a "five" on the second, and so forth. Explain the relationship of this observation to the expression for $f_{\tilde{N}+1}(X)$ in terms of $f_{\tilde{N}}(X - X')$ and $f_1(X')$ as given in the Example, and derive the latter expression.

21.1-2. Associate the value $+1$ with one side of a coin ("head") and the value -1 with the other side ("tail"). Plot the probability of finding a given "value" when throwing one, two, three, four, and five coins. (Note that the probability is discrete—for two coins the plot consists of just three points, with probability $= \frac{1}{4}$ for $X = \pm 1$ and probability $= \frac{1}{2}$ for $X = 0$.) Calculate σ for the case $n = 5$, and roughly sketch the Gaussian distribution for this value of σ.

21-2 SYMMETRY[2]

As a basis of thermostatistics the role of symmetry is less evident than the role of statistics. However, we first note that a basis in symmetry does rationalize the peculiar nonmetric character of thermodynamics. The results of thermodynamics characteristically relate apparently unlike quantities, yielding relationships such as $(\partial T/\partial P)_V = (\partial V/\partial S)_T$, but providing no numerical evaluation of either quantity. Such an emphasis on *relationships,* as contrasted with *quantitative evaluations,* is appropriately to be expected of a subject with roots in symmetry rather than in explicit quantitative laws.

Although symmetry considerations have been seen as basic in science since the dawn of scientific thought, the development of quantum mechanics in 1925 elevated symmetry considerations to a more profound level of power, generality, and fundamentality than they had enjoyed in classical physics. Rather than merely *restricting* physical possibilities, symmetry was increasingly seen as playing the fundamental role in *establishing the*

[2] H. Callen, *Foundations of Physics* **4**, 423 (1974).

form of physical laws. Eugene Wigner, Nobel laureate and great modern expositor of symmetry laws, suggested[3] that the relationship of symmetry properties to the laws of nature is closely analogous to the relationship of the laws of nature to individual events; the symmetry principles "provide a structure or coherence to the laws of nature just as the laws of nature provide a structure and coherence to the set of events." Contemporary "grand unified theories" conjecture that the very existence and strength of the four basic force fields of physical theory (electromagnetic, gravitational, "strong," and "weak") were *determined* by a symmetry genesis a mere 10^{-35} seconds after the Big Bang.

The simplest and most evident form of symmetry is the geometric symmetry of a physical object. Thus a sphere is symmetric under arbitrary rotations around any axis passing through its center, under reflections in any plane containing the center, and under inversion through the center itself. A cube is symmetric under fourfold rotations around axes through the face centers and under various other rotations, reflections, and inversion operations.

Because a sphere is symmetric under rotations through an angle that can take *continuous* values the rotational symmetry group of a sphere is said to be continuous. In contrast, the rotational symmetry group of a cube is discrete.

Each geometrical symmetry operation is described mathematically by a coordinate transformation. Reflection in the x–y plane corresponds to the transformation $x \to x'$, $y \to y'$, $z \to -z'$, whereas fourfold (90°) rotation around the z-axis is described by $x \to y'$, $y \to -x'$, $z \to z'$. The symmetry of a sphere under either of these operations corresponds to the fact that the equation of a sphere ($x^2 + y^2 + z^2 = r^2$) is identical in form if reexpressed in the primed coordinates.

The concept of a geometrical symmetry is easily generalized. A transformation of variables *defines* a symmetry operation. A function of those variables that is unchanged in form by the transformation is said to be symmetric with respect to the symmetry operation. Similarly a law of physics is said to be symmetric under the operation if the functional form of the law is invariant under the transformation.

Newton's law of dynamics, $\mathbf{f} = m(d^2\mathbf{r}/dt^2)$ is symmetric under time inversion ($\mathbf{r} \to \mathbf{r}'$, $t \to -t'$) for a system in which the force is a function of position only. Physically this "time-inversion symmetry" implies that a video tape of a ball thrown upward by an astronaut on the moon, and falling back to the lunar surface, looks identical if projected backward or forward. (On the earth, in the presence of air friction, the dynamics of the ball would not be symmetric under time inversion).

The symmetry of the dynamical behavior of a particular system is governed by the dynamical equation and by the mechanical potential that

[3] E. Wigner, "Symmetry and Conservation Laws," *Physics Today*, March 1964, p. 34.

determines the forces. For quantum mechanical problems the dynamical equation is more abstract (Schrödinger's equation rather than Newton's law), but the principles of symmetry are identical.

21-3 NOETHER'S THEOREM

A far reaching and profound physical consequence of symmetry is formulated in "Noether's theorem[4]". The theorem asserts that *every continuous symmetry of the dynamical behavior of a system* (i.e., of the dynamical equation and the mechanical potential) *implies a conservation law for that system*.

The dynamical equation for the motion of the center of mass point of any material system is Newton's law. If the external force does not depend upon the coordinate x, then both the potential and the dynamical equation are symmetric under spatial translation parallel to the x-axis. The quantity that is conserved as a consequence of this symmetry is the x-component of the momentum. Similarly the symmetry under translation along the y or z axes results in the conservation of the y or z components of the momentum. Symmetry under rotation around the z axis implies conservation of the z-component of the angular momentum.

Of enormous significance for thermostatistics is the *symmetry of dynamical laws under time translation*. That is, the fundamental dynamical laws of physics (such as Newton's law, Maxwell's equations, and Schrödinger's equation) are unchanged by the transformation $t \rightarrow t' + t_0$ (i.e., by a shift of the origin of the scale of time). If the external potential is independent of time, Noether's theorem predicts the existence of a conserved quantity. That conserved quantity is called the energy.

Immediately evident is the relevance of time-translation symmetry to what is often called the "first law of thermodynamics"—the existence of the energy as a conserved state function (recall Section 1.3 and Postulate I).

It is instructive to reflect on the profundity of Noether's theorem by comparing the conclusion here with the tortuous historical evolution of the energy concept in mechanics (recall Section 1.4). Identification of the conserved energy began in 1693 when Leibniz observed that $\frac{1}{2}mv^2 + mgh$ is a conserved quantity for a mass particle in the earth's gravitational field. As successively more complex systems were studied it was found that additional terms had to be appended to maintain a conservation principle,

---

[4] See E. Wigner, ibid. The physical content of Noether's theorem is implicit in Emmy Noether's purely mathematical studies. A beautiful appreciation of this brilliant mathematician's life and work in the face of implacable prejudice can be found in the introductory remarks to her collected works: Emmy Noether, *Gesammelte Abhandlungen*, (*Collected Papers*), Springer–Verlag, Berlin–New York, 1983.

but that in each case such an ad hoc addition was possible. The development of electromagnetic theory introduced the potential energy of the interaction of electric charges, subsequently to be augmented by the electromagnetic field energy. In 1905 Albert Einstein was inspired to alter the expression for the mechanical kinetic energy, and even to associate energy with stationary mass, in order to maintain the principle of energy conservation. In the 1930s Enrico Fermi postulated the existence of the neutrino solely for the purpose of retaining the energy conservation law in nuclear reactions. And so the process continues, successively accreting additional terms to the abstract concept of energy, *which is defined by its conservation law*. That conservation law was evolved historically by a long series of successive rediscoveries. It is now based on the assumption of time translation symmetry.

The evolution of the energy concept for macroscopic thermodynamic systems was even more difficult. The pioneers of the subject were guided neither by a general a priori conservation theorem nor by any specific analytic formula for the energy. Even empiricism was thwarted by the absence of a method of direct measurement of heat transfer. Only inspired insight guided by faith in the simplicity of nature somehow revealed the interplay of the concepts of energy and entropy, even in the absence of a priori definitions or of a means of measuring either!

21-4 ENERGY, MOMENTUM, AND ANGULAR MOMENTUM: THE GENERALIZED "FIRST LAW" OF THERMOSTATISTICS

In accepting the existence of a conserved macroscopic energy function as the first postulate of thermodynamics, we anchor that postulate directly in Noether's theorem and in the time-translation symmetry of physical laws.

An astute reader will perhaps turn the symmetry argument around. There are *seven* "first integrals of the motion" (as the conserved quantities are known in mechanics). These seven conserved quantities are the energy, the three components of linear momentum, and the three components of the angular momentum; and they follow in parallel fashion from the translation in "space–time" and from rotation. Why, then, does energy appear to play a unique role in thermostatistics? Should not momentum and angular momentum play parallel roles with the energy?

In fact, the energy is *not* unique in thermostatistics. The linear momentum and angular momentum play precisely parallel roles. *The asymmetry in our account of thermostatistics is a purely conventional one that obscures the true nature of the subject.*

We have followed the standard convention of restricting attention to systems that are macroscopically stationary, in which case the momentum and angular momentum *arbitrarily* are required to be zero and do not appear in the analysis. But astrophysicists, who apply thermostatistics to rotating galaxies, are quite familiar with a more complete form of thermostatistics. In that formulation the energy, linear momentum, and angular momentum play fully analogous roles.

The fully generalized canonical formalism is a straightforward extension of the canonical formalism of Chapters 16 and 17. Consider a subsystem consisting of N moles of stellar atmosphere. The stellar atmosphere has a particular mean molar energy (U/N), a particular mean molar momentum $(\mathbf{P}/N)$, and a particular mean molar angular momentum $(\mathbf{J}/N)$. The fraction of time that the subsystem spends in a particular microstate i (with energy E_i, momentum $\mathbf{P}_i$, and angular moment $\mathbf{J}_i$) is $f_i(E_i, \mathbf{P}_i, \mathbf{J}_i, V, N)$. Then f_i is determined by maximizing the disorder, or entropy, subject to the constraints that the average energy of the subsystem be the same as that of the stellar atmosphere, and similarly for momentum and angular momentum. As in Section 17.2, we quite evidently find

$$f_i = \frac{1}{Z} \exp\left(-\beta E_i - \boldsymbol{\lambda}_p \cdot \mathbf{P}_i - \boldsymbol{\lambda}_J \cdot \mathbf{J}_i \right) \tag{21.1}$$

The seven constants β, λ_{px}, λ_{py}, λ_{pz}, λ_{Jx}, λ_{Jy}, and λ_{Jz} all arise as Lagrange parameters and they play completely symmetric roles in the theory (just as $\beta\mu$ does in the grand canonical formalism).

The proper "first law of thermodynamics," (or the first postulate in our formulation) is the symmetry of the laws of physics under space–time translation and rotation, and the consequent existence of conserved energy, momentum, and angular momentum functions.

21-5 BROKEN SYMMETRY AND GOLDSTONE'S THEOREM

As we have seen, then, the entropy of a thermodynamic system is a function of various coordinates, among which the energy is a prominent member. The energy is, in fact, a surrogate for the seven quantities conserved by virtue of space–time translations and rotations. But other independent variables also exist—the volume, the magnetic moment, the mole numbers, and other similar variables. How do these arise in the theory?

The operational criterion for the independent variables of thermostatistics (recall Chapter 1) is that they be *macroscopically observable*. The low temporal and spatial resolving powers of macroscopic observations require that thermodynamic variables be essentially time independent on the atomic scale of time and spatially homogeneous on the atomic scale of distance. The time independence of the energy (and of the linear and angular momentum) has been rationalized through Noether's theorem.

The time independence of other variables is based on the concept of *broken symmetry* and *Goldstone's theorem*. These concepts are best introduced by a particular case and we focus specifically on the volume.

For definiteness, consider a crystalline solid. As we saw in Section 16.7, the vibrational modes of the crystal are described by a wave number $k(= 2\pi/\lambda$, where λ is the wavelength) and by an angular frequency $\omega(\mathbf{k})$. For very long wavelengths the modes become simple sound waves, and in this region the frequency is proportional to the wave number; $\omega = ck$ (recall Fig. 16.1). The significant feature is that $\omega(\mathbf{k})$ vanishes for $k = 0$ (i.e., for $\lambda \to \infty$). Thus, the very mode that is spatially homogeneous has zero frequency. Furthermore, as we have seen in Chapter 1 (refer also to Problem 21.5-1), the volume of a macroscopic sample is associated with the amplitude of the spatially homogeneous mode. Consequently the volume is an acceptably time independent thermodynamic coordinate.

The vanishing of the frequency of the homogeneous mode is not simply a fortunate accident, but rather it is associated with the general concept of broken symmetry. The concept of broken symmetry is clarified by reflecting on the process by which a crystal may be formed. Suppose the crystal to be solid carbon dioxide ("dry ice"), and suppose the CO_2 initially to be in the gaseous state, contained in some relatively large vessel ("infinite" in size). The gas is slowly cooled. At the temperature of the gas–solid phase transition a crystalline nucleus forms at some point in the gas. The nucleus thereafter grows until the gas pressure falls to that on the gas–solid coexistence curve (i.e., to the vapor pressure of the solid). From the point of view of symmetry the condensation is a quite remarkable development. In the "infinite" gas the system is symmetric under a continuous translation group, but the condensed solid has a lower symmetry! It is invariant only under a discrete translation group. Furthermore the location of the crystal is arbitrary, determined by the accident of the first microscopic nucleation. In that nucleation process the symmetry of the system suddenly and spontaneously lowers, and it does so by a nonpredictable, random event. The symmetry of the system is "broken."

Macroscopic sciences, such as solid state physics or thermodynamics, are qualitatively different from "microscopic" sciences because of the effects of broken symmetry, as was pointed out by P. W. Anderson[5] in an early but profound and easily readable essay which is highly recommended to the interested reader.

At sufficiently high temperature systems always exhibit the full symmetry of the "mechanical potential" (that is, of the Lagrangian or Hamiltonian functions). There do exist permissible microstates with lower symmetry, but these states are grouped in sets which *collectively* exhibit the full symmetry. Thus the microstates of a gas do include states with crystal-like spacing of the molecules—in fact, among the microstates all manner of different crystal-like spacings are represented, so that collec-

[5]P. W. Anderson, pp. 175–182 in *Concepts in Solids* (W. A. Benjamin Inc., New York, 1964).

tively the states of the gas retain no overall crystallinity whatever. However, as the temperature of the gas is lowered the molecules select that particular crystalline spacing of lowest energy, and the gas condenses into the corresponding crystal structure. This is a *partial* breaking of the symmetry. Even among the microstates with this crystalline periodicity there are a continuum of possibilities available to the system, for the incipient crystal could crystallize with any arbitrary position. Given one possible crystal position there exist infinitely many equally possible positions, slightly displaced by an arbitrary fraction of a "lattice constant". Among these possibilities, all of equal energy, the system chooses one position (i.e., a nucleation center for the condensing crystallite) arbitrarily and "accidentally".

An important general consequence of broken symmetry is formulated in the Goldstone theorem[6]. It asserts that *any system with broken symmetry* (*and with certain weak restrictions on the atomic interactions*) *has a spectrum of excitations for which the frequency approaches zero as the wavelength becomes infinitely large.*

For the crystal discussed here the Goldstone theorem ensures that a phonon excitation spectrum exists, and that its frequency vanishes in the long wavelength limit.

The proof of the Goldstone theorem is beyond the scope of this book, but its intuitive basis can be understood readily in terms of the crystal condensation example. The vibrational modes of the crystal oscillate with sinusoidal time dependence, their frequencies determined by the masses of the atoms and by the restoring forces which resist the crowding together or the separation of those atoms. But in a mode of very long wavelength the atoms move very nearly in phase; for the infinite wavelength mode the atoms move in unison. Such a mode does not call into action any of the interatomic forces. The very fact that the original position of the crystal was arbitrary—that a slightly displaced position would have had precisely the same energy—guarantees that no restoring forces are called into play by the infinite wavelength mode. Thus the vanishing of the frequency in the long wavelength limit is a direct consequence of the broken symmetry. The theorem, so transparent in this case, is true in a far broader context, with far-reaching and profound consequences.

In summary, then, the volume emerges as a thermodynamic coordinate by virtue of a fundamental symmetry principle grounded in the concept of broken symmetry and in Goldstone's theorem.

PROBLEMS

21.5-1. Draw a longitudinal vibrational mode in a one-dimensional system, with a node at the center of the system and with a wavelength twice the nominal length

---

[6]P. W. Anderson, *ibid.*

of the system. Show that the instantaneous length of the system is a linear function of the instantaneous amplitude of this mode. What is the order of magnitude of the wavelength if the system is macroscopic and if the wavelength is measured in dimensionless units (i.e., relative to interatomic lengths)?

21-6 OTHER BROKEN SYMMETRY COORDINATES—ELECTRIC AND MAGNETIC MOMENTS

In the preceding two sections we have witnessed the role of symmetry in determining several of the independent variables of thermostatistical theory. We shall soon explore other ways in which symmetry underlies the bases of thermostatistics, but in this section and in the following we continue to explore the nature of the extensive parameters. It should perhaps be noted that the choice of the variables in terms of which a given problem is formulated, while a seemingly innocuous step, is often the most crucial step in the solution.

In addition to the energy and the volume, other common extensive parameters are the magnetic and electric moments. These are also properly time independent by virtue of broken symmetry and Goldstone's theorem. For definiteness consider a crystal such as HCl. This material crystallizes with an HCl molecule at each lattice site. Each hydrogen ion can rotate freely around its relatively massive Cl partner, so that each molecule constitutes an electric dipole that is free to point in any arbitrary direction in space. At low temperatures the dipoles order, all pointing more or less in one common direction and thereby imbueing the crystal with a net dipole moment.

The direction of the net dipole moment is the residue of a random accident associated with the process of cooling below the ordering temperature. Above that temperature the crystal had a higher symmetry; below the ordering temperature it develops one unique axis—the direction of the net dipole moment.

Below the ordering temperature the dipoles are aligned generally (but not precisely) along a common direction. Around this direction the dipoles undergo small dynamic angular oscillations ("librations"), rather like a pendulum. The librational oscillations are coupled, so that librational waves propagate through the crystal. These librational waves are the Goldstone excitations. The Goldstone theorem implies that the librational modes of infinite wavelength have zero frequency[7]. Thus the electric

---

[7]In the interests of clarity I have oversimplified slightly. The discussion here overlooks the fact that the crystal structure would have already destroyed the spherical symmetry even above the ordering temperature of the dipoles. That is, the discussion as given would apply to an amorphous (spherically symmetric) crystal but not to a cubic crystal. In a cubic crystal each electric dipole would be coupled by an "anisotropy energy" to the cubic crystal structure, and this coupling would (naively) appear to provide a restoring force even to infinite wavelength librational modes. However, under these circumstances librations and crystal vibrations would couple to form mixed modes, and these coupled "libration–vibration" modes would again satisfy the Goldstone theorem.

dipole moment of the crystal qualifies as a time independent thermody-namic coordinate.

Similarly ferromagnetic crystals are characterized by a net magnetic moment arising from the alignment of electron spins. These spins par-ticipate in collective modes known as "spin waves." If the spins are not coupled to lattice axes (i.e., in the absence of "magnetocrystalline ani-sotropy") the spin waves are Goldstone modes and the frequency vanishes in the long wavelength limit. In the presence of magnetocrystalline ani-sotropy the Goldstone modes are coupled phonon–spin-wave excitations. In either case the total magnetic moment qualifies as a time independent thermodynamic coordinate.

21-7 MOLE NUMBERS AND GAUGE SYMMETRY

We come to the last representative type of thermodynamic coordinate, of which the mole numbers are an example.

Among the symmetry principles of physics perhaps the most abstract is the set of "gauge symmetries." The representative example is the "gauge transformation" of Maxwell's equations of electromagnetism. These equa-tions can be written in terms of the observable electric and magnetic fields, but a more convenient representation introduces a "scalar poten-tial" and a "vector potential." The electric and magnetic fields are derivable from these potentials by differentiation. However the electric and magnetic potentials are not unique. Either can be altered in form providing the other is altered in a compensatory fashion, the coupled alterations of the scalar and vector potentials constituting the "gauge transformation." The fact that the observable electric and magnetic fields are invariant to the gauge transformation is the "gauge symmetry" of electromagnetic theory. The quantity that is conserved by virtue of this symmetry is the electric charge[8].

Similar gauge symmetries of fundamental particle theory lead to con-servation of the numbers of leptons (electrons, mesons, and other particles of small rest mass) and of the numbers of baryons (protons, neutrons, and other particles of large rest mass).

In the thermodynamics of a hot stellar interior, where nuclear transfor-mations occur sufficiently rapidly to achieve nuclear equilibrium, the numbers of leptons and the numbers of baryons would be the appropriate "mole numbers" qualifying as thermodynamic extensive parameters.

In common terrestrial experience the baryons form long-lived associa-tions to constitute quasi-stable atomic nuclei. It is then a reasonable

---

[8] The result is a uniquely quantum mechanical result. It depends upon the fact that the phase of the quantum mechanical wave function is arbitrary ("gauge symmetry of the second kind"), and it is the interplay of the two types of gauge symmetry that leads to charge conservation.

approximation to consider atomic (or even molecular) species as being in quasi-stable equilibrium, and to consider the atomic mole numbers as appropriate thermodynamic coordinates.

21-8 TIME REVERSAL, THE EQUAL PROBABILITIES OF MICROSTATES, AND THE ENTROPY PRINCIPLE

We come finally to the essence of thermostatistics—to the principle that an isolated system spends equal fractions of the time in each of its permissible microstates. Given this principle it then follows that the number of occupied microstates is maximum consistent with the external constraints, that the logarithm of the number of microstates is also maximum (and that it is extensive), and that the entropy principle is validated by interpreting the entropy as proportional to $\ln \Omega$.

The permissible microstates of a system can be represented in an abstract, many-dimensional state space (recall Section 15.5). In the state space every permissible microstate is represented by a discrete point. The system then follows a random, erratic trajectory in the space as it undergoes stochastic transitions among the permissible states. These transitions are guaranteed by the random external perturbations which act on even a nominally "isolated" system (although other mechanisms may dominate in particular cases—recall Section 15.1).

The evolution of the system in state space is guided by a set of transition probabilities. If a system happens at a particular instant to be in a microstate i then it may make a transition to the state j, with probability (per unit time) f_{ij}. The transition probabilities $\{f_{ij}\}$ form a network joining pairs of states throughout the state space.

The formalism of quantum mechanics establishes that, at least in the absence of external magnetic fields[9]

$$f_{ij} = f_{ji} \tag{21.2}$$

That is, a system in the state i will undergo a transition to the state j with the same probability that a system in state j will undergo a transition to the state i.

The "*principle of detailed balance*" (equation 21.2) *follows from the symmetry of the relevant laws of quantum mechanics under time inversion* (i.e., under the transformation $t \rightarrow -t'$).

---

[9] The restriction that the external magnetic field must be zero can be dealt with most simply by including the source of the magnetic field as part of the system. In any case the presence of external magnetic fields complicates intermediate statements but does not alter final conclusions, and we shall here ignore such fields in the interests of simplicity and clarity.

Although we merely quote the principle of detailed balance as a quantum mechanical theorem, it is intuitively reasonable. Consider a system in the microstate i, and imagine a video tape of the dynamics of the system (a hypothetical form of video tape that records the microstate of the system!). After a brief moment the system makes a transition to the microstate j. If the video tape were to be played backwards the system would start in the state j and make a transition to the state i. Thus the interchangeability of future and past, or the time reversibility of physical laws, associates the transitions $i \rightarrow j$ and $j \rightarrow i$ and leads to the equality (21.2) of the transition probabilities.

The principle of equal probabilities of states in equilibrium ($f_i = 1/\Omega$) follows from the principle of detailed balance ($f_{ij} = f_{ji}$). To see that this is so we first observe that f_{ij} is the conditional probability that the system will undergo a transition to state j *if* it is initially in state i. The number of such transitions per unit time is then the product of f_{ij} and the probability f_i that the system is initially in the state i. Hence the total number of transitions per unit time out of the state i is $\sum_j f_i f_{ij}$. Similarly the number of transitions per unit time into the state i is $\sum_j f_j f_{ji}$. However in equilibrium the occupation probability f_i of the ith state must be independent of time; or

$$\frac{df_i}{dt} = -\sum_{j \neq i} f_i f_{ij} + \sum_{j \neq i} f_j f_{ji} = 0 \tag{21.3}$$

With the symmetry condition $f_{ij} = f_{ji}$, a general solution of equation 21.3 is $f_i = f_j$ for all i and j. That is, *the configuration $f_j = 1/\Omega$ is an equilibrium configuration for any set of transition probabilities $\{f_{ij}\}$ for which $f_{ij} = f_{ji}$.*

As the system undergoes random transitions among its microstates some states are "visited" frequently (i.e., $\sum_j f_{ji}$ is large), and others are visited only infrequently. Some states are tenacious of the system once it does arrive (i.e., $\sum_j f_{ij}$ is small), whereas others permit it to depart rapidly. Because of time reversal symmetry, however, those states that are visited only infrequently are tenacious of the system. Those states that are visited frequently host the system only fleetingly. By virtue of these compensating attributes the system spends the same fraction of time in each state.

The equal probabilities of permissible states for a closed system in equilibrium is a consequence of time reversal symmetry of the relevant quantum mechanical laws[10].

---

[10] In fact a weaker condition, $\sum_j (f_{ij} - f_{ji}) = 0$, which follows from a more abstract requirement of "causality," is also sufficient to ensure that $f_i = 1/\Omega$ in equilibrium. This fact does not invalidate the previous statement.

21-9 SYMMETRY AND COMPLETENESS

There is an additional, more subtle aspect of the principle of equal a priori probabilities of states. Consider the schematic representation of state space in Fig. 21.2. The boundary B separates the permissible states ("inside") from the nonpermissible states ("outside"). The transition probabilities f_{ij} are symmetric for all states i and j inside the boundary B.

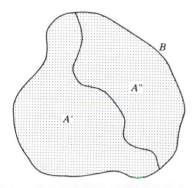

FIGURE 21.2

Suppose now that the permissible region in state space is divided into two subregions (denoted by A' and A'' in Fig. 21.2) such that all transition probabilities f_{ij} vanish if the state i is in A' and j is in A'', or vice versa. Such a set of transition probabilities is fully consistent with time reversal symmetry (or detailed balance), but it does *not* lead to a probability uniform over the physically permissible region ($A' + A''$). If the system were initially in A' the probability density would diffuse from the initial state to eventually cover the region A' uniformly, but it would not cross the internal boundary to the region A''.

The "accident" of such a zero transition boundary, separating the permissible states into nonconnected subsets, would lead to a failure of the assumption of equal probabilities throughout the permissible region of state space.

It is important to recognize how incredibly stringent must be the rule of vanishing of the f_{ij} between subregions if the principle of equal probabilities of states is to be violated. It is not sufficient for transition probabilities between subregions to be very small—*every* such transition probability must be absolutely and rigorously zero. If even one or a few transition probabilities were merely very small across the internal boundary it would take a very long time for the probability density to fill both A' and A'' uniformly, but eventually it would.

The "accident" that we feared might vitiate the conclusion of equal probabilities appears less and less likely—unless it is not an accident at all, but the consequence of some underlying principle. Throughout quantum physics the occurrence of outlandish accidents is disbarred; physics is neither mystical nor mischievous. If a physical quantity has a particular value, say 4.5172... then a second physical quantity will not have *precisely* that same value unless there is a compelling reason that ensures equality. Degeneracy of energy levels is the most familiar example—when it occurs it always reflects a symmetry origin. Similarly, transition probabilities do not accidentally assume the precise value zero; when they do vanish they do so by virtue of an underlying symmetry based reason. The vanishing of a transition probability as a consequence of symmetry is called a "selection rule."

Selection rules that divide the state space into disjoint regions do exist. They always reflect symmetry origins and they imply conservation principles. An already familiar example is provided by a ferromagnetic system. The states of the system can be classified by the components of the total angular momentum. States with different total angular momentum components have different symmetries under rotation, and the selection rules of quantum mechanics forbid transitions among such states. These selection rules give rise to the conservation of angular momentum.

More generally, then, the state space *can* be subdivided into disjoint regions, not connected by transition probabilities. These regions are never accidental; they reflect an underlying symmetry origin. Each region can be labeled according to the symmetry of its states—such labels are called the "characters of the group representation." The symmetry thereby gives rise to a conserved quantity, the possible values of which correspond to the distinguishing labels for the disjoint regions of state space.

In order that thermodynamics be valid it is necessary that the set of extensive parameters be complete. Any conserved quantity, such as that labelling a disjuncture of the state space, must be included in the set of thermodynamic coordinates. Specifying the value of that conserved quantity then restricts the permissible state space to a single disjoint sector (A' alone, or A'' alone, in Fig. 21.2). The principle of equal probabilities of states is restored only when all such symmetry based thermodynamic coordinates are recognized and included in the theory.

Occasionally the symmetry that leads to a selection rule is not evident, and the selection rule is not suspected in advance. Then conventional thermodynamics leads to conclusions discrepant with experiment. Puzzlement and consternation motivate exploration until the missing symmetry principle is recognized. Such an event occurred in the exploration of the properties of gaseous hydrogen at low temperatures. Hydrogen molecules can have their two nuclear spins parallel or antiparallel, the molecules then being designated as "ortho-hydrogen" or "para-hydrogen," respectively. The symmetries of the two types of molecules are quite different. In one case the molecule is symmetric under reflection in a plane perpendicular to the molecular axis, in the other case there is symmetry with respect to inversion through the center of the molecule. Consequently a selection rule prevents the conversion of one form of molecule to the other. This unsuspected selection rule led to spectacularly incorrect predictions of the thermodynamic properties of H_2 gas. But when the selection rule was at last recognized, the resolution of the difficulty was straightforward. Ortho- and para-hydrogen were simply considered to be two distinct gases, and the single mole number of "hydrogen" was replaced by separate mole numbers. With the theory thus extended to include an additional conserved coordinate, theory and experiment were fully reconciled.

Interestingly, a different "operational" solution of the ortho-H_2, para-H_2 problem was discovered. If a minute concentration of oxygen gas or water vapor is added to the hydrogen gas the properties are drastically changed. The oxygen atoms are paramagnetic, they interact strongly with the nuclear spins of the hydrogen molecules, and they destroy the symmetry that generates the selection rule. In the presence of a very few atoms of oxygen the ortho- and parahydrogen become interconvertible, and only a single mole number need be introduced. The original "naive" form of thermodynamics then becomes valid.

To return to the general formalism, we thus recognize that *all symmetries must be taken into account in specifying the relevant state space of a system*.

As additional symmetries are discovered in physics the scope of thermo-statistics will expand. Perhaps all the symmetries of an ideal gas at standard temperatures and pressures are known, but the case of ortho- and para-hydrogen cautions modesty even in familiar cases. Moreover thermodynamics has relevance to quasars, and black holes, and neutron stars and quark matter and gluon gases. For each of these there will be random perturbations, and symmetry principles, conservation laws, and Goldstone excitations,—and therefore thermostatistics.

SOME RELATIONS
INVOLVING PARTIAL
DERIVATIVES

A-1 PARTIAL DERIVATIVES

In thermodynamics we are interested in continuous functions of three (or more) variables

$$\psi = \psi(x, y, z) \tag{A.1}$$

If two independent variables, say y and z, are held constant, ψ becomes a function of only one independent variable x, and the derivative of ψ with respect to x may be defined and computed in the standard fashion. The derivative so obtained is called the *partial derivative* of ψ with respect to x and is denoted by the symbol $(\partial\psi/\partial x)_{y,z}$ or simply by $\partial\psi/\partial x$. The derivative depends upon x and upon the values at which y and z are held during the differentiation; that is $\partial\psi/\partial x$ is a function of x, y, and z. The derivatives $\partial\psi/\partial y$ and $\partial\psi/\partial z$ are defined in an identical manner.

The function $\partial\psi/\partial x$, if continuous, may itself be differentiated to yield three derivatives which are called the *second partial derivatives* of ψ

$$\frac{\partial}{\partial x}\left(\frac{\partial\psi}{\partial x}\right) \equiv \frac{\partial^2\psi}{\partial x^2}$$

$$\frac{\partial}{\partial y}\left(\frac{\partial\psi}{\partial x}\right) \equiv \frac{\partial^2\psi}{\partial y\,\partial x} \tag{A.2}$$

$$\frac{\partial}{\partial z}\left(\frac{\partial\psi}{\partial x}\right) \equiv \frac{\partial^2\psi}{\partial z\,\partial x}$$

By partial differentiation of the functions $\partial\psi/\partial y$ and $\partial\psi/\partial z$, we obtain other second partial derivatives of ψ

$$\frac{\partial^2\psi}{\partial x\,\partial y} \qquad \frac{\partial^2\psi}{\partial y^2} \qquad \frac{\partial^2\psi}{\partial z\,\partial y} \qquad \frac{\partial^2\psi}{\partial x\,\partial z} \qquad \frac{\partial^2\psi}{\partial y\,\partial z} \qquad \frac{\partial^2\psi}{\partial z^2}$$

It may be shown that under the continuity conditions that we have assumed for ψ and its partial derivatives the order of differentiation is immaterial, so that

$$\frac{\partial^2 \psi}{\partial x\, \partial y} = \frac{\partial^2 \psi}{\partial y\, \partial x}, \qquad \frac{\partial^2 \psi}{\partial x\, \partial z} = \frac{\partial^2 \psi}{\partial z\, \partial x}, \qquad \frac{\partial^2 \psi}{\partial y\, \partial z} = \frac{\partial^2 \psi}{\partial z\, \partial y} \quad (A.3)$$

There are therefore just six nonequivalent second partial derivatives of a function of three independent variables (three for a function of two variables, and $\frac{1}{2}n(n+1)$ for a function of n variables).

A-2 TAYLOR'S EXPANSION

The relationship between $\psi(x, y, z)$ and $\psi(x + dx, y + dy, z + dz)$, where dx, dy, and dz denote arbitrary increments in x, y, and z, is given by Taylor's expansion

$$\psi(x + dx, y + dy, z + dz)$$

$$= \psi(x, y, z) + \left(\frac{\partial \psi}{\partial x}\, dx + \frac{\partial \psi}{\partial y}\, dy + \frac{\partial \psi}{\partial z}\, dz \right) + \frac{1}{2}\left[\frac{\partial^2 \psi}{\partial x^2}(dx)^2 + \frac{\partial^2 \psi}{\partial y^2}(dy)^2 \right.$$

$$\left. + \frac{\partial^2 \psi}{\partial z^2}(dz)^2 + 2\frac{\partial^2 \psi}{\partial x\, \partial y}\, dx\, dy + 2\frac{\partial^2 \psi}{\partial x\, \partial z}\, dx\, dz + 2\frac{\partial^2 \psi}{\partial y\, \partial z}\, dy\, dz \right] + \cdots$$

$$(A.4)$$

This expansion can be written in a convenient symbolic form

$$\psi(x + dx, y + dy, z + dz) = e^{(dx(\partial/\partial x) + dy(\partial/\partial y) + dz(\partial/\partial z))}\psi(x, y, z)$$

$$(A.5)$$

Expansion of the symbolic exponential according to the usual series

$$e^x = 1 + x + \frac{1}{2!}x^2 + \cdots + \frac{1}{n!}x^n + \cdots \qquad (A.6)$$

then reproduces the Taylor expansion (equation A.4)

A-3   DIFFERENTIALS

The Taylor expansion (equation A.4) can also be written in the form

$$\psi(x + dx, y + dy, z + dz) - \psi(x, y, z)$$

$$= d\psi + \frac{1}{2!}d^2\psi + \cdots + \frac{1}{n!}d^n\psi \cdots \qquad (A.7)$$

where

$$d\psi \equiv \frac{\partial\psi}{\partial x}\,dx + \frac{\partial\psi}{\partial y}\,dy + \frac{\partial\psi}{\partial z}\,dz \qquad (A.8)$$

$$d^2\psi = \frac{\partial^2\psi}{\partial x^2}(dx)^2 + \frac{\partial^2\psi}{\partial y^2}(dy)^2 + \frac{\partial^2\psi}{\partial z^2}(dz)^2 + 2\frac{\partial^2\psi}{\partial x\,\partial y}\,dx\,dy$$

$$+2\frac{\partial^2\psi}{\partial x\,\partial z}\,dx\,dz + 2\frac{\partial^2\psi}{\partial y\,\partial z}\,dy\,dz \qquad (A.9)$$

and generally

$$d^n\psi = \left(dx\frac{\partial}{\partial x} + dy\frac{\partial}{\partial y} + dz\frac{\partial}{\partial z}\right)^n \psi(x, y, z) \qquad (A.10)$$

These quantities $d\psi, d^2\psi, \ldots, d^n\psi, \ldots$ are called the *first-, second-, and nth-order differentials* of ψ.

A-4   COMPOSITE FUNCTIONS

Returning to the first-order differential

$$d\psi = \left(\frac{\partial\psi}{\partial x}\right)_{y,z}dx + \left(\frac{\partial\psi}{\partial y}\right)_{x,z}dy + \left(\frac{\partial\psi}{\partial z}\right)_{x,y}dz \qquad (A.11)$$

an interesting case arises when x, y, and z are not varied independently but are themselves considered to be functions of some variable u. Then

$$dx = \frac{dx}{du}\,du \qquad dy = \frac{dy}{du}\,du \qquad \text{and} \qquad dz = \frac{dz}{du}\,du$$

whence

$$d\psi = \left[\left(\frac{\partial\psi}{\partial x}\right)_{y,z}\frac{dx}{du} + \left(\frac{\partial\psi}{\partial y}\right)_{x,z}\frac{dy}{du} + \left(\frac{\partial\psi}{\partial z}\right)_{x,y}\frac{dz}{du}\right]du \qquad (A.12)$$

or

$$\frac{d\psi}{du} = \left(\frac{\partial\psi}{\partial x}\right)_{y,z}\frac{dx}{du} + \left(\frac{\partial\psi}{\partial y}\right)_{x,z}\frac{dy}{du} + \left(\frac{\partial\psi}{\partial z}\right)_{x,y}\frac{dz}{du} \tag{A.13}$$

If x and y are functions of two (or more) variables, say u and v, then

$$dx = \left(\frac{\partial x}{\partial u}\right)_v du + \left(\frac{\partial x}{\partial v}\right)_u dv, \qquad \text{etc.}$$

and

$$d\psi = \left[\left(\frac{\partial\psi}{\partial x}\right)_{y,z}\left(\frac{\partial x}{\partial u}\right)_v + \left(\frac{\partial\psi}{\partial y}\right)_{x,z}\left(\frac{\partial y}{\partial u}\right)_v + \left(\frac{\partial\psi}{\partial u}\right)_{x,y}\left(\frac{\partial z}{\partial u}\right)_v\right]du$$

$$+ \left[\left(\frac{\partial\psi}{\partial x}\right)_{y,z}\left(\frac{\partial x}{\partial v}\right)_u + \left(\frac{\partial\psi}{\partial y}\right)_{x,z}\left(\frac{\partial y}{\partial v}\right)_u + \left(\frac{\partial\psi}{\partial z}\right)_{x,y}\left(\frac{\partial z}{\partial v}\right)_u\right]dv \tag{A.14}$$

or

$$d\psi = \left(\frac{\partial\psi}{\partial u}\right)_v du + \left(\frac{\partial\psi}{\partial v}\right)_u dv \tag{A.15}$$

where

$$\left(\frac{\partial\psi}{\partial u}\right)_v = \left(\frac{\partial\psi}{\partial x}\right)_{y,z}\left(\frac{\partial x}{\partial u}\right)_v + \left(\frac{\partial\psi}{\partial y}\right)_{x,z}\left(\frac{\partial y}{\partial u}\right)_v + \left(\frac{\partial\psi}{\partial z}\right)_{x,y}\left(\frac{\partial z}{\partial u}\right)_v \tag{A.16}$$

and similarly for $(\partial\psi/\partial v)_u$.

It may happen that u is identical to x itself. Then

$$\left(\frac{\partial\psi}{\partial x}\right)_v = \left(\frac{\partial\psi}{\partial x}\right)_{y,z} + \left(\frac{\partial\psi}{\partial y}\right)_{x,z}\left(\frac{\partial y}{\partial x}\right)_v + \left(\frac{\partial\psi}{\partial z}\right)_{x,y}\left(\frac{\partial z}{\partial x}\right)_v \tag{A.17}$$

Other special cases can be treated similarly.

A-5   IMPLICIT FUNCTIONS

If ψ is held constant, the variations of x, y, and z are not independent, and the relation

$$\psi(x, y, z) = \text{constant} \tag{A.18}$$

gives an implicit functional relation among x, y, and z. This relation may be solved for one variable, say z, in terms of the other two

$$z = z(x, y) \qquad \text{(A.19)}$$

This function can then be treated by the techniques previously described to derive certain relations among the partial derivatives. However, a more direct method of obtaining the appropriate relations among the partial derivatives is merely to put $d\psi = 0$ in equation A.8.

$$0 = \left(\frac{\partial \psi}{\partial x} \right)_{y,z} dx + \left(\frac{\partial \psi}{\partial y} \right)_{x,z} dy + \left(\frac{\partial \psi}{\partial z} \right)_{x,y} dz \qquad \text{(A.20)}$$

If we now put $dz = 0$ and divide through by dx, we find

$$0 = \left(\frac{\partial \psi}{\partial x} \right)_{y,z} + \left(\frac{\partial \psi}{\partial y} \right)_{x,z} \left(\frac{\partial y}{\partial x} \right)_{\psi,z} \qquad \text{(A.21)}$$

in which the symbol $(\partial y / \partial x)_{\psi,z}$ appropriately indicates that the implied functional relation between y and x is that determined by the constancy of ψ and z. Equation A.21 can be written in the convenient form

$$\left(\frac{\partial y}{\partial x} \right)_{\psi,z} = \frac{-(\partial \psi / \partial x)_{y,z}}{(\partial \psi / \partial y)_{x,z}} \qquad \text{(A.22)}$$

This equation plays a very prominent role in thermodynamic calculations. By successively putting $dy = 0$ and $dx = 0$ in equation A.20, we find the two similar relations

$$\left(\frac{\partial z}{\partial x} \right)_{\psi,y} = \frac{-(\partial \psi / \partial x)_{y,z}}{(\partial \psi / \partial z)_{x,y}} \qquad \text{(A.23)}$$

and

$$\left(\frac{\partial z}{\partial y} \right)_{\psi,x} = \frac{-(\partial \psi / \partial y)_{x,z}}{(\partial \psi / \partial z)_{x,y}} \qquad \text{(A.24)}$$

Returning to equation A.20 we again put $dz = 0$, but we now divide through by dy rather than by dx

$$0 = \left(\frac{\partial \psi}{\partial x} \right)_{y,z} \left(\frac{\partial x}{\partial y} \right)_{\psi,z} + \left(\frac{\partial \psi}{\partial y} \right)_{x,z} \qquad \text{(A.25)}$$

whence

$$\left(\frac{\partial x}{\partial y} \right)_{\psi,z} = \frac{-(\partial \psi / \partial y)_{x,z}}{(\partial \psi / \partial x)_{y,z}} \qquad \text{(A.26)}$$

and, on comparison with equation A.21, we find the very reasonable result that

$$\left(\frac{\partial x}{\partial y}\right)_{\psi,z} = \frac{1}{(\partial y/\partial x)_{\psi,z}} \tag{A.27}$$

From equations A.22 to A.24 we then find

$$\left(\frac{\partial x}{\partial y}\right)_{\psi,z}\left(\frac{\partial y}{\partial z}\right)_{\psi,x}\left(\frac{\partial z}{\partial x}\right)_{\psi,y} = -1 \tag{A.28}$$

Finally we return to our basic equation, which defines the differential $d\psi$, and consider the case in which x, y, and z are themselves functions of a variable u (as in equation A.12)

$$d\psi = \left[\left(\frac{\partial \psi}{\partial x}\right)_{y,z}\frac{dx}{du} + \left(\frac{\partial \psi}{\partial y}\right)_{x,z}\frac{dy}{du} + \left(\frac{\partial \psi}{\partial z}\right)_{x,y}\frac{dz}{du}\right]du \tag{A.29}$$

If ψ is to be constant, there must be a relation among x, y, and z, hence also among dx/du, dy/du, and dz/du. We find

$$0 = \left(\frac{\partial \psi}{\partial x}\right)_{y,z}\left(\frac{dx}{du}\right)_{\psi} + \left(\frac{\partial \psi}{\partial y}\right)_{x,z}\left(\frac{dy}{du}\right)_{\psi} + \left(\frac{\partial \psi}{\partial z}\right)_{x,y}\left(\frac{dz}{du}\right)_{\psi} \tag{A.30}$$

If we further require that z shall be a constant independent of u we find

$$0 = \left(\frac{\partial \psi}{\partial x}\right)_{y,z}\left(\frac{\partial x}{\partial u}\right)_{\psi,z} + \left(\frac{\partial \psi}{\partial y}\right)_{x,z}\left(\frac{\partial y}{\partial u}\right)_{\psi,z} \tag{A.31}$$

or

$$\frac{(\partial y/\partial u)_{\psi,z}}{(\partial x/\partial u)_{\psi,z}} = -\frac{(\partial \psi/\partial x)_{y,z}}{(\partial \psi/\partial y)_{x,z}} \tag{A.32}$$

Comparison with equation A.22 shows that

$$\left(\frac{\partial y}{\partial x}\right)_{\psi,z} = \frac{(\partial y/\partial u)_{\psi,z}}{(\partial x/\partial u)_{\psi,z}} \tag{A.33}$$

Equations A.22, A.27, and A.33 are among the most useful formal manipulations in thermodynamic calculations.

APPENDIX **B**

MAGNETIC
SYSTEMS

If matter is acted on by a magnetic field it generally develops a magnetic moment. A description of this magnetic property, and of its interaction with thermal and mechanical properties, requires the adoption of an additional extensive parameter. This additional extensive parameter X and its corresponding intensive parameter P are to be chosen so that the *magnetic work* dW_{mag} is

$$dW_{\mathrm{mag}} = P\,dX \tag{B.1}$$

where

$$dU = dQ + dW_M + dW_c + dW_{\mathrm{mag}} \tag{B.2}$$

Here dQ is the heat $T\,dS$, dW_M is the mechanical work (e.g., $-P\,dV$), and dW_c is the chemical work $\Sigma\mu_j\,dN_j$. We consider a specific situation that clearly indicates the appropriate choice of parameters X and P.

Consider a solenoid, or coil, as shown in Fig. B.1. The wire of which the solenoid is wound is assumed to have zero electrical resistance (superconducting). A battery is connected to the solenoid, and the electromotive force (emf) of the battery is adjustable at will. The thermodynamic system is inside the solenoid, and the solenoid is enclosed within an adiabatic wall.

If no changes occur within the system, and if the current I is constant, the battery need supply no emf because of the perfect conductivity of the wire.

Let the current be I and let the local magnetization of the thermodynamic system be $\mathbf{M}(\mathbf{r})$. The current I can be altered at will by controlling the battery emf. The magnetization $\mathbf{M}(\mathbf{r})$ then will change also. We assume that the magnetization at any position $\mathbf{r}$ is a single-valued function of the current

$$\mathbf{M}(\mathbf{r}) = \mathbf{M}(\mathbf{r}; I) \tag{B.3}$$

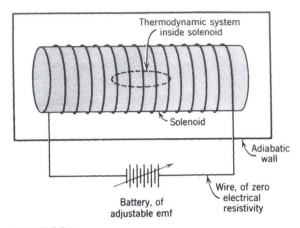

FIGURE B.1

Systems for which $\mathbf{M}(\mathbf{r}; I)$ is not single valued in I are said to ex-hibit *hysteresis*; most ferromagnetic systems have this property. Hysteresis generally is associated with a magnetic heterogeneity of the sample, the separate regions being known as domains. The analysis we shall develop is generally applicable *within* a ferromagnetic domain, but for simplicity we explicitly exclude all hysteretic systems. Paramagnetic, diamagnetic, and antiferromagnetic systems satisfy the requirement that $\mathbf{M}(\mathbf{r}; I)$ is single valued in I.

If the thermodynamic system were *not* within the solenoid, the current I would produce a magnetic field (more accurately, a magnetic *flux density*) $\mathbf{B}_e(I)$. This external "field" may be a function of position within the solenoid, but it is linear in I. That is

$$\mathbf{B}_e = \mathbf{b}I \qquad (\text{B.4})$$

where $\mathbf{b}$ is a vector function of position.

We suppose that the current is increased, thereby increasing the exter-nal field $\mathbf{B}_e$. The magnetic moment changes in response. In order to accomplish these changes, the battery must deliver work, and we seek the relationship between the work done and the changes in $\mathbf{B}_e$ and $\mathbf{M}$.

The rate at which work is done by the battery is given by

$$\frac{dW_{\text{mag}}}{dt} = I \times (\text{voltage}) \qquad (\text{B.5})$$

in which (voltage) denotes the back emf induced in the solenoid windings by the changes that occur within the coil.

The induced emf in the solenoid arises from two sources. One source is independent of the thermodynamic system and results from a change in the flux associated with the field $\mathbf{B}_e$. Rather than compute this flux and voltage, we can write the resultant contribution to dW_{mag} directly. For an *empty* solenoid the work is just the change in the energy of the magnetic field, or

$$dW_{mag} = d\left(\frac{1}{2\mu_0} \int B_e^2 \, dV \right) \tag{B.6}$$

in which $\mu_0 = 4\pi \times 10^{-7} T \cdot m/A$, and in which the integral is taken over the entire volume of the solenoid.

The second contribution to dW_{mag} results from the thermodynamic system itself and consequently is of more direct interest to us. It is evident that the change of magnetic moment of each infinitesimal element of the system contributes separately and additively to the total induced emf, and furthermore that the induced emf produced by any change in dipole moment depends not on the nature of the dipole but only on the rate of change of its moment and on its position in the solenoid. Consider then a particular model of an elementary dipole at the position r: a small current loop of area $\mathbf{a}$ and current i, with a magnetic moment of $\mathbf{m} = i\mathbf{a}$. If the current in the solenoid is I, the field produced by the solenoid at the point $\mathbf{r}$ is $\mathbf{B}_e(\mathbf{r}) = \mathbf{b}(\mathbf{r})I$. This field produces a flux linkage through the small current loop of magnitude $\mathbf{b}(\mathbf{r}) \cdot \mathbf{a}I$. Thus the mutual inductance between solenoid and current loop is $\mathbf{b}(\mathbf{r}) \cdot \mathbf{a}$. If the current in the current loop changes, it consequently induces a voltage in the solenoid given by

$$(\text{voltage}) = [\mathbf{b}(\mathbf{r}) \cdot \mathbf{a}] \frac{di}{dt} \tag{B.7}$$

$$= \mathbf{b}(\mathbf{r}) \cdot \frac{d\mathbf{m}}{dt} \tag{B.8}$$

$$= \frac{1}{I} \mathbf{B}_e(\mathbf{r}) \cdot \frac{d\mathbf{m}}{dt} \tag{B.9}$$

Thus the work done by the battery is

$$\frac{dW_{mag}}{dt} = \mathbf{B}_e(\mathbf{r}) \cdot \frac{d\mathbf{m}}{dt} \tag{B.10}$$

Although this result has been obtained for a particular model of an elementary dipole, it holds for any change in elementary dipole moment. In particular if $\mathbf{M}(\mathbf{r})$ is the *magnetization*, or the dipole moment per unit volume in the system at the point $\mathbf{r}$, we set

$$\mathbf{m} = \int \mathbf{M}(\mathbf{r}) \, dV \tag{B.11}$$

To obtain the total work, we sum over all elementary dipoles, or integrate over the volume of the sample

$$\frac{dW_{\text{mag}}}{dt} = \int \mathbf{B}_e \cdot \frac{d\mathbf{M}}{dt} \, dV \tag{B.12}$$

Adding the two contributions to the magnetic work, we find

$$dW_{\text{mag}} = d\left(\frac{1}{2\mu_0} \int B_e^2 \, dV \right) + \int (\mathbf{B}_e \cdot d\mathbf{M}) \, dV \tag{B.13}$$

This is the fundamental result on which the thermodynamics of magnetic systems is based.

In passing we note that the *local* field $\mathbf{H}$ can be introduced in place of the *external* field $\mathbf{H}_e$ by noting that the difference $\mathbf{H} - \mathbf{H}_e$ is just the field produced by the magnetization $\mathbf{M}(\mathbf{r})$ acting as a magnetostatic source. In this way it can be shown[1] that

$$dW_{\text{mag}} = \int \mathbf{H} \cdot d\mathbf{B} \, dV \tag{B.14}$$

where $\mathbf{H}$ and $\mathbf{B}$ are *local* values. However the form of the magnetic work expression we shall find most convenient is the first derived (equation B.13).

In the general case the magnetization $\mathbf{M}(\mathbf{r})$ will vary from point to point within the system, even if the external field $\mathbf{B}_e$ is constant. This variation may arise from inherent inhomogeneities in the properties of the system, or it may result from demagnetization effects of the boundaries of the system. We wish to develop the theory for homogeneous systems. We therefore assume that $\mathbf{B}_e$ is constant and that the intrinsic properties of the system are homogeneous. We further assume that the system is ellipsoidal in shape. For such a system the magnetization $\mathbf{M}$ is independent of position, as shown in any text on magnetostatics.

The magnetic work equation can now be written as

$$dW_{\text{mag}} = d\left(\frac{1}{2\mu_0} \int B_e^2 \, dV \right) + \mathbf{B}_e \cdot d\mathbf{I} \tag{B.15}$$

where $\mathbf{I}$ is the total magnetic dipole moment of the system

$$\mathbf{I} = \int \mathbf{M} \, dV = \mathbf{M}V \tag{B.16}$$

[1] See V. Heine, *Proc. Cambridge Phil. Soc.*, **52**, 546 (1956).

The energy differential is

$$d(\text{Energy}) = T\,dS - P\,dV + d\left(\frac{1}{2\mu_0}\int B_e^2\,dV\right) + \mathbf{B}_e \cdot d\mathbf{I} + \sum_1^r \mu_j\,dN_j$$

$$(B.17)$$

The third term on the right of the foregoing equation does not involve the thermodynamic system itself but arises only from the magnetostatic energy of the empty solenoid. Consequently it is convenient to absorb this term into the definition of the energy. We define the energy U by

$$U \equiv \text{Energy} - \frac{1}{2\mu_0}\int B_e^2\,dV \qquad (B.18)$$

so that U is the total energy contained within the solenoid relative to the state in which the system is removed to its field free fiducial state and the solenoid is left with the field $\mathbf{B}_e$. This redefinition of the internal energy does not alter any of the formalism of thermodynamics. Thus we write

$$dU = T\,dS - P\,dV + B_e\,dI_B + \sum_1^r \mu_j\,dN_j \qquad (B.19)$$

where I_B is the component of $\mathbf{I}$ parallel to $\mathbf{B}_e$.

The extensive parameter descriptive of the magnetic properties of a system is $\mathbf{I}_B$, *the component of the total magnetic moment parallel to the external field. The intensive parameter in the energy representation is* $\mathbf{B}_e$.

The fundamental equation is

$$U = U(S, V, I_B, N_1, \ldots, N_r) \qquad (B.20)$$

and

$$\left(\frac{\partial U}{\partial I_B}\right)_{S,V,N_1,\ldots,N_r} = B_e \qquad (B.21)$$

GENERAL REFERENCES

THERMODYNAMICS

R. Kubo, *Thermodynamics*, Wiley, 1960. Concise text with many problems and explicit solutions.

K. J. Laidler and J. F. Meiser, *Physical Chemistry*, Benjamin/Cummings, 1982. Chemical applications of thermodynamics.

A. B. Pippard, *Elements of Classical Thermodynamics*, Cambridge University Press, 1966. A scholarly and rigorous treatment.

R. E. Sonntag and G. J. Van Wylen, *Introduction to Thermodynamics, Classical & Statistical*, 2nd edition, Wiley, 1982. Very thorough thermodynamic treatment. Engineering viewpoint.

G. Weinreich, *Fundamental Thermodynamics*, Addison-Wesley, 1968. Idiosyncratic, insightful, and original.

M. W. Zemansky and R. H. Dittman, *Heat and Thermodynamics, An Intermediate Textbook*, 6th edition, McGraw-Hill, 1981. Contains careful and full discussions of empirical data, experimental methods, practical thermometry, and applications.

STATISTICAL MECHANICS

R. P. Feynman, *Statistical Mechanics, A Set of Lectures*, W. A. Benjamin, 1972. Advanced-level notes with the unique Feynman flair. Particularly strong emphasis on the Bogoliubov variational theorem.

R. J. Finkelstein, *Thermodynamics and Statistical Physics-A Short Introduction*, W. H. Freeman and Co., 1969. A brief and unconventional formulation of the logic of thermostatistics.

J. W. Gibbs, *The Scientific Papers of J. Willard Gibbs, Volume 1, Thermodynamics*, Dover, 1961. Gibbs not only invented modern thermodynamics and statistical mechanics, but he also anticipated, explicitly or implicitly, almost every subsequent development. His exposition is not noted for its clarity.

C. Huang, *Statistical Mechanics*, Wiley, 1963. Classic graduate text.

C. Kittel and H. Kroemer, *Thermal Physics*, 2nd edition, W. H. Freeman, 1980. Introductory treatment. Large number of interesting illustrative applications.

R. Kubo, *Statistical Mechanics*, Wiley, 1965. Concise text with many problems and explicit solutions.

L. D. Landau and E. M. Lifshitz, *Statistical Physics*, 3rd edition, Part 1, by E. M. Lifshitz and L. P. Pitaevskii, Pergamon Press, 1980.

E. M. Lifshitz and L. P. Pitaevskii, *Statistical Physics*, Part 2 of reference above, Pergamon Press, 1980. Advanced treatment.

P. T. Landsburg, *Thermodynamics and Statistical Mechanics*, Oxford Univ. Press, 1978. Contains many novel observations and 120 fully-solved problems.

F. Reif, *Fundamentals of Statistical and Thermal Physics*, McGraw-Hill, 1965. Classic text, with an immense collection of excellent problems.

M. Tribus, *Thermostatics and Thermodynamics*, Van Nostrand, 1961. A development based on the information-theoretic approach of E. T. Jaynes.

CRITICAL PHENOMENA

D. J. Amit, *Field Theory, the Renormalization Group, and Critical Phenomena*, McGraw-Hill, 1978. Advanced theory.

Shang-Keng Ma, *Modern Theory of Critical Phenomena*, Benjamin, 1976.

P. Pfeuty and G. Toulouse, *Introduction to the Renormalization Group and to Critical Phenomena*, Wiley, 1977.

H. E. Stanley, *Introduction to Phase Transitions and Critical Phenomena*, Oxford University Press, 1971. Excellent introduction. Pre-dates Wilson renormalization theory.

CONCEPTUAL OVERVIEWS

P. W. Anderson, *Basic Notions of Condensed Matter Physics*, Benjamin/Cummings, 1984. A profound and pentrating analysis of the role of symmetry in the general theory of properties of matter. Although the level is quite advanced, Anderson's interest in underlying principles of universal generality, rather than in mathematical techniques of calculation, make the book a treasure for the reader at any technical level.

R. D. Rosenkrantz, editor, *E. T. Jaynes: Papers on Probability, Statistics and Statistical Physics*, Reidel, 1983. An unconventional conceptualization of statistical mechanics as an information–theoretic exercise in prediction. Jaynes' point of view is reflected (in a rather pale form) in Chapter 17 of this text.

INDEX

13792213R00293